TRANSBOUNDARY
WATER MANAGEMENT

TRANSBOUNDARY
WATER MANAGEMENT

Principles and Practice

Edited by
Anton Earle, Anders Jägerskog and Joakim Öjendal

publishing for a sustainable future
London • New York

First published in 2010 by Earthscan

SIWI – Independent, Leading-Edge Water Competence for Future-Oriented Action
The Stockholm International Water Institute (SIWI) is a policy institute that contributes to international efforts to combat the world's escalating water crisis. SIWI advocates future-oriented, knowledge-integrated water views in decision making, nationally and internationally, that lead to sustainable use of the world's water resources and sustainable development of societies.

 Sida

This document has been financed by the Swedish
International Development Cooperation Agency,
Sida. Sida does not necessarily share the views
expressed in this material. Responsibility for its
contents rests entirely with the author.

Earthscan

2 Park Square, Milton Park, Abingdon, Oxon OX14 4RN

Simultaneously published in the USA and Canada by Earthscan

711 Third Avenue, New York, NY 10017

Earthscan in an imprint of the Taylor & Francis Group, an informa business

Earthscan publishes in association with the International Institute for Environment and Development

ISBN   978-1-84971-137-1 hardback
            978-1-84971-138-8 paperback

Typeset by MapSet, Gateshead, UK
Cover design by Dan Bramall

A catalogue record for this book is available from the British Library

Library of Congress Cataloging-in-Publication Data
   Transboundary water management : principles and practice / edited by Anton Earle, Anders Jägerskog, and Joakim Öjendal. — 1st ed.
        p. cm.
   Includes bibliographical references and index.
      ISBN 978-1-84971-137-1 (hardback) — ISBN 978-1-84971-138-8 (pbk.) 1. Water resources development. 2. Water-supply. I. Earle, Anton. II. Jägerskog, Anders. III. Öjendal, Joakim.
   HD1691.T733 2010
   333.91—dc22

                                                                              2010015267

At Earthscan we strive to minimize our environmental impacts and carbon footprint through reducing waste, recycling and offsetting our CO_2 emissions, including those created through publication of this book.

Contents

PART 3 – CHALLENGES AND OPPORTUNITIES

List of Figures and Tables

Figures

Tables

Foreword

Water knows no political, economic or social boundaries as it passes through the various phases of the hydrological cycle. By linking plants, animals, people and ecological processes, water physically embodies our 'common future'. The sound use and management of this precious resource can lead to benefits for humanity and ecosystems; just as mismanagement of it will be to their detriment.

The challenge facing water managers today is how to unlock the benefits from the water cycle and to make them available to a broad range of social and economic sectors in a sustainable way. Increasing competition for transboundary water resources between economic sectors, ecosystems, communities and nations is often assumed to lead necessarily to conflict. The additional pressures of climate change, economic development and demographic shifts would seem only to increase the likelihood of such a negative outcome.

Yet, historically, water has more often been a source of cooperation between countries, leading to economic development for those involved, than a catalyst for conflict. Since the earliest years of the nation state, institutions have been set up to develop infrastructure for water storage and supply, as well as to provide protection from hazards such as flooding.

As governments all over the world are coming to recognize, transboundary waters can be the object of peaceful cooperation and sustainable socioeconomic development, provided sufficient institutional capacity exists to manage the challenges. Investments need to be made in data monitoring and exchange, in the creation of legal frameworks, in human resources capacity-building, in education programmes for a range of stakeholders, and in the generation of scientific knowledge.

United Nations agencies are actively involved in a number of transboundary water management initiatives. These include UNESCO's 'Potential Conflict to Cooperation Potential' and 'Internationally Shared Aquifer Resource Management' programmes. Both flagships facilitate multilevel and interdisciplinary dialogues in order to foster peace, cooperation and development related to the management of transboundary water resources. Through research and capacity-building activities, the programmes have brought together players engaged in transboundary surface water and aquifers management, and have helped to increase the opportunities for actual cooperation and development.

Addressing transboundary water issues is a real and pressing challenge. With the backing of the policy makers who represent them and the water users whom they serve, water resource managers are able to take proactive steps towards achieving a common vision and development outcomes.

It is essential that the international community support these efforts by strengthening all facets of institutional capacity for effective water management. This volume represents a tangible contribution to this effort by placing the current knowledge and experience of water management professionals and researchers within reach of a broad audience.

Irina Bokova

Irina Bokova
Director-General of UNESCO
June 2010

List of Contributors

Contributors

Editors

Anton Earle is Project Director – Capacity Building, at the Stockholm International Water Institute (SIWI), Sweden. He is a geographer with an academic background in environmental management and specializes in transboundary water resource management, facilitating the interaction between governments, basin organizations and other stakeholders in international river and lake basins. He has experience in developing capacity-building initiatives for the various groups involved in water management and water use at the inter-state level in the southern and East African regions, the Middle East and internationally, and has published extensively on transboundary water management. For more information on SIWI's capacity-building programmes in the water sector see www.siwi.org/capacitybuilding.

Anders Jägerskog is Project Director – Applied Research, at the Stockholm International Water Institute (SIWI) and Associate Professor, School of Global Studies (Peace and Development Research), Gothenburg University. His work is focused on transboundary waters with particular emphasis on the MENA regions and Africa. Previously he has worked at the secretariat for the Expert Group on Development Issues (EGDI) at the Swedish Ministry for Foreign Affairs; at the Swedish International Development Cooperation Agency (Sida), on water resources management in eastern and southern Africa; and at the Stockholm International Peace Research Institute (SIPRI), on Middle Eastern security issues. In 2003 he finished his PhD on the water negotiations in the Jordan River Basin at the Department of Water and Environmental Studies at the Linköping University, Sweden. He has published over 60 scientific articles, book chapters, debate articles and reports.

Joakim Öjendal is Professor at the School of Global Studies (SGS), Peace and Development Research, University of Gothenburg, Sweden. He has been researching water management for 15 years at SGS, where he also wrote his PhD thesis *Sharing the Good* (2000), and has subsequently published widely, including *Transboundary Water Co-operation as a Tool for Conflict Prevention and Broader Benefit Sharing* (with Phillips, Daoudy, McCaffrey and Turton). He is currently involved in the research project *Reconstruction of War-Torn Societies*, as well as another on *Democratization and Decentralization* in Cambodia. Other fields of research for Öjendal have been the scope of globalization in Southeast Asia, in *Deepening Democracy and Restructuring Governance: Responses to globalization in Southeast Asia* (Nias Press; co-edited with Francis Loh), and regionalization as a global phenomenon in *Regionalization in a Globalizing World* (Zed Press; co-edited with Schulz and Söderbaum).

Chapter and Case Study Authors

J. A. (Tony) Allan, Professor Emeritus at King's College London and at the School of Oriental and African Studies, University of London, UK.

Cate Brown, Southern Waters, Freshwater Research Unit, Zoology Department, University of Cape Town, South Africa.

Ana Cascão, Project Manager – Capacity Building, at the Stockholm International Water Institute, Sweden.

Marwa Daoudy, Departmental Lecturer, St Anthony's College and Department of Politics and International Relations, University of Oxford, UK.

Malin Falkenmark, Senior Scientific Adviser at the Stockholm International Water Institute, Sweden, and Guest Professor at Stockholm Resilience Centre, Sweden.

Jakob Granit, Project Director – Advisory Services, at the Stockholm International Water Institute, Sweden.

Virginia Hooper, Water Consultant at Arup, UK.

Todd Jarvis, Associate Director at the Institute for Water and Watersheds, Oregon State University, USA.

Jackie King, Water Matters, Freshwater Research Unit, Zoology Department, University of Cape Town, South Africa.

Nicole Kranz, Senior Fellow at the Ecologic Institute Berlin, Germany.

Owen McIntyre, Senior Lecturer, Faculty of Law, University College Cork, National University of Ireland.

Michael McWilliams, former Communications Director at Stockholm International Water Institute, currently a private consultant.

Naho Mirumachi, a PhD student at the Department of Geography, School of Social Science and Public Policy, King's College London, UK.

Erik Mostert, Managing Director of the Centre for Research on River Basin Administration, Analysis and Management at Delft University of Technology, The Netherlands.

Shaminder Puri, Secretary General of the International Association of Hydrogeologists and a Senior Consultant to UNESCO–IHP serving as global coordinator of the ISARM Programme.

Léna Salamé, Programme Specialist (Water Conflicts and Cooperation) at the Division of Water Sciences, UNESCO.

Wilhelm Struckmeier, President of the International Association of Hydrogeologists and a Director and Professor of the BGR (German Federal Institute for Geosciences and Natural Resources) where he heads the information unit on groundwater and soils.

Pieter van der Zaag, Professor of Water Resources Management at UNESCO-IHE Institute for Water Education, Delft, The Netherlands.

Aaron Wolf, Professor and Chair of the Department of Geosciences, and Director: Program in Water Conflict Management and Transformation (PWCMT) at Oregon State University, USA.

Mark Zeitoun, Senior Lecturer at the School of International Development and founder of the UEA Water Security Research Centre, University of East Anglia, UK.

Acknowledgements

This volume seeks to support water management professionals, researchers, politicians and others in unleashing the contribution transboundary waters can make to sustainable socio-economic development. The need for such a book was identified while running a series of international training programmes on transboundary water management (TWM) aimed at mid-career professionals from around the world. Although much has been written on the topic, little has been assimilated in a systematic way, requiring interested readers to consult a variety of sources to understand key concepts.

The editors hope that this volume provides a readily accessible guide to the main issues that those tasked with the management of transboundary waters deal with. The production and dissemination of such a book is a collective effort, with the authors contributing their time, energy and creativity to the project and various organizations providing funds to support the process.

The training programmes on TWM provided a forum for testing new ideas and approaches, and for gaining valuable feedback from practitioners on their suitability. The Swedish International Development Cooperation Agency (Sida), by providing long-term funding for the training programmes, facilitated the experimental space for many of the concepts introduced in this book to be further developed. In addition, Sida has financially supported the production of this book – for which the editors, as well as SIWI, are grateful.

The Swedish Water House, as part of their commitment to strengthening communication between research, Swedish policy development, international processes and practical implementation, contributed funding to the development of the book. The governance of transboundary waters is an important theme which the SWH is focused on and we are grateful for the support provided for producing the book.

The TWM Programme in the Southern African Development Community (SADC), funded by the German and UK Governments and implemented by GTZ, committed support to the book project while still in its inception phase. Specifically the insight and willingness to nurture new ideas, shown by Peter Qwist-Hoffmann of GTZ, contributed to the viability of the book. He recognized that the book is in confluence with the aims of their TWM Programme – 'developing capacities of the SADC Secretariat, river basin organizations and national organizations to plan and implement transboundary water infrastructure'.

The editors would also like to thank the co-publisher of the book, the Stockholm International Water Institute, under the leadership of Anders Berntell, for their support throughout the project. Specific mention goes to our colleague Jakob Granit, whose tireless efforts resulted in securing core support for the production of the book. We trust that the book lives up to the SIWI credo of 'independent and leading-edge water competence for future-oriented action'.

We also thank the team at Earthscan for their encouragement, dedication and professional attention to detail, all of which contributed to a successful finished product. However, any errors or omissions remain the responsibility of the editors and authors alone.

List of Acronyms and Abbreviations

ACWR	African Centre for Water Research
ARIA	antagonism, resonance, invention and action
ASEAN	Association of Southeast Asian Nations
AWIRU	African Water Issues Research Unit
BAR	Basins at Risk
BATNA	best alternative to a negotiated agreement
BGR	Bundesanstalt für Geowissenschaften und Rohstoffe (Federal Institute for Geosciences and Natural Resources, Germany)
BMZ	Bundesministerium für wirtschaftliche Zusammenarbeit und Entwicklung (German Federal Ministry for Economic Cooperation and Development)
BWF	Basin-Wide Forum
CFA	cooperative framework agreement
CGA	Critical Groundwater Area
CIC	Comité Intergubernamental Coordinador de los Países de la Cuenca del Plata (Intergovernmental Coordinating Committee of the La Plata Basin)
CIDA	Canadian International Development Agency
CPR	common pool resources
CSIR	Council for Scientific and Industrial Research
CTUIR	Confederated Tribes of the Umatilla Indian Reservation
CWRC	Changjiang (Yangtze) Water Resources Commission
DEF	Danube Environmental Forum
DFID	Department for International Development
DRPC	Danube River Protection Convention
DSS	decision support system
EAC	East African Community
EAP	Environmental Action Plan
ECO-Asia	Environmental Cooperation-Asia
ECO-Tiras	International Association of the Dniester River Keepers
EF	environmental flow
EFA	environmental flow assessment
EGDI	Expert Group on Development Issues
ENVSEC	Environment and Security (Initiative)
ERJC	Estonian-Russian Joint Commission (on the protection and rational use of transboundary waters)
ESCWA	Economic and Social Commission for Western Asia
ETIC	Euphrates-Tigris Initiative for Cooperation

FAO	Food and Agriculture Organization of the United Nations
FRIEND	Flow Regimes from International Experimental and Network Data
GAP	Güneydoğu Anadolu Projesi
GAS	Guarani Aquifer System
GBM	Ganges–Brahmaputra–Meghna Basin
GEF	Global Environment Facility
GTZ	Deutsche Gesellschaft für Technische Zusammenarbeit
HH	hydrogen-hegemon
IAH	International Association of Hydrogeologists
IBFA	integrated basin flow assessment
ICJ	International Court of Justice
ICP	international cooperating partners
ICPDR	International Commission for the Protection of the Danube River
IFI	international financial institution
IFIM	Instream Flow Incremental Methodology
IHE	Institute for Water Education
IHP	International Hydrological Programme
IJC	International Joint Commission
ILA	International Law Association
ILC	International Law Commission
IMTA	Instituto Mexicano de Tecnología del Agua (Mexican Water Technology Institute)
ISARM	Internationally Shared Aquifer Resources Management
ITP	International Training Programme
IUCN	International Union for Conservation of Nature
IWMI	International Water Management Institute
IWR	Institute for Water Resources
IWRM	integrated water resources management
JFF	joint fact finding
JWC	Joint Water Committee
KCS	Kalahari Conservation Society
LCC	local coordination committee
LVBC	Lake Victoria Basin Commission
MAR	managed aquifer recharge
MAR	mean annual runoff
MENA	Middle East and North Africa
MRC	Mekong River Commission
NBCBN-RE	Nile Basin Capacity Building Network for River Engineering
NBI	Nile Basin Initiative
NNF	Namibia Nature Foundation
NSAS	Nubian Sandstone Aquifer System
OAS	Organization of American States
OKACOM	Permanent Okavango Basin River Water Commission
OMVS	Organisation Pour la Mise en Valeur du Fleuve Sénégal (Senegal River Development Organization)
ORASECOM	Orange-Senqu River Commission
OSCE	Organization for Security and Cooperation in Europe
OSU	Oregon State University
PCIJ	Permanent Court of International Justice

PWCMT	Program in Water Conflict Management and Transformation
3Rs	recharge–retention–reuse
RAK	River Awareness Kit
RBO	river basin organization
REC	regional economic communities
SAARC	South Asian Association for Regional Cooperation
SADC	Southern African Development Community
SAP	strategic action plan
SEA	strategic environmental assessment
SGS	School of Global Studies, Peace and Development Research, University of Gothenburg, Sweden
SI	Sensitivity Index
Sida	Swedish International Development Cooperation Agency
SIPRI	Stockholm International Peace Research Institute
SIWI	Stockholm International Water Institute
SOAS	School of Oriental and African Studies
SUMTACA	Sustainable Use and Management of the Transboundary Aquifers in Central Asia
TARM	Transboundary Aquifer Resource Management
TECCONILE	Technical Cooperation Committee for the Promotion of the Development and Environmental Protection of the Nile Basin
TRBI	transboundary river basin institution
TRBO	transboundary river basin organization
TWINS	transboundary waters interaction nexus
TWM	transboundary water management
TWO	Transboundary Water Opportunity
UCGT	Umatilla County Critical Groundwater Task Force
UEA	University of East Anglia
UNDP	United Nations Development Programme
UNECE	United Nations Economic Commission for Europe
UNEP	United Nations Environment Programme
UNESCO	United Nations Educational, Scientific and Cultural Organization
UNESCO PCCP	United Nations Educational, Scientific and Cultural Organization Potential Conflict to Cooperation Potential
USAID	United States Agency for International Development
WERRD	Water and Ecosystem Resources in Rural Development
WFD	Water Framework Directive
WHYMAP	Worldwide Hydrogeological Mapping and Assessment Project
ZRA	Zambezi River Authority

1

Introduction: Setting the Scene for Transboundary Water Management Approaches

Anton Earle, Anders Jägerskog and Joakim Öjendal

- Transboundary waters can make a contribution to regional development and peace if the institutional capacity exists to manage them cooperatively to the benefit of all basin states.
- The three major groupings in transboundary water management (TWM) initiatives are the water resource community (including water managers from government, as well as water users from the private sector and civil society), the research and academic community (including international financial institutions (IFIs) and development partners) and politicians.
- The three broad groups interact in a variety of ways – influencing each other and learning from each other, but the overall pace and direction of TWM processes is set by the politicians.

Setting the Scene

The more than 263 surface water basins shared between two or more states account for roughly 60 per cent of global freshwater flows and cover almost half the earth's land surface area (Carius et al, 2004; Wolf et al, 2005). Added to this is the large number of transboundary aquifers, constituting the primary source of water for over two billion people in the world (Puri and Struckmeier, Chapter 6). The availability, distribution and control of freshwater resources have been at the centre of the human story since the start of the Neolithic revolution roughly 12,000 years ago. With the advent of the modern nation state and its attendant emphasis on sovereignty, self-sufficiency and rivalry, it comes as no surprise that interactions between states over shared watercourses have at times been tense and conflictual. Water, as a fugitive resource, respects neither political boundaries nor commonly accepted notions of fairness or equity. Variable in both time and space, water has defied the efforts of politicians, economists and engineers to tame its capricious nature. No wonder much of the literature has focused on the possibility of disputes over water spilling over into outright conflict between states (Wolf et al, 2005). Water is an indispensible input to almost all human activity – manufacturing products, delivering services, producing food, transporting goods and sustaining life itself. The fact that water cannot readily be substituted by other resources leads, in part, to the long-term cooperation between states over its management and development.

Work carried out over the past decade by scientists such as Aaron Wolf (1998; Wolf et al, 2003b), Anthony Allan (1998a, 1998b, 1999, 2000, 2002) and Anthony Turton (Turton, 2003; Turton and Earle, 2005), among many others, has demonstrated that issues of national identity, cultural values and world view are more likely to lead to conflict between states, than are disputes over water (Kalpakian, 2004).

Disputes over water do occur, but they very rarely develop into greater conflicts as this would jeopardize the use of the resource itself (Wolf, 1998). Instead, states either reach a stalemate or deadlock over their shared waters, or manage to cooperate to some limited degree. This has been described by John Galtung as a 'negative peace' – merely the absence of violence, without further constructive collaboration (Galtung, 1996). In cases of water scarcity, where the likelihood of disputes between states over shared waters may be greater, the corollary is that there is also more evidence of cooperation (Wolf et al, 2003a). This cooperation is promoted and enhanced by institutions, such as laws or agreements, organizations and customary practices, which have been developed either on a bilateral or a multilateral level between states. These institutions (formal or informal) offer a forum where disputes can be discussed and amicably settled and may lead to the sustainable development of shared water resources, making a contribution to national and regional socioeconomic development. In the context of global change these institutions are bulwarks against the pressures introduced through natural climatic variability, resource degradation from socioeconomic development and climate change. Indeed, assuming sufficient institutional development, transboundary waters can become avenues of cooperation between countries, contributing to socioeconomic development and regional integration.

In the Basins at Risk (BAR) study carried out by Wolf and colleagues it is proposed that 'the likelihood and intensity of conflict rises as the rate of change within a basin exceeds the institutional capacity to absorb that change' (Wolf et al, 2003a). Rapid changes in the institutional framework (such as key staff members leaving) or in physical factors (climate, water demand, demographics etc.) which outpace the institutional capacity to absorb such change are at the heart of most water conflicts (UNEP, 2005). Thus, where a well-capacitated institution is in place (encompassing human resources, legislative framework, financial sustainability and political will, among other factors), the chances of being able to withstand the pressures of global change are enhanced. This can improve the potential for cooperation between basin states and even lead to the co-management and co-development of shared water resources, shifting the balance towards cooperative management and away from the lack of development associated with the stalemate situation of a 'negative peace'. This is an important aspect considering that various parts of the developing world are in the early stages of creating water infrastructure (dams, inter-basin transfers etc.) which will have transboundary impacts. Hence, in our view, there is nothing predetermined about the outcome of TWM processes – neither 'war' nor 'cooperation' – but rather we view it as 'malleable', measures taken being dependent on context and interests.

The rationale for developing this book on TWM stems from the Swedish International Development Cooperation Agency (Sida)-supported International Training Programmes (ITPs) on TWM, implemented jointly by SIWI and the international consultants Ramboll Natura, in collaboration with regional partners. The ITPs were aimed at mid-career professionals involved in the management of transboundary waters and represent sectors such as government (national, regional, local), river basin organizations (RBOs), NGOs, academia, the media and the private sector. Through the programmes, the participants were introduced to some of the core elements involved in TWM, challenges and emergent solutions, using a range of inputs including lectures, role-plays, individual projects, panel discussions and technical site visits.

While collating reading material for the programmes it soon became apparent that no single-volume work covered the range of topics associated with TWM. Many individual papers, journal articles, policy briefs and reports exist and, for the purposes of the programme, were collated into a body of reading for the participants. There are several books dealing with a specific element of TWM, such as negotiations, water law or conflict management, but none provided an overall picture. The current volume represents a partial solution to fulfilling this need. Partial in that it cannot hope to cover all the topics necessary to manage transboundary waters in any specific setting, thus the respective chapters provide an introduction and overview of their respective subjects and not an in-depth analysis. But also partial in the sense that the editors believe transboundary waters can and in most cases do act as catalysts of sustainable socioeconomic development.

The nature of the book is to some extent dictated by the intended audience – the people involved in managing transboundary waters. These are the same types of *practitioners* at whom the Sida ITPs were aimed and who need to refer to a book to give them input on specific challenges they may face in performing their duties. Most of these practitioners are skilled water managers of some sort, working in the public, private and non-profit sectors – what they may lack is the specific knowledge needed in the transboundary context. For them the section on *TWM Polity and Practice* should provide a practical overview of the skills, tools, and mechanisms used in managing transboundary waters of various types.

The second intended audience of the book is the broader research and *academic community*. In common with the practitioners mentioned above, they also come from a specific epistemic background and may lack an overview of the skills and knowledge needed to conduct research on transboundary issues. In addition, the chapters in the section on the *Analytical Approaches to TWM* are more theoretical in nature, providing the opportunity of developing the discourse further and contributing to solutions to TWM challenges. It is believed that the theoretical section would also be of use to the practitioners, as this would assist them in developing appropriate responses to the specific set of challenges encountered in their basin situation.

A group frequently neglected in TWM initiatives is that of the *politicians*. Arguably, politicians

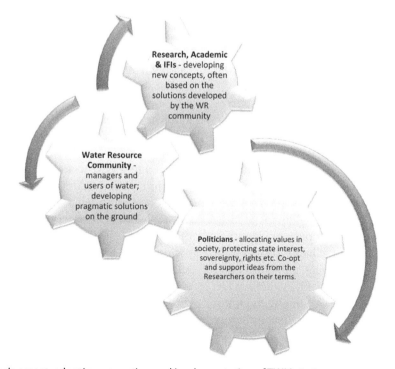

Figure 1.1   Development, adoption, co-option and implementation of TWM strategy

have the greatest impact on the outcome, successful or otherwise, of any TWM initiative. With ultimate responsibility for the 'allocation of values in society', they have great influence over water management domestically (Allan, 2001). Coupled with the responsibility to protect the sovereignty and rights of their state, they emerge collectively as the group with most influence over the direction, speed and quality of transboundary interactions over water (see Figure 1.1).

The three broad groups involved in TWM strategy can be seen as cogs turning in a system, with greater leverage being exerted by the largest cog. The *water resource community* – including water managers from government, as well as water users from the private sector and civil society, implements TWM strategies. They do this within the framework developed by the politicians, while also having a role in the formation of that framework. It is within this community that some of the innovative and pragmatic solutions to TWM challenges develop. In cases where the broader political environment does not allow politicians to enter into formal cooperative structures, it is often the water resource community which manages to transcend boundaries and cooperate for a greater good, often based on common professional understanding or water-use needs across political borders. Whether this is formalized by the political structures depends on the relations, at a political level, between the states concerned (Jägerskog, 2003).

There is a limit to what the water resources community can achieve – in the absence of the required political will it becomes difficult for the functional cooperation evident at this level to become more institutionalized. It often remains at the level where it is driven by and dependent on specific individuals. The broader *research community*, including academics, international financial institutions (IFIs) and development partners (donors), plays an active role developing theory and generalized practice aimed at explaining, influencing and improving TWM. Much of what they develop has its genesis in the action observed among the water resource community. The research community would like to see uptake of its ideas in the broader TWM strategy framework for specific basins or regions; indeed this may be the prime objective of

development partners. They believe that approaches such as benefit sharing, equitable share allocations, decentralized decision making and institutional cooperation can improve a given water management situation. Typically, they seek to introduce these ideas into the water resource community, hoping that by strengthening the desire to cooperate across international borders at this level some form of durable institutional structure is developed. In essence they are outsiders, observing and commenting on processes from a distance. Some members of this community enjoy better access to the other two communities. This is specifically the case for development partners, who often work closely with governmental entities. Their power to influence stems primarily from the fact that they support many of the TWM processes taking place in countries and regions.

The reality is that approaches developed by the research community will be adopted or co-opted by the politicians in accordance with the goals and pressures of the latter. It should be recognized that the political community, visualized in Figure 1.1 as a large cog, is not heterogeneous. There are domestic realities and pressures on a national scale, from within as well as outside of the water sector, in addition to power relations on the basin or international scale. For any given basin the rate and direction of change in TWM strategy will be highly influenced by the hegemonic power in that situation, with a range of cooperative and non- or less cooperative actions evident (Zeitoun and Mirumachi, 2008).

The 'set of cogs' metaphor is useful in visualizing the interactions between the various groups involved in TWM strategy and management as it recognizes that the direction as well as the speed of rotation of the cogs is linked. The inertia of the most powerful cog determines the rotation of the others. What it does not show is how individual communities interact directly with each other – if all the cogs were to engage with each other the system would jam. The reality is that there are direct linkages between the politicians and the research community, some formal and others less so. The key point to consider is that in the context of TWM strategy and management, the ideas and approaches emerging from the research community, as well as from the water resource community, are only implementable in so far as they receive support from the politicians. This approach places emphasis on the role of the state in TWM, driving the range of possible management options. A more effective way of developing and sharing approaches and solutions to TWM challenges between the three groups is needed.

Moving the Research Frontier

The reality of the emerging global water crisis (Gleick, 2009) is gradually pushing the broader water sector – water resource community, research community and politicians – to seek more efficient methods for managing available water, as well as ways to access other sources sustainably. These trends point towards the demand for improved transboundary governance. First, as we saw above, research and knowledge production, water policy and management, and political spheres do not interact optimally. Second, historically, utilization of transboundary waters – and especially in the light of increasing water scarcity – has been associated with diverging interests, conflict, violence and, as it was commonly predicted in the 1990s, war (see Philips et al, 2006).

However, what is now well established through the groundbreaking research led by Oregon State University (Wolf et al, 2003b), 'war' is unusual as a result of water rivalry (only), and although 'conflict' – the pursuance of incompatible interests – is common over any scarce resource, cooperation is also common, to the extent that it constitutes the dominant pattern at international transboundary scales. What is less commonly observed is that cooperation as it typically occurs in these cases is shallow, reluctant and piecemeal: it may serve to avoid violent conflicts, but it does not markedly assist development efforts, or even serve to guarantee long-term sustainability of the tentative agreements. Cooperation will tend to be on the terms of the hegemon in the basin, with approaches and initiatives co-opted by them as needed. The lack of a more nuanced understanding of cooperation stems from the under-theorized nature of cooperation as a concept (Axelrod and

Keohane, 1985; Oye, 1985).

Put differently, the fact is that 'cooperation' is common and necessary, but insufficient for dealing with accelerating water crises and the underlying development demands. Crucially, in international systems, interests, power and hegemony are structural features that cannot be wished away (Warner and Zeitoun, 2008). How should this cooperation be deepened and subsequently transformed into actions that are conducive to development? We have chosen to approach the field with the view that transboundary waters can be properly managed, leading to peace and sustainable development, only if a number of conditions are present, and if good practices on various levels in the system are applied. This book seeks to illuminate systematically these 'ifs'.

TWM carries some specific difficulties, which in combination constitute a field that needs to be comprehensively addressed. These include:

- scale (emanating from the size of a basin-wide approach);
- fragmentation (from its division into several judicial systems);
- complexity (from the range of cross-sectoral issues in large-scale water management);
- interests (vested in the basin states).

These combine to make up a formidable field of issues that need to be addressed. Overall, this volume serves to display the key issues involved with transboundary management and review the best-practice responses to the dilemmas at hand, while it draws on, and synthesizes from, these experiences.

This volume will address these issues in three sections. First, the section on *Analytical Approaches to TWM* will present three analytical perspectives illustrating the difficulties and/or the possibilities of enhancing the effectiveness and efficiency of TWM. This section aims to explain why wars over water are unlikely, explain some of the interactions between states from a hydropolitical perspective and describe how expanding the focus of negotiations to include benefits associated with a water system can ease possible disputes over shared waters.

The second section on *TWM Polity and Practice*, forming the core of the volume, presents an experience-based inventory of issues typically associated with the governance of transboundary waters. These chapters form a link between the theoretical approaches introduced in the first section and their further refinement through empirical observation and the actions associated with their implementation. All these explicitly include a 'user-friendly' section on 'good practices' as well as providing an overview of the key challenges faced.

Third, in its concluding section on *Challenges and Opportunities*, it will draw on both the analytical and the empirical reviews in this volume and commence a construction of an analytical framework for transboundary waters. This will be complemented by forward-looking chapters on facing environmental change, and the role of science and education in building capacity to manage effectively the range of changes creating pressure in transboundary basins.

The Chapters

As described above, much of the content of this book originated from the Sida ITPs on TWM. These training programmes sought to introduce participants to the range of management approaches needed to manage shared waters. The chapters chosen for this book provide an overview of the various types of issues which would be encountered when managing the resource, covering topics which are theoretical as well as those of a more practical nature. The authors have discussed the importance of their topics to TWM, sketched the theoretical background, provided examples of good practice and reflected on possible future developments in their conclusions. Included as a chapter is a set of 14 short case studies, illustrating specific elements of TWM as outlined in the chapters. These cases were chosen to complement the issues discussed by the authors in their chapters and explicit links are made – as well as cross-references between the various chapters. A sequential overview of the chapters in the book follows.

Analytical approaches to TWM

Anthony Allan and *Naho Mirumachi* opens the theoretical debate by debunking the premise of 'water wars', instead arguing that the ability of a state to tap into the international markets for water-intensive products allows it to ignore increasing water scarcity. He demonstrates how flows of virtual water alleviate physical water scarcity and in so doing allow politicians to maintain technically sub-optimal water management solutions because they are politically inexpensive. The predictors of the degree of water security a state can achieve are data on socioeconomic development, and not on hydrology. An economically diversified economy can thus trade its way to water security. Furthermore they describe how the capacity of a state to project power on its neighbours or other basin states determines the degree and type of cooperation in a specific context. Precisely because the management of transboundary waters falls into the realm of international relations, it becomes difficult for outsiders – such as researchers, development partners and financiers (as well as technical insiders such as water managers) – to know fully and to influence water management decisions. They end contentiously by concluding that this makes the topic of transboundary water management 'un-researchable' – precisely because so much of what goes on is done in secret.

The chapter by *Ana Elisa Cascão* and *Mark Zeitoun* focuses on issues related to power and TWM. Taking as their point of the departure the old political science question of 'who gets what, when, where and why?' (Lasswell, 1936) they encourage the reader to take a fresh look at TWM practices through the 'power lens'. The authors argue that power asymmetries in transboundary water settings are a more important determinant of outcomes in terms of water allocations then mere riparian position. Indeed, power in all its forms – from the more obvious economic and military to the less obvious related to ability to control discourse and expertise – is argued to be determining for how cooperation (including its quality) is developed. Further, the authors also argue that not all cooperation is good – sometimes cooperation is 'domination dressed up as cooperation' (Selby, 2003) and not what it may appear to be at first glance. The chapter is based on theoretical research as well as interaction with practitioners.

In an attempt to extend our thinking beyond the basin, *Marwa Daoudy* outlines the challenges associated with developing a 'benefit-sharing' approach to TWM. The author identifies three key dilemmas underlying the debate on conflict and cooperation on international watercourses. First is that optimal water-usage solutions may not be congruent with the principle of equitable utilization; a technically optimal solution might not be viewed by the parties as politically feasible. The second dilemma is the perennial belief that common pool resources, such as international watercourses, are inevitably destined for environmental degradation. This is potentially exacerbated by the various power asymmetries present in all such basins. The third flows as a consequence of the increasing levels of water stress prevalent in almost all transboundary watercourses today. States perceive this as a threat to their national interests and respond accordingly, placing water management decisions firmly in the realm of security issues. By shifting the focus to look at a range of benefits associated with watercourses it just may be possible to achieve positive-sum outcomes for all parties concerned.

TWM polity and practice

Opening the more practically oriented section of the book, in Chapter 5 *Owen McIntyre* focuses on the development of international water law and its central features. It provides an overview of the 150 years of development of the body of rules that is supposed to guide transboundary waters. This includes in-depth discussion of the early Harmon Doctrine and moves all the way through the Helsinki Rules on to the UN convention on the Law of the Non-navigational Uses of International Watercourses and the recent addition of the draft articles for a UN convention on transboundary aquifers from 2008. It usefully discusses various transboundary cases from a legal perspective and considers how states have used the international water law rather diversely in order to

support their respective practices. It also discusses how certain principles are more popular with upstream states while others are more fashionable in downstream states. Recent inclusions into the body of international water law, including concepts such as the ecosystems approach and water as a human right are also discussed.

Shifting our attention underground, *Shammy Puri* and *Wilhelm Struckmeier* focus on the management of transboundary aquifers. Taking a clear practitioner's perspective the authors present the reader with a set of tools to use the 'hidden resource' that aquifer water represents. Laying out the basics of aquifers (discussing issues such as inflow, throughflow, storage, abstraction and replenishment) combined with a global overview of transboundary aquifers, the authors move on to discuss current policy practices and policy options for transboundary aquifer management. Included is a discussion of the usefulness of the UN International Law Commission's (ILC) draft articles on the Use of Transboundary Aquifers for improved management of transboundary aquifers as well the need for increased monitoring and mapping of existing aquifer resources.

A theoretical discussion is followed by practical examples of stakeholder participation in TWM processes in the chapter by *Nicole Kranz* and *Eric Mostert*. The authors highlight the crucial role that stakeholders play in management activities, and stress the need to bring the appropriate stakeholders into the process as early as possible, recognizing that this is made more complex in the transboundary environment. Five key questions are discussed in the theoretical component:

1   Should participation be conducted as an add-on or as an integral part of planning processes?
2   Which stakeholders should be involved and how should their involvement be stimulated?
3   At what level or levels should they be involved: the river basin, the national or the local level?
4   Who should initiate the participatory process: an international commission, a national institution or perhaps one of the stakeholders themselves?
5   Which participation methods should be used?

In an effort to illustrate and map responses to the above questions the authors draw on experiences gained from work in three international transboundary basins – the Danube, the Okavango and the Dniester.

The backbone upon which all human activity relies is the integrity of ecosystems – no less so in the world of freshwater resources. These ecosystems play a vital role in cleaning water which passes through them, providing products for sustainable livelihoods as well as essential services such as flood protection and navigation. In their chapter *Cate Brown* and *Jackie King* introduce the concept of environmental flows for ecosystem maintenance, how they can be assessed, and their perceived role in the development and management of transboundary river basins. Increasingly the environment is viewed as a 'legitimate' user of water in its own right, but in situations of competition over scarce supplies it becomes important for parties to agree on a common methodology for assessing this need. In recognition that this is an essentially political process the authors introduce the belief that freshwater ecosystems can be managed at different levels of condition, from pristine, when they provide a range of natural services of benefit to humans; through various stages of change from pristine, when the original services disappear and others appear; to serious degradation, when virtually all natural services essentially disappear.

Striving for consensus, *Todd Jarvis* and *Aaron Wolf* focus on water negotiations and various techniques and practices for dispute resolutions in transboundary water settings. Taking stock of the fact that water management is by definition a way of managing conflict, the chapter leads the reader through conceptual discussions on water and conflict, the causes of conflicts over water and the interplay between the technical and political levels of water management. Based on those discussions the chapter moves on to discuss the role of experts in water dispute resolution and joint learning in transforming a water conflict. The authors develop a matrix combining stages of water conflict with a model for various stages of conflict transformation. They conclude on a positive note, arguing that although the world has seen (and sees)

many water conflicts, most of the outcomes have been predominantly on the cooperative side. They hold that riparians have shown both resilience and creativity, through employing, for example, preventive diplomacy and the benefit-sharing approach (allowing a move from zero-sum to positive-sum), to arrive at cooperative outcomes.

Defining (TWM) business models as 'frameworks for creating economic, social and environmental value in a transboundary watercourse context', *Jakob Granit* presents a functionalist overview of the various designs available for the architecture of TWM institutions. Different approaches to management present a range of costs and benefits, with advantages and disadvantages associated with pursuing unilateral or cooperative actions. The author identifies three drivers of collaborative water management institutions: the water supply gap (communities, countries, basins or regions running out of water resources to support socioeconomic development); the water quality gap (most notably the low quality of or destruction of many aquatic environments globally); and the current under-investment in the entire water value chain (investments needed in the water resource management as well as water supply sectors to address the above gaps). Transboundary waters become an avenue for promoting wider development and regional integration objectives, with several regional economic communities now playing an active role in TWM. By drawing on the language used by business management strategists, the author conveys the point that good TWM is not an end in itself – instead cooperation should be aimed at the achievement of a mutually beneficial outcome, however defined by the populations of the watercourse states concerned.

Challenges and opportunities

Legal agreements between states form the bedrock of cooperative water management internationally. In the first chapter of the section investigating challenges and opportunities associated with TWM, *Malin Falkenmark* and *Anders Jägerskog* address the need for water agreements to be resilient to global change. Such change, whether climatic, demographic or developmental, places extra pressure on water resources, thus increasing the need for effective institutions to promote cooperative management. If agreements do not possess the needed flexibility or durability to respond to pressure there is a real risk of conflict between parties. The authors start by analysing the implications of developing forward-looking agreements at the physical basin scale. This views the watercourse as a unitary whole which needs to be managed through a cohesive process. When allocating water shares between countries it is important to work with relative, rather than absolute shares, allowing the flexibility to adjust volumes during drought years. Next they extend our thinking towards looking at how such agreements could in the future include more elements of benefit sharing between parties – maximizing the opportunities associated with the resource. To this end they discuss mechanisms such as paying compensation to a country impacted negatively by a water use or by identifying and sharing benefits associated with watercourses within agreements.

TWM is knowledge- as well as data-intensive, leading *Léna Salamé* and *Pieter van der Zaag* to conclude that the commitment of financial resources and political will alone is not enough to ensure effective outcomes. The authors argue that there is a distinct role for the development of knowledge and capacity to strengthen TWM institutions, irrespective of the water body being managed. Institutional capacity is a vital ingredient for developing the resilience needed to face global change or deterioration in relations between states. The authors contend that both the specialists (hydrologists, engineers, economists etc.) as well as the generalists involved in TWM tend to focus on 'things', instead of on interests and people. The development of knowledge and capacity among water managers and other stakeholders can play a role in promoting good relations between countries, as it helps in overcoming the language, cultural, religious or political divides associated with transboundary basins. Thus the process of developing capacity is as important as the type of knowledge introduced. To illustrate the role that knowledge and data play in TWM the authors analyse four cases – two rivers, an aquifer and a lake, distilling the key factors which contributed to an improved management outcome.

The bulk of this book is aimed at developing a conceptual framework and describing good TWM policy. Authors have been encouraged to keep their chapters generic enough for readers to apply the concepts introduced to their own situation – no matter the differences in climate, economic development or political system. To some degree this has meant compromising the degree of detail that authors could incorporate in their respective chapters – the emphasis being placed on introducing generically applicable concepts. Recognizing that no two watercourses or TWM situations are exactly the same, the editors have included a chapter devoted to case studies of TWM processes. Some of the cases have been contributed by chapter authors and several of them are based on work first developed by Aaron Wolf and Joshua Newton, but all 14 cases have been edited by *Virginia Hooper* and *Michael McWilliams* into a standardized format – short, pithy and presenting the key lessons learned. The cases were chosen so as to represent a range of issues (based on those raised in chapters in the book), geographical areas, climates and types of water bodies. Most of them display examples of good practices – however several are useful despite being unsuccessful TWM processes; in most cases the judgement is left to the reader.

Finally, the editors conclude the book by discussing some of the key emergent themes identified in the preceding chapters. In part this chapter is a retrospective overview of some of the main debates in TWM over the past two decades, showing how the understanding of the field has grown more nuanced and complex – how it is maturing. However, there is still some way to go before several paradoxes can be resolved. In the final chapter *Joakim Öjendal, Anton Earle* and *Anders Jägerskog* question several of the core assumptions of TWM and specifically critique the exclusive implementation of the hydrographic management unit approach to the transboundary context, in essence arguing that scale matters. Naturally as various strands are woven together and others are broken what emerges is a set of new questions, leading into avenues not necessarily covered by this volume. It is hoped that these questions inspire the research community to keep pushing the state of knowledge in this young and dynamic field.

References

Allan, J. A. (1998a) 'Virtual water: an essential element in stabilizing the political economies of the Middle East', *Yale University Forestry & Environmental Studies Bulletin*, vol 103, pp141–149

Allan, J. A. (1998b) 'Virtual water: a strategic resource: global solutions to regional deficits', *Ground Water*, vol 36, no 4, pp545–546

Allan, J. A. (1999) 'Avoiding war over natural resources', in S. Fleming (ed.) *War and Water*, ICRC Publication Division, Geneva

Allan, J. A. (2000) *The Middle East Water Question: Hydropolitics and the Global Economy*, I. B. Tauris, London

Allan, J. A. (2001) *The Middle East Water Question: Hydropolitics and the Global Economy*, I. B. Tauris, London

Allan, J. A. (2002) 'Water resources in semi-arid regions: real deficits and economically invisible and politically silent solutions', in A. R. Turton and R. Henwood (eds) *Hydropolitics in the Developing World: A Southern African Perspective*, African Water Issues Research Unit (AWIRU), Pretoria, South Africa, pp 23–36

Axelrod, R. and Keohane, R. (1985) 'Achieving cooperation under anarchy: strategies and institutions', in K. A. Oye (ed.) *Cooperation Under Anarchy*, Princeton University Press, Princeton, NJ

Carius, A., Dabelko, G. D. and Wolf, A. T. (2004) 'Water, conflict, and cooperation', background paper for the UN Global Security Initiative expert workshop in cooperation with the Environmental Change & Security Project at the Woodrow Wilson Center for International Scholars, 2 June 2004, Washington, DC

Galtung, J. (1996) *Peace by Peaceful Means: Peace and Conflict, Development and Civilisation*, PRIO, Oslo, Norway

Gleick, P. (2009) *The World's Water 2008–2009: The Biennial Report on Freshwater Resources*, Pacific Institute, San Francisco, CA

Jägerskog, A. (2003) *Why States Cooperate over Shared Water: The Water Negotiations in the Jordan River Basin*, Department of Water and Environmental Studies, Linköping University, Linköping, Sweden

Kalpakian, J. (2004) *Identity, Conflict and Cooperation in International River Systems*, Ashgate, Aldershot, UK

Lasswell, H. D. (1936) *Politics: Who Gets What, When, How*, McGraw-Hill, New York, NY

Oye, K. A. (ed.) (1985) *Cooperation Under Anarchy*, Princeton University Press, Princeton, NJ

Phillips, D. J. H., Daoudy, M., Öjendal, J., Turton, A. and McCaffrey, S. (2006) *Trans-boundary Water Cooperation as a Tool for Conflict Prevention and for Broader Benefit-sharing*, Ministry for Foreign Affairs, Stockholm, Sweden

Selby, J. (2003) 'Dressing up domination as "co-operation": the case of Israeli–Palestinian water relations', *Review of International Studies*, vol 29, no 1, pp121–138

Turton, A. R. (2003) 'The hydropolitical dynamics of cooperation in Southern Africa: a strategic perspective on institutional development in international river basins', in A. R. Turton, P. Ashton and T. E. Cloete (eds) *Transboundary rivers, sovereignty, and development: Hydropolitical drivers in the Okavango River Basin*, AWIRU and Green Cross International, Pretoria and Geneva, pp83–103

Turton, A. R. and Earle, A. (2005) 'Post-apartheid institutional development in selected Southern African international river basins', in C. Gopalakrishnan, C. Tortajada and A. K. Biswas (eds) *Water Resources Management – Structure, Evolution and Performance of Water Institutions*, Springer, Heidelberg, pp154–168

UNEP (2005) *Hydropolitical Vulnerability and Resilience along International Waters: Africa*, UNEP, Nairobi, Kenya

Warner, J. F. and Zeitoun, M. (2008) 'International relations theory and water do mix: a response to Furlong's troubled waters, hydro-hegemony and international relations', *Political Geography*, vol 27, pp802–810

Wolf, A. T. (1998) 'Conflict and cooperation along international waterways', *Water Policy*, vol 1, no 2, pp51–65

Wolf, A. T., Stahl, K. and Macomber, M. F. (2003a) 'Conflict and cooperation within international river basins: the importance of institutional capacity', *Water Resources Update*, vol 125, pp31–40

Wolf, A. T., Yoffe, S., and Giordano, M. (2003b) 'International waters: identifying basins at risk', *Water Policy*, vol 5, no 1, pp29–60

Wolf, A. T., Kramer, A., Carius, A. and Dabelko, G. D. (2005) 'Managing water conflict and cooperation', in *State Of The World: Redefining Global Security*, World Watch Institute, Washington, DC

Zeitoun, M. and Mirumachi, N. (2008) 'Transboundary water interaction I: reconsidering conflict and cooperation', *Journal of International Environmental Agreements*, vol 8, pp297–316

PART I

ANALYTICAL APPROACHES TO
TRANSBOUNDARY WATER MANAGEMENT

2

Why Negotiate? Asymmetric Endowments, Asymmetric Power and the Invisible Nexus of Water, Trade and Power That Brings Apparent Water Security

J. A. (Tony) Allan and Naho Mirumachi

- International relations over transboundary waters take place in a context of conflict and cooperation. *Conflict and cooperation coexist.*
- Conflict over transboundary waters is normal. There are four levels of conflict: *non-politicized, politicized, securitized and violized.*
- Violent conflict can occur at sub-national levels *but there has been no international armed (violent) conflict over transboundary waters since the mid-1960s.* International trade in water-intensive food commodities mitigates such conflict very effectively.
- As transboundary water debates become more intense and politicized, ministries of foreign affairs take on responsibility for the topic. The data, information and analysis about the status of the water resources and their management become their concern. Their priority is sovereignty.
- At the point at which transboundary water relations become securitized, responsibility for transboundary water affairs passes to the 'shadow state' and its security services. These relations progressively disappear from the public domain as the intensity of conflict increases.
- Analysts need to be aware of: (1) the conflict-reducing impacts of virtual water 'trade' which is economically invisible and politically silent; (2) the challenges of researching the quiet securitized processes at the level of shadow states.

The need to keep politically hazardous conditions invisible normally overwhelms the option of recognizing underlying fundamentals.

Introduction

The purpose of this chapter is to show that there are international processes which very effectively reduce conflict over transboundary water resources. It is a precautionary message, in that it draws the attention of water scientists, water professionals and those analysing international transboundary water relations, to the importance of invisible and hard-to-research issues that should not be ignored. These processes are *trade* and the *economic diversification* of water-scarce economies. Economic diversification and trade enable them to be water secure by accessing volumes of water in water-intensive commodities in global markets that can never be available via negotiation. The chapter also identifies the conditions that make the *operationalization* of integrated management and the *implementation of cooperative basin initiatives* very difficult and sometimes impossible. The first of these conditions is the power relations of the riparians in a river or aquifer basin. These relations are usually asymmetric. The second condition is – again – the existence of *international trade* in water-intensive commodities, especially the 'low-value' water embedded in food commodities.

It is usual to take three issues into account when analysing transboundary water relations:

1   The *relative endowments of freshwater* are always considered, but soil water, which can be the much greater water endowment compared with that of surface and groundwater is not.
2   It is usual to examine the different levels of riparian conflict, examining how some riparians *project power* and can *securitize* access to shared water resources, and how others can only deploy limited power to advance their interests.
3   There are numerous reviews of the evolution of international customary water law, and of *cooperative approaches* to the development of river basin institutions, which examine whether international customary water law has been observed.

Deep and significant engagements over international transboundary waters are invisible and take place out of the public realm. In addition the water-insecure achieve a version of water security by engaging in commodity trade, the benefits of which are also economically invisible and politically silent. As a consequence, the subordinate players who contribute knowledge on underlying fundamentals or nurture cooperation and basin initiatives play very minor roles.

The chapter concludes with a short critique of the difficulties of researching and analysing transboundary international water resource relations. It cautions that in most cases – and always in contexts where the riparians have non-diverse and weak economies – researchers, water resource professionals, donor water resources specialists and international water lawyers do not have access to the major decision-making forums where significant engagement could or does take place.

Contexts and Points of Departure

Politicized and securitized relations over transboundary water disappear first into ministries of foreign affairs and then into what has become known as the shadow state.

The usual ways to research and analyse transboundary water relations involve:

1   The examination of *relative riparian endowments* of surface freshwaters. Freshwater endowments have proved to be a very partial and incomplete definition of riparian water endowments. It is also unfortunate that transboundary water relations started their history in disputes over navigation. Contemporary high politics over shared transboundary waters tend to be over freshwater – increasingly known as *blue water* – abstractions and rights to abstract. Transboundary water studies have progressed in the past three decades from focusing on the water resources in the water-

course to embracing the notion that all the surface and groundwater resources of an international river basin should be the relevant focus of analysis (McCaffrey, 2007; Cullet, 2009; Cullet et al, 2010). But whether the *soil or green water* in the basin should be considered part of the water endowment is certainly not yet assumed or agreed. It will take some time for *green water* to be seen to be politically acceptable as part of the calculus of international transboundary relations.

The idea that the levels of economic diversification of the riparian partners and their capacities to trade their way to water security has made its way – albeit awkwardly – on to the transboundary waters discourse a decade or so ago. But nowhere is the political economy of water resources taken as a point of departure. There are examples in this publication of the first approach in the chapters focusing on underlying hydrological fundamentals of groundwater and minimum flows. Most of the case studies also make reference to underlying fundamentals. As usual soil water is a neglected issue.

2   It is usual to examine the different levels of riparian conflict, looking at how some riparians *project power* and can *securitize* access to shared water resources, and how others can only deploy limited power to advance their interests and respond to such securitization. The second issue, conflict and the projection of power, figures in the chapters in this volume on conflict and the evidence of power projection.

3   It is common to look at *cooperative approaches* and the development of river basin institutions. An important element of the cooperative approach is advocacy of the introduction of, and operation of, *international water law* and in its absence the observance of *international customary water law* (McCaffrey, 2007; McIntyre, this book Chapter 5). This book is rich in material on this third issue, with many chapters on law, cooperative processes and institutions.

The purpose of the chapter will be to highlight the international processes that are very much more

effective in the reduction of conflict over shared waters than customary international law and river basin cooperation and institution building. These processes are *trade* and *economic diversification*. Together they enable economies to trade so that they can access and pay for water-intensive commodities in global markets. Water security for almost all of 210 or so economies of the world will be determined by the way water is moved from water-surplus regions to water-scarce regions – via trade. These processes also help the water scarce to have the option to protect the environmental services of water.

The analysis will also highlight that there are *two conditions* that make the *research and analysis* of transboundary water relations especially challenging. These conditions also make the *operationalization of integrated management* and the *implementation of cooperative basin initiatives* very difficult and frequently impossible. The first condition is the *power relations* of the riparians in a river or aquifer basin. These relations are usually asymmetric; often extremely asymmetric. The second is the inaccessibility of international transboundary water issues as soon as they become politicized. Politicized and securitized relations over transboundary water disappear first into ministries of foreign affairs and then into what has become known as the *shadow state* (Tripp, 2007).

Political and Political Economy Landscapes That Determine Transboundary Water Relations and the Challenges of Researching Them

Riparians enjoy different water resource endowments, co-evolve asymmetric riparian power relations and may develop diversified economies. The greatest of these – and the condition that determines water security – is the possession of a diversified economy.

Investing effectively in human resources has proved to be the best way to develop a diversified economy and in turn achieve water security at all levels of social organization and especially at the national level.

There are numerous dangers in focusing on the river basin in any analysis of the water security of individual riparians. It is a little-recognized truth that the extent to which a riparian has to share a surface or groundwater resource with other riparians is almost always a minor source of explanation of the degree to which it is water secure.

Water resource endowments are significant but not determining. Power relations with other riparians are usually more significant but again they are not determining. It is the possession of a diversified and strong economy that determines water security. Of the three conditions enjoyed by a riparian – first, its water resource endowments; second, its relative power *vis-à-vis* other riparians; and third, the possession of a diversified economy – the greatest of them is the possession of a diversified economy.

Those who focus on water resource endowments and the way they are used by riparians have too narrow a focus. In addition, those who expect legal principles to be a sound basis for sharing international waters reach for tools that have proved to be wanting as a principled basis for the achievement of water security. The sound and enlightened principles such as *reasonable and equitable use* as well as the awkward ones shaped by *sovereignty* or *prior use* have proved to be inadequate. Often these apparently obvious ways of engaging with the problem of sharing water resources play a minor role in the actual achievement of strategic water security. Major water scarcity problems are usually very effectively solved outside the river basin, the water sector and the realm of international law (Allan, 2000) by *economic diversification* and *international trade*. The experience of Singapore is an extreme example of these non-intuitive solutions.

Singapore has only about 5 per cent of the water it needs for water self-sufficiency – that is for its food security based on local water, for the water needed by its very diverse industrial sector and for its domestic water use. Singapore's industrial and domestic uses have always needed every drop of local freshwater. Half of these freshwater needs have had to be imported from Malaysia. In this decade they are being partly provided by desalination as the costs of desalination have fallen to levels as low as those for treating and importing water across the straits from the peninsula of Malaya. The Singapore case is useful because it does not have the complicating element of transboundary waters. It has no transboundary waters and scarcely any water resources of its own.

Singapore has no green water and no freshwater to spare for food production. Its only option to achieve food security for the current population of 4.5 million is to buy food produced in regions that have surplus soil water, or spare surface and groundwaters. Its future population of 6 million can only be food and water secure on the basis of even higher levels of food imports in the future. Singapore's experience shows that a national economy can be water secure by 'accessing' virtual water in food from the global economy. Singapore has very successfully avoided conflictual international transboundary water politics. It has proved that fruitless and tortured international hydro-politics need not concern advanced industrial economies even in extremely water-scarce regions (Long, 2002). So successful is the Singapore political economy in handling water resource investment as well as in allocating and managing its tiny water resources that the world beats its way to its door to attend its annual international water conference (Low, 1998; Soon, 2009). The almost waterless Singapore is a global hydro-hub that makes part of its living out of exporting water-managing and water-treating technologies, water science and radical water policy approaches. It has certainly made the most of water scarcity (World Bank, 2007).

The secret of Singapore's success in dealing with water scarcity is its political leadership. Water security has been one of the strategic issues addressed by Singapore's leadership since independence in the 1960s. Policies have evolved to address water allocation and management in a sublime environment of enterprise-friendly public sector management (Low, 1998, Lee, 2000; Tongzon, 2002). Its public investment decisions have been inspired. At the same time a long-term treaty was made with Malaysia to pipe water from Johor across the straits. Singapore invested in water treatment plants in Johor that served both Singapore and Johor with high-quality water for domestic use and industry. Water relations have been very effectively desecuritized (Long, 2002).

Even more important has been the priority given to enhancing the skills of its workforce. Investing heavily in human resources has proved to be a spectacularly successful way to ensure water security. Singapore's leadership focused investment on improving skills in engineering, science, finance, health services, administration, business and marketing (Lee, 2000). As a consequence Singapore is water secure because its increasingly skilled human resources accelerated the diversification of the economy.

Singapore will always be sustainably water secure in a world at peace. Skilled human resources generate goods and services which the world needs. Revenues from the sale of these goods and services enable Singapore to trade its way to water security. It accesses virtual water embedded in water-intensive food commodities and in other goods in the global economy. Singapore's well-ordered society is also of crucial importance. A well-ordered society can generate tax revenues which enable effective regulation and secure circumstances for local enterprising commerce which in turn engages effectively in global trade.

Diversified, but water-deficit, economies everywhere – 26 out of the 27 EU economies,[1] a number of East Asian economies – Japan, Korea, Taiwan and Hong Kong before it was integrated into China – have been well able to be food and water secure and to enjoy low-intensity hydropolitics in the peaceful late 20th century. This security was made easy by international 'trade' of the big volumes of virtual water needed to meet regional food deficits. Global systems can also meet the water needs of the world's economies, although there is no space to develop the argument here (Allan, 2010).

Trade in food and livestock commodities was made very attractive to the water and food scarce – albeit at the same time as being very damaging for many weak economies in Africa – as a consequence of the low food commodity prices that have prevailed for the past four decades. Low food commodity prices have been especially important in very virtual water 'trade'-dependent regions

such as the Middle East. Many non-diversified economies, such as those of the Middle East, have also benefited greatly in the last quarter of the 20th century from reductions in international transboundary water tension as a consequence of the low food commodity prices. Importing half-cost food because of production and trade subsidies in the EU and the USA greatly eased budgetary stress in all food-importing economies. Full-cost food imports would have fiercely concentrated the minds of those negotiating transboundary waters. High global food prices would on the other hand have also greatly intensified the very uncomfortable politics of reforming water use and allocation within national economies, as happened during the commodity price spikes of late 2008 (Rice, 2009). They would also have hugely intensified the hydropolitics of transboundary relations in all economies except those with very high levels of GDP per head.

The *capacity to trade* internationally and the *price levels of food commodities* influence the intensity of riparian relations over the big volumes of water needed for food security. It is water technology, however, that has had the major impact on water security for the small volumes of water needed for domestic and industrial uses. The challenge of achieving water security for these *small volumes of water* – about 10 per cent of total water needed by an individual or economy – has been very much reduced by the fall in desalination costs in the late 1990s. Singapore will always be *small water secure* because it is surrounded by seawater. Seventy per cent of the world's population also lives beside the sea or beside a major river. These societies will always be able to meet their *small water* needs – that is for domestic uses and for industry and services – by manufacturing freshwater from seawater and from other local brackish water.

The technological solution to Singapore's *small water* needs is intuitive and easily grasped both by politicians, policy makers and society. On the other hand, the spectacular success of investing in human resources to diversify an economy to secure all its water needs – *big and small* – for food, industry and domestic uses – is non-intuitive for most voter–consumers and for most policy makers. Water users and politicians expect to achieve water security by investing in the water sector. Singapore has invested extremely effectively in its tiny water resources by protecting and optimizing their use. Its approach to managing water in this domain is better than any other political economy in the world. It has also developed and invested in technology to augment its other freshwater needs by importing water from Malaysia. The 90 per cent of its water needs to grow its food has been 'imported' as virtual water.

It is evident that Singapore's water security has been achieved outside the water sector. It is an extreme case. But all economies that achieve their food security by 'importing' virtual water – that is about 190 out of 210 world economies – gain this form of water security partly by using their limited available water. But their strategic water security is achieved via international trade which has proved to be an extremely flexible and effective remedy for water scarcity.

This dependence on international trade to achieve water security is normal. Most economies are net food importers. About 20 of the world's economies enable the other 190 to be food and water secure. By contrast the number of economies in the world that achieve water security by negotiating transboundary water entitlements is extremely small. Even where a powerful riparian captures almost all the shareable water, actual hydro-security is achieved in the world's distorted markets rather than in fraught transboundary hydropolitics. Egypt has 'captured' three-quarters of the Nile flow. But even if it could negotiate to have all of the flow it would still need more water. There is not enough water in the Nile system to meet Egypt's present needs, never mind those of a future doubled population. It is an illusion – albeit an unshiftable illusion – that Egypt's water security lies in the $10km^3$ of water of *negotiable* annual Nile flow. It is already '*importing*' $40km^3$ of virtual water annually and it will need at least twice as much as this to meet its future food needs. This uncomfortable fact is understandably kept out of the public discourse.

The most important feature of the *political economy* explanation of the nature and level of hydro-conflict in international transboundary relations is the *invisibility* of the political economy processes. There are a number of reasons for their invisibility:

1   The role of the political economy in achieving water security is invisible and non-intuitive. In addition people do not know how much water is needed to raise their food or that of the rest of the nation. Neither do they realize that 90 per cent of the daily water use is for their food, nor do they have any notion of how much of their food is imported.

2   They are also blind to the existence of the massive volumes of *green water* – that is the water in soil profiles used in rainfed crop production – that accounts for 70 per cent of global food production. This green water also accounts for 90 per cent of the *virtual water* 'traded' internationally that ensures worldwide local water security for the water scarce.

3   The virtual water embedded in traded commodities is also *invisible*. The concept of virtual water has only been in scientific currency for about 17 years, and in international and national political discourse for about 5 (Hoekstra and Chapagain, 2008; Allan, 2010). Future transboundary hydropolitics that take into account the political economy of water as well as the role of virtual water will operate differently from current transboundary international relations.

All these underlying invisibles are very unwelcome for societies that have accommodated unwittingly to dependence on food imports. They are even more worrying for political leaders supposedly responsible for a nation's water security, who pretend that this lies in negotiating allocations with neighbouring riparians. Drawing attention to the linked insecurities of food and water dependence threatens the stability of a state. Society and political élites easily align to reject the scary message that the invisibles reveal. These illusions – often constructed over millennia – have so far proved to be very durable. As already noted, long-held beliefs weave the illusion that national water security is dependent on locally accessible water resources, and shared transboundary waters are easily kept in place.

It is more politically feasible for national leaders to appear to engage in hydropolitical contention than to reveal the extent of the national dependence on imported food. In practice the hydropolitical contention is usually fruitless while the invisible remedy of a diversified economy enabling food imports is comprehensively effective. Paradoxically the invisibility of the development/trade remedy makes it possible to appear to be in legitimate contention over the shared waters. The economic security (including water security) enabled by the development and trade remedy allows the strong and the weak riparians to enjoy having the political space to contend, and simultaneously cooperate, over shared waters (Mirumachi, 2007; Mirumachi and Warner, 2008). They can promote their different preferred principles – sovereignty, integrity, and reasonable and equitable use – for decades, because the problem of water security is being invisibly addressed.

Power Asymmetries in Transboundary Relations: Another Invisible

The analysis so far has demonstrated that very effective solutions to water security operate outside the water sector. These solutions operate invisibly in the individual riparian economies and in the global economy.

In addition to these economic processes there are also a number of political conditions that determine the terms of engagement of riparians over shared transboundary waters. Riparians are all members of the international systems of nations. They are supposed to engage as equals, but in practice *power relations are asymmetric* in all river basins. There is always a hegemon. In river basins where all the economies have advanced industrialized economies these riparians can engage as equals. They do not even need a legal framework to be able to adopt outcomes that are equitable and reasonable.

For example, in the river basins of Europe the riparians on the Rhine – Switzerland, France, Germany, The Netherlands and Belgium and the international corporations that utilize the Rhine

waters – have cooperated to remedy pollution and protect the resource. International water law would have accelerated the improved joint management but its absence has not prevented the introduction of improved water quality standards and other environmentally appropriate measures. EU economies and many other industrialized economies have put in place environmental regulation, including the regulation of international transboundary waters that is many times more binding and intrusive than those evolving in the UN system.

Figure 2.1 illustrates the TWINS concept and shows how the transboundary waters interaction nexus for some different basins can be compared. The Jordan is characterized by protracted high conflict and low cooperation. The Ganges by relatively low conflict and relatively low cooperation. The EU basins and those of North America have experienced low conflict and high cooperation. Advanced economies with high GDP levels appear to be able to enjoy high levels of cooperation. An individual economy, such as that of Singapore, can avoid both conflict and the need to cooperate by developing an advanced and strong economy. Singapore does not need to contend over scarce water as it has the capacity through its economic strength to 'import' virtual water.

Figure 2.2 shows that the Jordan Basin which has three economies with very different levels of GDP per head is located in the part of TWINS space where there is high conflict and low cooperation over shared waters. Riparians that enjoy relatively high levels of GDP per head – for example the economies of Singapore and Malaysia, as well as those of Europe and North America – can cooperate and avoid conflict.

Other basins are characterized by major differences in the levels of strength and diversification of the riparian economies. For example, the economies of the Jordan riparians range from the highly diversified Israeli economy with high levels of GDP per head, to the Palestinian Authority where GDP/capita is currently less than US$2000 per capita (Table 2.1). These indicators of economic asymmetry are partly the result of the international power asymmetries – including in hydropolitics.

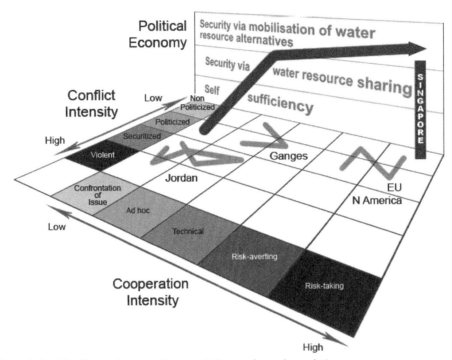

Figure 2.1   Conflict and cooperation coexist in transboundary relations

Note: The diagram illustrates the differing intensities of conflict and cooperation through time in different basins. It also shows how diverse and strong economies can avoid the risks and transaction costs of conflict and cooperation over transboundary waters.

Source: adapted from Mirumachi, 2007

Every basin is different in its water endowments and in its economic and its political asymmetries. The Nile Basin has ten riparians, and Egypt – in this case a downstream riparian – has enjoyed the position of hegemon for the past two centuries. Egypt has always had a GDP per capita many times that of the other riparians. An oil-enriched Sudan is beginning to catch up and as a consequence the relations over water in the Nile Basin are changing rapidly. But while the conflictual politics may intensify, the way to remedy the water deficits in both Egypt and the Sudan will be the same as it has been for the last three or more decades. Virtual water 'imports' from outside the basin will meet the increasing water deficits. The conflictual position taken by the contending parties will be prominent for a period. But the invisible and silent virtual water 'trade' will quietly solve the problem. Global water resources will continue to be 'importable' because of the diversifying economies or the oil revenues of the 'importing' economies.

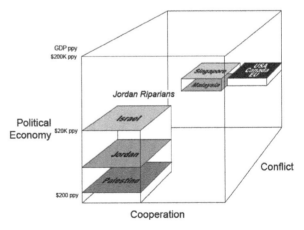

Figure 2.2   Diverse and strong political economies have the option to adopt alternative water resource management approaches via technology and economic diversification and trade

Source: Allan, based on Mirumachi, 2007

The Impossibility of Researching International Transboundary Waters and the Reasons for the Low Impact of Such Research

It has proved to be possible to establish apparently significant confidence-building processes such as river basin initiatives at the same time as these cooperative processes are being vetoed in the world of disappeared politics.

The power relations that feature in the politics of transboundary relations, and the invisible processes which feature in the international economy, make researching transboundary water resource allocation and the joint international management of such waters virtually impossible. In addition the contentious politics have effectively prevented the adoption of the 1997 UN Convention on the Non-navigational Uses of International Water Courses (United Nations, 1997).

The reason that transboundary politics are complex to analyse and international water law has proved to be non-implementable can be partially explained by theories of securitization and international conflict (Buzan, 1995; Buzan et al, 1998; Neumann, 1998; Warner, 2004a, b). The Copenhagen School (Buzan et al, 1998) identified three levels of intensity in international conflictual politics. In the case of water resources when water is not perceived to be scarce, relations are *non-conflictual*. Once shared water resource scarcity is on the political agenda, transboundary relations over shared international water resources become *politicized*. When awareness of water scarcity is

Table 2.1   Economies of Jordan riparians

Country/administrative area	GDP per capita, US$ (approximate)
Israel	30,000
Lebanon	13,000
Jordan	5500
Syria	4700
Palestinian Authority	<2000

Source: Israel, Lebanon, Jordan, Syria (UN data); Palestinian Authority

perceived as an existential threat that evidently requires a remedy to protect the resource – that is to *securitize* it in the national interest – then measures beyond usual politics are taken by the hegemonic riparian to capture water resources. Note that securitization of water does not necessarily have to be based on actual scarcity and that perceptions matter. The extraordinary measures taken by the hegemonic riparian to capture water resources can include constructing dams; filling reservoirs; groundwater pumping; or the blocking of such interventions by other riparians, or threats to do any of these. Riparians would go to war over water but the invisible and silent remedy of virtual water 'trade' is so invisibly and silently effective that riparians take the economically attractive and politically costless route to water security. They import water-intensive commodities.

Beyond *securitization* the hegemon has the option of *violence*; of launching a water war. Neumann (1998) conceptualized this phase of conflict intensification as *violized* relations. Warner (2004a, b) first linked these four conflictual levels to the international relations over transboundary waters (see the conflict axis of Figure 2.1). In Chapter 9 Jarvis and Wolf have provided evidence of the absence of water wars and armed conflict over transboundary water resources. The explanation of hydro-peace (Allan, 2002) lies in the political economy and the capacity of contending riparians for the *theatre of illusion*.

The level of difficulty in carrying out research on transboundary waters is directly proportional to the intensity of the contention over shared waters. In basins where water is abundant, research into the hydrology and the economics of water is non-controversial. Hydropolitics will be both non-conflictual and researchable. Extensive experience and insight of the first author show that it is very difficult to research transboundary water relations where awareness of water scarcity has shifted hydropolitics beyond the non-politicized and where at the same time inter-riparian power asymmetries exist. Where power asymmetries are extreme it is impossible to research transboundary relations effectively.

The institutions and procedures of nation states explain why researching transboundary international water relations is difficult and sometimes impossible. At the point where transboundary relations move from the non-politicized to the politicized, responsibility for international water relations move out of the departments of state *responsible* for *managing* water resources – namely the water, agriculture, planning and finance ministries responsible for *investing* in them – and from the ministries responsible for *protecting* them to the Ministry of Foreign Affairs. Officials in the Ministries of Foreign Affairs, or Attorneys General are given the responsibility for the issue and for negotiations over transboundary waters (McCaffrey, 2007). Water is then assumed to be a *sovereignty* issue. Some riparians have the resources and the capacity to augment the staff of their Ministry of Foreign Affairs. They can appoint professionals with a deep understanding of the technical features of water resources management and with expertise in the governance and legal aspects of water management and regulation, law and international law. Most do not.

A consequence of moving relations and negotiation over water resources into the remit of Ministries of Foreign Affairs is that the issues and priorities on the agenda change from the possibilities of *hydrological and economic mutuality and compensation* to *sovereignty*. In the absence of international water law the Ministry of Foreign Affairs of upstream states insists on principles of ownership of – that is *sovereignty* over – the water resource in its territory and over the water that flows out of it. Downstream states insist on continuing access to water resources on which they have become economically dependent. They argue for the principle of *prior use*. These approaches, based on a very poor understanding of both hydrology and economics, have proved to be a recipe for enduring deadlock and the tendency for the hegemon to accumulate water.

At the next level of conflictual intensity when water relations become an issue of threat to nation-state survival – although the perception of insecurity is just as potent – they can disappear altogether, into what is increasingly called the *shadow state* (Tripp, 2001, 2007; Springborg, 2007). The process occurs in both hegemonic and non-hegemonic, non-democratic states.[2] Even in highly industrialized neo-liberal political economies the topic of sharing water transboundary water resources can get lost in inter-ministerial labyrinths.[3]

Engagement over shared water resources in intensely conflictual circumstances, where measures to securitize access to water have been put in place, is inaccessible to all but tiny inner groups of officials very close to power in each riparian. There are examples in all the Middle Eastern, South Asian and South-East Asian basins. These officials are very tightly networked. It will be only a very small number of the insider shadow state network in any riparian that will deal with transboundary issues when tensions periodically gain prominence. The difference between the *politicized* and the *securitized* phases of transboundary water relations is that those involved are generally much more informed in the securitized shadow state phase than in the politicized phase.

When the shadow state takes over transboundary water relations it has the capacity and the resources to get itself up to speed to deal with the hydropolitical complexities. Security service staff can be deployed to gather knowledge and expertise. They can at the same time monitor what is being discussed in the public domain. It has proved to be possible to establish apparently significant confidence-building processes such as river basin initiatives at the same time as these cooperative processes are being vetoed in the inaccessible politics determined by the secret processes of interstate relations. All these activities take place outside of the public realm where the hegemon can insist on the *status quo* and the non-hegemon has little or no counter-hegemonic purchase. Some international agencies are allowed occasional access to the parallel and non-engaged intentions of the hegemon and normally non-aligned counter-hegemons, but these unreportable activities have little influence as these agencies have complicated relations, and sometimes have deep dependencies especially on the basin hegemon.

Table 2.2 provides a version of the relationship between water scarcity and the intensity of contention over shared transboundary water resources. It also defines the levels of intensity of conflictual relations over shared water resources

Table 2.2 The relationship between water scarcity, the intensity of contention over shared transboundary water resources, levels of conflict relations, the mode of engagement and the degree of researchability

Level of water scarcity in the basin	Type of conflictual relations	Mode of engagement	Researchability of:	
			Public activities such as 'cooperative' river basin initiatives	**Actual interstate decision making and negotiations**
Low	Non-politicized	No need for negotiations	Researchers can participate	Researchers can examine little riparian engagement
Moderate	Politicized	Contention normally based on constructed knowledge and not on observed science or economics	Researchers can participate	Researchers have limited access to these processes
High	Securitized	Hegemon riparians can determine outcomes	Researchers can participate but the public processes may be of minor relevance to actual decision making	Researchers have only retrospective access to data on these processes
Very high	Intuitively associated with potential violence (Neumann, 1998, coined the term 'violized')	Violent measures are *prevented by* processes in the international political economy – virtual water 'trade'		

Note: Shaded areas are those where it is possible for researchers to play a role.

according to the Copenhagen School (Buzan, 1995; Buzan et al, 1998; Neumann, 1998) and how these relate to the degree of engagement of researchers and of the engagement and the researchability of international hydropolitics in river basins. Researchers can participate in 'cooperative' river basin initiatives. It should be noted that even if there are cooperative water management initiatives it does not mean that essential issues such as water allocation are being addressed. Zeitoun and Mirumachi (2008) highlighted that cooperation is not always pretty and that the value of cooperation needs to be scrutinized. Outsider water professionals, outsider researchers, outsider international water lawyers and outsider officials in international agencies can impair their analysis by assuming that cooperation is sensible, neutral and inevitably a public good. Outsider analysts need to be aware that the scientific knowledge of the outsider is not easily accepted and integrated by the hegemon (Zeitoun and Allan, 2008).

This bleak analysis of the inaccessible processes integral to international transboundary water resources relations has been included to ensure that the difficulties of analysing the inaccessible are at least noted. It is in this region of shadow state politics beyond the public world that hegemons decide when and how to securitize water resources, prevent the recognition of the interests of non-hegemons and manipulate the constructive intentions of major and minor international agencies, thus determining water allocation outcomes. However, the challenges of researching shadow state decision making over transboundary waters is not unique. Other highly politicized and securitized issues such as the war on terror, trade relations, financial regulation and climate change negotiations have processes of decision making that are also partially accessible, or totally inaccessible, to the average researcher. Those wishing to research international transboundary issues or actually engage in them should not expect that the non-politicized levels of engagement, which take place in a transparent or relatively transparent mode, in any way reflect the activities that take place beyond the public realm once relations move into politicized and securitized modes. The message is to be aware of the politics of power asymmetry and that there exists determining engagement beyond the public realm.

Conclusion

The extent to which an economy can mobilize socioeconomic development that enables international trade determines its water security and the nature of inter-riparian contention over transboundary waters in a world at peace.

... this trade in water-intensive commodities reduces conflict over water with neighbouring riparians and makes armed conflict unnecessary.

Remaining in power can depend on keeping the roles of power and trade off the public agenda.

The concept of embedded water was coined in the late 1980s by the first author, and the more engaging terms of *virtual water* and *virtual water 'trade'* in 1992, to explain the absence of armed conflict over transboundary waters in the Middle East. During the 1970s and 1980s the promised water wars over shared transboundary waters had not materialized despite the water resource position of all the economies in the region having very significantly and progressively worsened. It was embarrassing for a regional water resources specialist long uncomfortable with the spectre of environmental determinism. It was an encounter with trends in Egypt's grain imports in the 1970s that provided the answer to the absence of armed conflict over shared transboundary waters. Egypt could silently and invisibly import grain. No need for impossibly expensive and ineffective war.

The virtual water explanation not only explained the transboundary water relations of the 1970s and 1980s. It has continued to explain them for the next three decades. The future role of the invisible process is essential if regional and global security and peace are to endure.

The main message of this chapter is that both transboundary water research and attempts to operationalize cooperative basin initiatives have overemphasized the underlying fundamentals of hydrology, engineering and capturing new water in the region. There has also been a tendency to expect too much of legal principles.

The close relationship between achieving *high levels of economic diversity and strength* which enable forms of cooperation that fall short of equitable and reasonable water sharing arrangements has been demonstrated. It has been shown that the *invisibles – trade/development* and *politics* – actually underpin and shape the complex political economy of water resources and of international relations over water. Understanding these political landscapes is immeasurably more important than knowing the underlying hydrology and economics, when trying to operationalize cooperative and integrated transboundary water management (TWM).

It has been suggested that the most seriously water-scarce economies have brought water scarcity upon themselves. Their societies have increased their water needs as a consequence of demographic pressures and consumption practices. Populations have increased and the consumption of scarce natural resources has been exacerbated by modes of consumption that are excessive and often very wasteful in the manner of their utilization.

Governments managing increasingly water scarce economies face impossible political challenges in addressing their water supply predicaments. They cannot mobilize significant volumes of 'big water' for food production. The political challenge of changing the consumption preferences of their own societies is just as impossible, at least in the short term. Getting societies to live within their water means has proved to be too difficult for almost all political economies. Politicians and societies facing water scarcity, in basins with equally water-scarce neighbours, have unwittingly encountered an unforeseen and politically and socially feasible solution to their water scarcity. The economically invisible and politically silent international political economy of water has been very effective indeed in providing virtual water that makes water scarcity invisible. Possibly more important, this trade in water-intensive commodities reduces conflict over water with neighbouring riparians and makes armed conflict unnecessary. Water-scarce economies have found solutions in global markets. The extent to which an economy can mobilize socioeconomic development that enables international trade determines its water security and the nature of inter-riparian contention over transboundary waters in a world at peace.

The analysis has also been shown that those with influence at the sub-national level (for example farming interests), as well as those responsible for transboundary water international relations, have incentives to reinforce the invisibility of these conditions. Remaining in power can depend on keeping the role of *hegemony* and *trade* off the public agenda. Keeping *politically hazardous conditions* invisible normally overwhelms the option of recognizing *underlying fundamentals*. Researchers and international agencies tend to be blind to these invisible but determining underlying fundamentals. The discussion concluded by highlighting the problems of researching transboundary international relations. It showed that, except in neo-liberal advanced political economies, transboundary riparian relations and engagement takes place in a world of disappeared hydropolitics where the interests of the hegemon normally prevail. Water scientists, water professionals and those analysing transboundary waters international relations need to note the precautionary message of this chapter. The quiet, even invisible, processes highlighted in the chapter reduce the politicization of riparian relations over water and the need for the securitization of water resources. They reduce the urgency of even engaging in transboundary water negotiations at all, especially from the perspective of the hegemonic riparian state, and from that of states that have achieved the status of diverse and strong economies.

Notes

1   France is the only EU economy out of the current 27 that is a net food exporter and does not therefore depend on other economies for its food, although even France imports tropical crops.

2   Associated with the process of:

de-democratization is a more sinister and probably more important development. This is the further expansion of invisible, informal executive power, known by the term 'shadow states' coined by SOAS' Charles Tripp or, in the case of Morocco, simply le pouvoir *or, as in Jordan, the* makhzan, *which is Arabic for treasury but refers to the* [court *that is the*] *king, his entourage, and the power spreading out from them.* (Springborg, 2007)

3   In the UK it has proved to be impossible to identify the department of state responsible for the UK position on signing the 1997 UN Convention on the Non-Navigational Uses of International Watercourses.

References

Allan, J. A. (2000) *The Middle East Water Question: Hydropolitics and the Global Economy*, I. B. Tauris, London

Allan, J. A. (2002) 'Hydro-Peace in the Middle East: why no water wars? A case study of the Jordan River Basin', *SAIS Review*, vol 22, no 2 (Summer–Fall), pp255–272

Allan, J. A. (2010) *Virtual Water*, I. B. Tauris, London

Buzan, B. (1995) 'The levels of analysis problem in international relations reconsidered', in K. Booth and S. Smith (eds) *International Relations Theory Today*, Polity Press, Cambridge, pp198–216

Buzan, B., Waever, O. and de Wilde, J. (1998) *Security: A New Framework for Analysis*, Lynne Rienner, Boulder, CO and London, pp5–7

Cullet, P. (2009) *Water Law and Water Sector Reforms: National and International Perspectives*, Oxford University Press, Oxford

Cullet, P., Gowlland-Gualtieri, A., Madhav, R. and Ramanathan, U. (2010) *Water Law for the Twenty-First Century: National and International Aspects of Water Law Reform in India*, Routledge, Oxford and New York, NY

Hoekstra, A. Y. and Chapagain, A. (2008) *The Globalization of Water*, Blackwell Publishing, Oxford

Hornidge, A.-K. (2007) *Knowledge Society: Vision and Social Construction of Reality in Germany and Singapore*, Lit, Berlin and Münster

Lee, K. Y. (2000) *From Third World to First: the Singapore Story: 1965–2000*, HarperCollins, New York, NY

Long, Y. (2002) 'On the threshold of self-sufficiency: towards the desecuritization of the water issue in Singapore–Malaysia relations', in C. G. Kwa (ed.) *Beyond Vulnerability? Water in Singapore–Malaysia Relations*, Institute of Defence and Strategic Studies, Singapore, pp97–140

Low, L. (1998) *The Political Economy of a City-State: Government-made Singapore*, Oxford University Press, Singapore

McCaffrey, S. (2007) *The Law of International Watercourses*, Oxford Monographs in International Law, Oxford University Press, Oxford

Mirumachi, N. (2007) 'Fluxing relations in water history: conceptualizing the range of relations in transboundary river basins. Pasts and futures of water', Proceedings of the *5th International Water History Association Conference*, Tampere, Finland, 13–17 June 2006

Mirumachi, N. and Warner, J. (2008) 'Co-existing conflict and cooperation in transboundary waters', paper presented at the *International Studies Association Annual Conference*, San Francisco, 26–30 March 2008

Neumann, I. B. (1998) 'Identity and the outbreak of war: or why the Copenhagen School of security studies should include the idea of "violization" in its framework of analysis', *International Journal of Peace Studies*, vol 3, no 1, pp1–10

Rice, A. (2009) Is there such a thing as agro-imperialism?, *New York Times*, 16 November, www.nytimes.com/2009/11/22/magazine/22land-t.html?_r=3&pagewanted=all, accessed 27 November 2009

Soon, T. Y. (2009) *Clean, Green and Blue: Singapore's Journey Towards Environmental and Water Sustainability*, Singapore: Institute of Southeast Asian Studies, Singapore

Springborg, R. (2007) 'De-democratisation and the apotheosis of Arab shadow states', *The Middle East in London*, vol 4, no 5, pp3–4

Tongzon, J. L. (2002) *The Economics of Southeast Asia: Before and After the Crisis*, 2nd ed., Edward Elgar, Cheltenham, UK

Tripp, C. H. (2001) 'States, elites and the "management of change"', in H. Hakimian and Z. Moshaver (eds) *The State and Global Change: The Political Economy of Transition in the Middle East and North Africa*, Curzon Press, Richmond, UK, pp211–231

Tripp, C. H. (2007) *A History of Iraq*, Cambridge University Press, Cambridge

United Nations (1997) *The UN Convention on the Non-navigational Uses of International Water Courses*, International Law Commission of the United Nations, New York, NY

Verweij, M. (1999) 'A watershed on the Rhine: changing approaches to international environmental cooperation', *GeoJournal*, vol 47, no 3, pp453–461

Warner, J. (2004a) *Plugging the GAP. Working with Buzan: the Ilisu Dam as a Security Issue*, Occasional Paper no 67, SOAS/King's College London Water Issues Group, London

Warner, J. (2004b) 'Water, wine, vinegar, blood: on politics, participation, violence and conflict over the hydro-social contract', in World Water Council, *Proceedings of the Workshop on Water and Politics: Understanding the Role of Politics in Water Management*, Marseilles, France, 26–27 February 2004, pp7–18

World Bank (2007) *Making the Most of Scarcity: Accountability for Better Water Management in the Middle East and North Africa*, Orientations in Development, MENA Development Reports on Water, The World Bank, Washington, DC

Zeitoun, M. and Allan, J. A. (2008) 'Applying hegemony and power theory to transboundary water analysis', *Water Policy*, vol 10, supplement 2, pp3–12

Zeitoun, M. and Mirumachi, N. (2008) 'Transboundary water interaction I: Reconsidering conflict and cooperation', *International Environmental Agreements: Politics, Law and Economics*, vol 8, pp297–316

3

Power, Hegemony and Critical Hydropolitics

Ana Elisa Cascão and Mark Zeitoun

- Hydro-hegemony is hegemony active at the basin scale, and occurs where control over transboundary flows is consolidated by the more powerful actor.
- Four forms of power can be used to evaluate hydro-hegemonic situations: geography; material power; bargaining power; and ideational power.
- Explicit consideration of the perspective of the non-hegemonic actor(s) can provide valuable insight into the process and outcome of transboundary water interaction.
- Testing of the theoretical concepts by mid-level water managers suggests that not all forms of power are equal, with material and bargaining power counting more than geographic position or ideational power, for instance.
- It was also found that what is labelled as 'cooperation' is not always as intended, and distinctions should be made between 'non-cooperation', limited or dominative-type 'cooperation', and comprehensive cooperation.

Introduction

This chapter may be considered a 'primer' on established and emerging critical hydropolitical theory. It intends to contribute to the theoretical underpinning necessary for the effective interpretation and implementation of transboundary water management (TWM). The chapter focuses on the role of power and hegemony in particular, as a complement to the wide range of issues covered in the following contributions to this volume (including international water law, groundwater science, stakeholder participation and negotiations). The theoretical developments reviewed are based on recent experience in developing, testing and debating aspects of critical hydropolitical theory with academics and practitioners from a wide range of disciplines, primarily in the Middle East and North Africa.

This chapter takes as explicit the base assumption that TWM is a political process. It is thus subject to the fundamentally political processes of control, utilization and allocation, just as it is to the laws of nature and the physical processes of the hydro-cycle. Earle et al (see Chapter 1) have it right, in other words. Policy makers and politicians have the greater leverage over the direction that transboundary water interaction will take (see

Figure 1.1, Chapter 1). Application of critical hydropolitics is thus useful for interpretation of the power plays that grease or block the cogs of the decision-making machinery. Observing interstate interaction from inside and outside these processes on the Nile, Jordan, and Tigris and Euphrates River Basins, we argue that various riparian states are endowed with highly asymmetric capacity to use both overt and covert forms of power. As we will see, overt and covert forms of power are also commonly understood as, but not directly analogous to, 'hard' and 'soft' forms of power. We assert that the power asymmetries determine to a significant (not total) extent the fundamentally political distributional issue of 'who gets what, when, where and why' (Lasswell, 1936).

The framework of 'hydro-hegemony' is used as an analytical approach by which to flesh out and test the assertion. The approach recognizes that a characteristic common to most international river basins is the existence of 'hydro-hegemonic' configurations based on power plays, including in the influence of transboundary institutions and regimes. Hydro-hegemony is taken as 'hegemony at the river basin level, achieved through water resource control strategies ... that are enabled by the exploitation of existing power asymmetries (Zeitoun and Warner, 2006). In other words, the more powerful basin state can exploit its advantage in a number of ways to ensure the configuration is in its favour. But an advantage in power does not necessarily mean an inequitable outcome. The hydro-hegemon (HH) also has the ability to lead on the basin, and create a more optimal outcome for all parties. This essentially occurs when this is perceived to bring gains to the hydro-hegemonic riparian, namely in terms of additional allocation of water resources. Furthermore, the so-called weaker ('non-hegemonic' is the preferred term) states are not always as weak or optionless as they are credited to be. Deeper examination of each case reveals evidence of counter-hegemonic mechanisms employed by the non-hegemonic states, with the aim to change the outcomes of water control and allocation towards a more equitable configuration (Cascão 2008a).

A more nuanced point that this chapter hopes to make is that power and hegemony also deter-

mine the perceptions and practices of those of us involved in transboundary interaction analysis or implementation. Perspective takes on even greater analytical importance when considering the fixed mindsets related to established hegemonic configurations on river basins. We contend that explicit consideration of the perspective of the non-hegemonic actor(s) offers valuable insight into the process and outcome of transboundary water interaction. A more interesting and revealing picture may be found by considering the case of the 'underdog', in other words.

The first part of this chapter reviews the literature to offer the reader an update on hydropolitics theory. The second part reviews the theoretical fundamentals of hydro-hegemony and counter-hydro-hegemony theory. The third part tests, refines and substantiates the theory, through extensive feedback from practitioners, chiefly during the TWM training programme for the Middle East and North Africa (MENA) region. The Jordan, Nile, and Tigris and Euphrates contexts are primarily drawn upon throughout, though a wider discussion also includes examples from North Africa. The chapter concludes with thoughts on the applicability of critical hydropolitics to the cases, issues and features of TWM described in the following chapters of the book.

Updating 'Hydropolitics'

This section intends to lead to a more robust understanding of theories associated with hydropolitics, and to demonstrate its utility to real-world situations. Attempts to define 'hydropolitics' are rare, and an accepted and clear definition is not likely to be established any time soon. Attempts even to define it are rare enough. Turton (2002) picks up on the distributive issue in his definition of hydropolitics as 'the authoritative allocation of values in society with respect to water'. We can add to this definition by explicit incorporation of the role that power plays as an essential feature of water conflict and cooperation in practice.

We take *politics* in its broadest sense: 'who gets what, when, where and why?' (Lasswell, 1936).

The question is answered only by considering how political decisions of distribution and allocation of resources (namely natural, political and financial) are made. Allocative politics are affected by their socioeconomic and political contexts, of course, at multiple scales (from the individual to the global). As the allocation of resources in each dimension of each context tends to be asymmetric, our analysis must also answer the crucial question of 'who gets left out?' of the equation (Markovitz, 1987).

The bulk of academic hydropolitical research has concentrated on basins in the Middle East and North Africa (e.g. Waterbury, 1979; Falkenmark 1989; Wolf, 1998; Elhance, 1999; Allan, 2001). Hydropolitics has also been strongly associated with the 'water wars' concept, wherein interstate armed conflicts were expected to occur in any number of 'hydropolitical security complexes' such as the Tigris and Euphrates (Schulz, 1995). The analytical dyads of 'water–conflict' and 'water–security' are among the major forms of bias in the hydropolitics literature. The consistent association of hydropolitics with conflict or security issues has led to an impoverished debate and hindered understanding of hydropolitics as a dynamic and ongoing process involving several other key dynamics – notably society, environment and culture. Sales-driven news media and books hyping water wars are as plentiful as they are unhelpful for any illumination or insight into the complex dynamics, yet they persist (e.g. Bullock and Darwish, 1993; Thomson, 2005; *The Independent*, 2006; Lewis, 2007).

In part a response to the hype surrounding water and violent conflict, students of hydropolitics have focused on transboundary water cooperation. The quantitative work led by Aaron Wolf on the Transboundary Freshwater Dispute Database (e.g. Wolf, 2002; Wolf et al, 2005) has been instrumental in demonstrating that cooperative forms of interaction over transboundary waters are not exceptional. Another strand of quantitative research has been dedicated to the prediction or examination of causal factors of international water conflicts, primarily by the Peace and Research Institute of Oslo (see, for example, Toset et al, 2000). The findings have shown few direct links between water and violent

conflict with such criteria as a state's physical size or level of democracy (Gleditsch et al, 2006; Hensel and Brochmann, 2007; Brochmann and Hensel, 2008).

The analytical tool of choice for many of these authors is the conflict–cooperation continuum (Delli-Priscoli, 1996; NATO, 1999). Knowing that conflict and cooperation are in fact ever-present and may be two sides of the same coin rather than opposing ends of a spectrum, we begin to gather that a continuum may be the wrong tool for the job. Not all conflict is necessarily negative, and some forms of 'cooperation' can be based on coercion, or temporary submissiveness.[1]

Understanding relations over transboundary waters in terms of *interaction* rather than conflict or cooperation has implications for policy makers and academics in a number of ways. Practitioners involved in stakeholder participatory processes, for example, may want to consider whether the inclusion of weaker parties in a process is genuine rather than token (see Krantz and Mostert, this volume Chapter 7). Analysts examining a non-violent international conflict over water issues may similarly wish to look deeper than the statements from officials of all sides as evidence of 'cooperation'. Understanding coexisting conflict and cooperation also facilitates the work of those from both groups involved in the design and execution of negotiation strategies at the multilateral level, or in the development of positive-sum solutions addressing water and benefit-sharing paradigms in a balanced manner (see Daoudy, this volume Chapter 4; Granit, this volume Chapter 10; and Jarvis and Wolf, this volume Chapter 9).

Water politics brings to political science theory many peculiarities of the resource itself, only two of which are touched upon here. First, in a very strong sense, the distributional issue for water – the 'what' question – is not only about quantity. The *quality* of water in many cases matters as much as the volumes concerned, though the issue is not addressed further here. A second unavoidable peculiarity of water that shapes its politics is its fluid nature. The movement of flows across static political boundaries in a sense stitches otherwise unwilling states – and their political economies – together. The 'who' question thus becomes

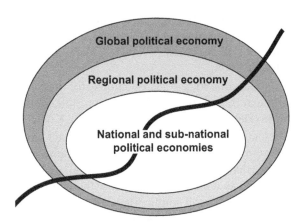

Figure 3.1   Rivers run through them: transboundary waters spanning all levels of political economies

complex quite quickly. As shown in Figure 3.1, the resource passes through sub-national to global levels. Reconciling the interests and contexts of two or more neighbouring states is as significant a challenge as it is to satisfy competing demands of various economic sectors and activities at the sub-national level. But reconciling interests at the global or regional/basin level is even more challenging than at national level, taking into account that clear transnational political tools are not yet available. The challenge is exponentially compounded when each entity is part of global or regional international systems, wherein parallel power games are also played.

The implications that the fluid nature of water have on notions of deeply held convictions and state sovereignty should not be under-emphasized. As with the implications of clear-cut harvesting half of a transboundary forest, it is insufficient for a state to claim territorial rights of exploitation without consideration of the effects – the possibility of spread of disease or loss of species habitat, in this case – on the neighbouring state. Transboundary waters are an even more direct challenge to sovereignty, for it is not only the effects of exploitation that must be considered, but physical sharing of the resource itself. Chapter 5 on TWM and law by McIntyre delves into this subject deeper.

With an appreciation of such insights in mind, our *critical* engagement with the 'who/what/when/where/how' question must address

asymmetric power relations. A strict realist interpretation would see power as the ultimate determinant of the outcome of a competition for a transboundary resource. However, another embedded idea of critical hydropolitics is that power relations are not an irreversible or a static reality; the *status quo* does not last forever. Power, and power asymmetry, are constantly being contested and challenged (Cascão, 2008a). While the basin hegemon is typically stronger in all fields of the various forms of power, it certainly does not follow that the non-hegemonic riparians are 'powerless'. As we will see, non-hegemonic riparians have capacities of bargaining power in particular which can be the main element of counter-hegemonic strategy. Through increased use of bargaining power, the non-hegemonic riparians can in theory 'level the playing field', influence the regional agendas and negotiations, and might contribute to change the hegemonic configuration in the basin.

Power and Hegemony over Transboundary Water Resources

Given the enabling role that power plays in the allocative politics of transboundary waters, it is necessary to be able to conceptualize power in a useful manner. The breadth of the subject at hand and the very peculiar nature of each basin mean it is not surprising that there is no common template that can be used to interpret all transboundary contexts. It is nonetheless helpful to group contexts according to specific criteria – in terms of the character of the control exerted, for instance, as shown in Figure 3.2.

At the right-hand extreme of the continuum, control over the transboundary waters is openly competed for, sometimes in somewhat hostile political environments. Wegerich (2008) suggests that this is the case in post-Soviet Central Asia, and in particular in the Amu Darya Basin, where downstream Uzbekistan is now obliged to confront actions both threatened and carried out by the relatively weaker upstream Tajikistan. At the opposite extreme is the cooperative form of interaction, based on the principle of full equality, and manifested in terms of

Figure 3.2   Range of forms of interaction over transboundary water resources

Source: based on Zeitoun and Warner, 2006

economic integration, equitable distributive politics and collective decision-making processes. The European Union serves to exemplify, where the EU Water Frame-work Directive envisions shared and equitable control of transboundary water resources. In such a cooperative context, for example, The Netherlands can hold sway over much 'bigger' upstream powers such as Germany, for instance (Warner, 2008; de Silva and van der Zaag, 2009).

The bulk of current transboundary water interaction lies conceptually between these two extremes, however – where control is shared in principle, but not in practice. For the purpose of this chapter, this middle area is characterized as hegemony, where the 'first among equals' carries more relative power. Control of water resources in this position is consolidated in favour of the most powerful riparian state. There may be overt or covert competition for the water, but it is generally stifled through a variety of strategies and tactics which rely in turn on both overt and covert expressions of power, discussed in Zeitoun and Warner (2006) and elaborated upon here.

Forms and fields of power

The original work on hydro-hegemony was developed from examination of three MENA transboundary water contexts – the Jordan, Nile, and Tigris and Euphrates Basins. The study revealed that open or 'overt' forms of power – such as 'material' power in the form of military force or economic juggernauts – are not as common as the water wars hype would make them out to be. The same and similar studies showed that in fact more hidden 'covert' forms of power were much more prevalent in transboundary water contexts. Both covert and overt forms of power have at least two aspects, or 'fields', associated with them:

- *Geographical power.* One of the most influential types of overt power is that of riparian position. The importance of a state's position on a river was conceptualized most robustly by Frey and Naff (1985). It relies on the distinct advantage that geography provides to an upstream state to manipulate the flows, i.e. to dam or divert them. In the Tigris Basin, various Turkish governments have been exploiting this position with the continued construction of the Southeastern Anatolia Güneydo u Anadolu Projesi (GAP) project. Consideration of Egypt's downstream yet dominative position on the Nile, however, shows that geographic position can be less influential and determining than other fields of power (see Nile case study, this volume Chapter 13).
- *Material power.* This most visible form of power includes economic power, military might, technological prowess and international political and financial support. India's ability to undertake a massive river interlinking programme, with little consideration of the upstream or downstream protestations from Nepal and Bangladesh, serves as an example. Asymmetries in material power can influence the control exercised over water, in particular when combined with bargaining and ideational dimensions of power.
- *Bargaining power.* This field of power refers to the capability of actors to control the rules of the game and set agendas, in the sense of their ability to define the political parameters of an agenda (Bachrach and Baratz, 1962; Lukes, 2004). It is also evident in the power of actors to influence the terms of negotiations and agreements, through their ability to provide and/or influence incentives that may encour-

age weaker parties to comply (Lustick, 2002; Zeitoun, 2006). The relations between actors are crucial in determining the applicability of bargaining power. If each is legitimate in the eyes of the other, an actor with much less capacity in the material dimension of power may still retain influence over the so-called stronger actor. Daoudy (2009), for instance, recognizes the leverage that Syria was able to generate from linking non-water issues (e.g. regional security, political alliances) in its Tigris and Euphrates negotiations with the much 'stronger' Turkey.

- *Ideational power.* This dimension refers to 'power over ideas' (Lukes, 2004) which represents the capacity of a riparian to impose and legitimize particular ideas and narratives. In sum, ideational power allows the basin hegemon to control the perceptions of the allocative config- uration of the societies both in its own country and in the neighbouring riparian countries, thereby reinforcing its legitimacy. An abstract conception, ideational power may be exercised through knowledge structures, sanctioned discourse and the imposition of narratives and storylines (see for example Hajer, 1997). Hegemonic riparians may manipulate the interaction with the neighbouring riparians through a number of tools, including lack of knowledge and data sharing, or the use of time, silence or ambiguity. A refusal by the Israeli side to share data on water use by Israeli settlers resident in the Palestinian West Bank (Tagar et al, 2004; World Bank, 2009), for example, can be understood as hegemonic use of ideational power. In persistently delaying negotiations over its preferred 1959 treaty on Nile allocations, Egypt is operating in the realm of ideational power by being present at the negotiations table and yet playing with time to maintain its hegemonic position in the Basin (Cascão 2008b). A further example of ideational power common to the Nile, Jordan, and Tigris and Euphrates Basins is based on the 'sanctioning of the discourse' (Allan, 2001). Arguments against equitable sharing based on state security have been so common that the governments transform the water issues into a

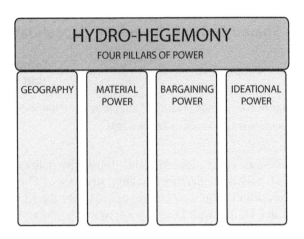

Figure 3.3   Revised pillars of hydro-hegemony

matter of top national security (securitization process) that can now be used interchangeably to silence and/or overplay certain issues, according to the hegemon's interests.

The fields of power reviewed are of course at use concurrently, and usually combine to determine the hegemonic configurations. The early work on hydro-hegemony provided a method to interpret this visually, by suggesting that the most powerful state may create and maintain a situation of hydro-hegemony through the development of three 'pillars' that support it. The concept has been informally refined to incorporate four foundational pillars shown in Figure 3.3 (Cascão 2008c, adapted from Zeitoun and Warner, 2006).

The revised pillars of hydro-hegemony suggest that a hegemonic situation on transboundary waters is built on the four fields of both covert and overt forms of power. Without going into too much detail, the suggested relative measures of each field for the Eastern Nile, Tigris and Euphrates, and Lower Jordan River Basins are given in Figure 3.4. The lengths of pillars are relative to other basin states, and not quantified. These plots are based in the current political context, but as we know power and hegemonic situations are not static, the plots in 3, 13 or 30 years would certainly look different. Just as the suggested plots of Figure 3.4 are based on the authors' experience, perspective certainly enters any evaluation. As such, the evaluation by different authors, for example by hegemonic and

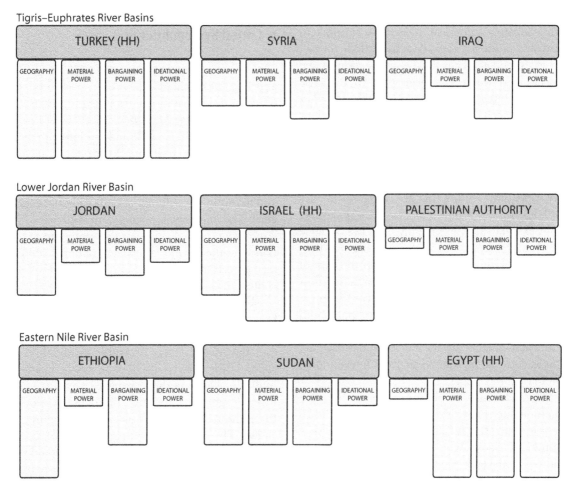

Figure 3.4   Suggested plots of hydro-hegemonic configurations in the Eastern Nile, Tigris and Euphrates, and Lower Jordan river basins (2009 estimates)

non-hegemonic actors, would certainly be different. Offering representatives of the states involved the opportunity to evaluate from their own perspective would not only provide three different diagrams, but also a critical opportunity to see power relations from another perspective – through the eyes of counterparts.

A quick glance at Figure 3.4 confirms what is commonly understood, for instance that Turkey, Egypt and Israel are the basin hegemons, despite their different riparian positions on the respective rivers. It also shows that the hegemons are usually very strong in all dimensions of power, at least in relation to their neighbours.

A deeper look reveals rather more interesting features, particularly when considering the

perspective of the non-hegemonic states. One feature of note is that non-hegemonic riparians all have a somewhat substantial measure of bargaining power available, and this can be or is used to counterbalance their weaknesses in other power dimensions. Ethiopia's bargaining power, for instance, is perhaps much greater than normally acknowledged. It is shored up by Ethiopia's geographic advantage as upstream riparian and provider of 85 per cent of the Nile flows, to be sure. The bargaining power is precisely what is being drawn upon in the ongoing multilateral legal negotiations (see Nile case study, this volume Chapter 13). Another example of bargaining power is the government of Syria's negotiations (see above), with the bargaining strategy informed through linking Turkey's multi-dam GAP project with the latter's interests in ending support to Kurdish separatist groups (Warner, 2008), as well as Syria's repeated calls for the application of the international water law principles to the case (Daoudy 2008). Bargaining power is thus noted as one of the mechanisms that holds key potential for the non-hegemon – and attempts to challenge an inequitable *status quo* on a basin (Cascão 2009a).

A further feature of note from Figure 3.4 is the restricted space of manoeuvre that afflicts particularly weak states. States that are relatively weaker in all dimensions of power (e.g. Iraq and Palestine, in our example above) evidently have less 'wriggle room' to begin to level the playing field or challenge inequity. Nonetheless, as in the case of Ethiopia and Syria, bargaining power is the most likely prospect for improved equity. Counter-hegemonic strategies driven from Baghdad could build on bargaining power by improving Iraqi technical and negotiation capacity, and anchoring of its negotiation position in terms of international water law, for example.

The relevance of the approach of 'hydro-hegemony' to real-world situations thus begins to become clear. But it is still underdeveloped in a number of ways. Academic critics have rightly pointed out the lack of consideration of hegemony at the sub-national level (Selby, 2007; Furlong, 2008), for instance. There is furthermore a possibility of policy mis-steps deriving from over-ascribing 'false consciousness' to the actors concerned (something which haunts scholars of hegemony from all disciplines). Assertions of hegemony such as those in Figure 3.4 imply, after all, that some actors are submitting either consciously or unconsciously to a situation over which they are deemed to have little control. The so-called weaker actors may in fact be engaged instead in some kind of 'strategic cooperation' that meets their short- or long-term interests, and to which analysts are not privy. Practitioners as well have helped to point out deficiencies with and limitations to the theory. The following section reviews these more substantially, and begins a refinement of the theory.

Practitioners' Contribution to Critical Hydropolitics

The concepts derived from critical hydropolitics and hegemony theory have been presented and tested in the international training programme on *Transboundary Water Management for the MENA Region* in 2008,[2] in which over 30 mid-level practitioners[3] from the Middle East and North Africa participated. The goal of the training was twofold: to introduce the participants to hydropolitics theory and to ground new thinking on field-based experience. The interactive process went both ways – the application of hydropolitics theory by the practitioners to their own regional and national contexts has served to enhance understanding of the context in which they were operating.

Freshwater resources in the MENA region are known to be relatively scarce physically, with low average river flows and groundwater recharges and the lowest levels of water availability per capita in the world (UNDP et al, 2000). There are multiple water bodies crossing over or under diverse and competing political economies throughout the MENA region, as Figure 3.5 shows. The region is characterized by the practitioners as a mix of physical scarcity coupled with complex regional political contexts. The political and geographical peculiarities that make each sub-context unique were equally emphasized, however.

The wide range of participants from the MENA region also permitted some comparison of limitations and similarities between basins in quite distinct regional or national political economies. The refinement can be considered through a number of 'lessons'.

Lesson 1: Not all transboundary waters are 'shared'

While MENA contexts may be transboundary by definition, the different contexts certainly displayed significant differences in terms of degree of 'sharing'. Water allocations varied vastly from basin to basin, in other words, and a clear message that emerged was for a clearer distinction between 'transboundary' and 'shared' water resources. The terms are often used interchangeably, but have quite different implications. It was suggested that

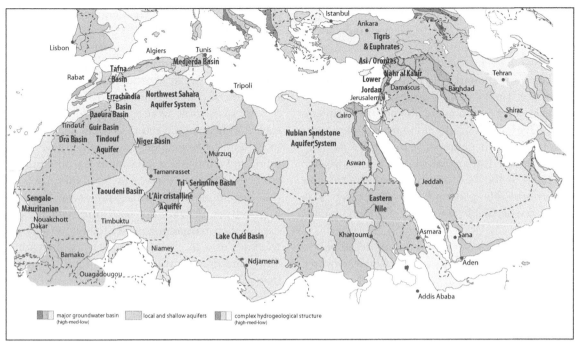

Figure 3.5   Select transboundary aquifers (shown) and rivers (named only) in the MENA region

Source: adapted from UNESCO, 2008

'transboundary water resources' should maintain a strict geography-based definition. Water bodies that cross the administrative borders between two or more countries should be considered transboundary. 'Shared water resources', on the other hand, should be defined by a political interpretation. Shared water resources would take into account how water is in fact allocated or used between the states (the 'who/what/when/where/how' question again). Bearing the distinction in mind, it was asserted that the Jordan or the Tigris–Euphrates Basins, for example, should be considered 'transboundary' basins. They could not be considered equitably 'shared', however, due to the prevalent extreme asymmetry in use and allocation between the non-hegemonic actors and the basin hegemons (Figure 3.4).

The semantic distinction refines hydropolitics theory in at least two ways. First, basing the difference in the terms on physical and political grounds serves to guide analysts in their examination of transboundary water relations. Just as water 'scarcity' has a political component (Turton, 1999; Mehta, 2001) to complement the more accepted understanding of physical scarcity, more precise (or selective) use of the term 'shared' should encourage clarity. Second, if the political definition of border-crossing water ('shared') is used, it becomes apparent to any reader that the term must be immediately qualified in terms of how water resources are (or are not) being actually shared. In the case of the Tigris and Euphrates just discussed, the implication of 'inequitable' was used to qualify 'shared'. Practitioner readers of this article could consider other qualifiers of basins they are familiar with, such as 'equitably shared', 'periodically shared', 'historically shared', or 'theoretically shared'.

Lesson 2: Not all power is equal

Application of the critical hydropolitics theory by the practitioners to their own contexts revealed further refinements. The visual ('pillars') component of hydro-hegemony theory suggests that each of the four fields of power carries equal weight, by virtue that each pillar is the same width. The testing of the theory suggests, however, that some dimensions of power are more influential than others. First of all, riparian position (geography-based power) was found to be relevant only under certain conditions. Position, it was found, can be an advantage to an upstreamer *primarily* if combined with material, financial and geopolitical power (e.g. Turkey in the Tigris–Euphrates Basin). An upstream position in the absence of advantage in the other pillars was found to be less relevant in various hydro-hegemonic configurations (e.g. Ethiopia and Equatorial riparians are weaker riparians in the Nile Basin). Furthermore, the importance of geography-based power was considered negligible in the case of transboundary aquifers.

Second, it was suggested that material power be considered more case-specific. Material power was considered a significant element of power in the Middle Eastern part of MENA, but much less so in the North African parts. The finding reflects the strong asymmetries that exist between Middle East states in terms of economic, military and geopolitical powers, and access to external support. Egypt, Israel and Turkey have clear material power supremacy over their neighbours. Power relations between the four North African countries – Morocco, Algeria, Tunisia and Libya – are relatively more symmetrical. More accurate analysis of transboundary contexts in North Africa would result, it was felt, if the 'material power' pillar of Figure 3.3 was not given the same 'weight' as the other pillars.

Third, greater nuance was suggested for the pillar of 'bargaining' power. Experience asserted that bargaining power was understood as a double-faced power resource, which both hegemonic and non-hegemonic riparians can make use of to strengthen their position. It was generally recognized on one hand that strong asymmetries in terms of access and production of data, information and knowledge weaken the non-hegemonic riparians in their ability to influence the political agenda or to bargain at the negotiations table. The bargaining gap between Palestinians and Israelis during previous and recent water negotiations was cited as the major example. On the other hand, participants enumerated several bargaining tools that are also available to non-hegemonic riparians to counter

the hegemons. These were identified as: a) claiming the moral high ground (application of the principles of international water law); b) public media and legal advocacy campaigns against unilateral projects; c) issue-linkage; and d) the formation of coalitions among weaker states.

All such tools were considered relevant to such non-hegemonic actors as Syria, Iraq, Jordan or Ethiopia. Once again, 'bargaining power' was found to be much less relevant to less contentious contexts in North Africa. The main message was: non-hegemonic riparians are not as weak as one might think, because they can (and typically do) exploit their bargaining power to counterbalance weaknesses in other power dimensions. The findings in this case reaffirm counter-hydro-hegemony theory. The case study on the Nile Basin (this volume Chapter 13) provides a clear example of how the non-hegemonic Nile riparians have been using collective bargaining power to counter the hegemonic power in the Basin, and how this has contributed to levelling the playing field among upstream and downstream riparians.

Finally, the practitioners also confirmed that the less visible forms of power – especially ideational power – are key explanatory factors for asymmetric power relations in transboundary basins. The ability to influence perceptions, the agenda, the discourse, or the timing of negotiations and projects are highly asymmetric among riparians. Several examples of co-option, issue-exclusion, stalling and securitization tactics related to ideational power were provided. Israel, for example, benefited from its superiority in ideational power to co-opt its Jordanian co-riparian through the terms of the 1994 Jordanian–Israeli Agreement, though whether it was a strategic or short-sighted move from the Jordanian perspective is debated. Israel also leveraged its ideational power – with limited success – by excluding the issue of Palestinian access to the Jordan River from the agenda, during the 2008 'Annapolis' round of negotiations. Seeking not to derail the negotiations, the chief Palestinian water negotiator initially consented to the exclusion (though it was later raised, but not resolved) (Al Attili, pers comm, 2008)

Table 3.1   Examples of transboundary water cooperative arrangements in the MENA region

Context	Initiative	Main achievements and limitations	Donors
Nile	Nile Basin Initiative (all 10 riparians)	• NBI – provisional cooperative mechanism (since 1999) • Ambitious goals/ involves all 10 riparians/strong involvement of donors • Shared vision and subsidiary action programmes • Capacity-building and trust achieved • Not yet a legal framework or projects on the ground • Nevertheless, seen as a good model of cooperation	World Bank, UNDP, African Development Bank, FAO, GEF, Canada, Denmark, European Commission, Finland, France, Netherlands, Norway, Sweden, UK, Germany, Japan, Switzerland, US
Niger	Niger Basin Authority (9 riparians)	• One of the oldest intergovernmentals in Africa (Convention signed in 1987) • Goal: integrated water management and economic development • Shared vision and several investment projects • Joint basin-wide hydrological monitoring system • Active involvement of donors, but also civil society and environmentalists	World Bank, UNDP, African Development Bank, Canada, European Commission, France, US
Lake Chad	Lake Chad Basin Commission (5 riparians) (Algeria, Sudan and Libya not members)	• Old organization – since 1964 (failed to prevent environmental catastrophe) • Goal: regulation and planning of the uses of water and natural resources • Still focusing primarily on surface water, and not groundwater • Ambitious project of water diversion from Congo River to Lake Chad	World Bank, UNDP, Denmark, European Commission, France
Nubian Aquifer	Nubian Sandstone Aquifer System Project (all 4 riparians)	• Goal: rational and equitable management of the NSAS • In the first stages of cooperation (setting) • No legal and institutional frameworks yet, nor projects	International Atomic Energy Agency, UNDP, GEF, UNESCO
Jordan Basin	EXACT (3 riparians)	• Database project • Forum whereby infrastructure and research are discussed • Small-scale projects (e.g. wastewater treatment) • Data on Israeli settler use of water in the West Bank not shared	United States, France, EU, the Netherlands, Canada
	Israeli–Palestinian Joint Water Committee (just aquifers, not Jordan River)	• Goal: rational water resource management and IWRM • JWC: decisions on water projects in the West Bank by consensus • Presented as model of cooperation • Criticism: domination dressed up as cooperation (Israel: veto & licensing)	Currently no donors support projects not licensed through the JWC
	Israeli–Jordanian Joint Water Committee (just Israel and Jordan)	• JWC goal: the implementation of the water clauses of the Peace Treaty • Continued to function despite conflicts between countries • Several problems in disentangling ambiguities embedded in the treaty	Not applicable
Tigris–Euphrates Basin	Euphrates-Tigris Initiative for Cooperation (ETIC) (not governmental, just civil society of 4 riparians)	• No cooperation at interstate level • Civil society engagement • Aims to facilitate cooperation • Goals: capacity-building and institutional strengthening	UNESCO, private sector, Universities

Benefiting from a bilateral volumetric water agreement signed in 1959 with Sudan, Egypt was generally considered to be in a comfortable pivotal position that often allows stalling or delaying tactics that prevent a basin-wide agreement being reached. Through securitization tactics, Turkish governments had been able to overplay the importance of the Tigris–Euphrates waters for its national security and political economy to the international community. At the same time it was seen to reject the idea that Syria and Iraq also have rights to develop the Tigris–Euphrates water resources for their own economic development.

Lesson 3: Transboundary water cooperation is not always a good thing

Prior to their exposure to the dual nature of coexisting conflict and cooperation, the bulk of participants considered a) that conflict in transboundary river basins is something to be avoided and must be (or is already being, in some cases) replaced by cooperation; and b) that cooperation is something desirable but difficult in a region where conflict (though not water conflicts) is pervasive. The views were challenged following a deeper exploration of the nature and effectiveness of the cooperative arrangements. The arrangements are provided in Table 3.1.

Commonly held notions of the nature and effectiveness of the cooperative arrangements were challenged when a critical hydropolitical lens was applied. Participants concurred that 'cooperation' efforts were also about enduring and/or changing the perceptions among the riparian states (including the hegemon) as well as those of third parties.

Various forms of cooperation are evident from Table 3.1. Examples of all-inclusive cooperation are found in the Niger Basin and the Nubian Sandstone Aquifer. The Niger Basin is considered an example of effective long-term cooperation, which included the establishment of a river basin organization which addresses a wide range of water-related and development issues in a manner that includes non-governmental stakeholders. The Nubian Sandstone Aquifer System Project is a case of a transboundary cooperation initiative at its early stage, and one that is widely supported by the scientific and donor community in the hopes of emulating the success on the Niger. The table also reveals a more 'narrow' type of cooperation. There is non-inclusive cooperation occurring at the inter-governmental level in the MENA region on the Nile, Jordan and Lake Chad. The extent of the cooperation is thus hobbled by exclusivity or non-comprehensiveness. The non-official cooperation through the academic cooperative forum Euphrates-Tigris Initiative for Cooperation (ETIC) in the Tigris and Euphrates Basin was considered in a similar light, though with the hope that it will bear influence in the high-level political echelons in Turkey, Syria and Iraq.

More specifically, three types of cooperation were identified from Table 3.1: a) non-cooperation (Tigris–Euphrates Basin); b) limited or domination-type cooperation (Jordan Basin); and c) comprehensive cooperation (Nile Basin). The findings confirm the earlier lesson learned about the importance of distinguishing between transboundary and shared resources. The practitioners found that the type of cooperation is related to the type of hegemonic configuration of each of the basins. The lack of discussion on permanent water allocations in the Tigris and Euphrates Basin can in part be explained by the fact that basin hegemon Turkey has benefited from its advantageous upstream position to develop unilateral hydraulic projects. Failing to even recognize the Tigris–Euphrates as an 'international river', Turkish authorities do not need or are even inclined to promote multilateral water cooperation. In the Jordan Basin case, 'cooperative' agreements were signed between the hegemon, Israel, and the non-hegemonic riparians (Kingdom of Jordan in 1994, the Palestine Liberation Organization in 1995). There was consensus among the Palestinian participants that the arrangement was one of 'domination dressed up as cooperation' (Selby, 2003), as the agreements are limited in scope and ultimately have only contributed to strengthen and legitimize Israel's hegemonic and inequitable position. The Jordanian participants did not reach a similar consensus, with some participants evaluating the form of cooperation established with Israel as coinciding with Jordanian interests. As such, the forms of cooperation occurring in the Jordan Basin were seen as both genuine as well as dominative.

Cooperation in the Nile Basin, on the other hand, was perceived by many of the participants as a good example of multilateral, all-inclusive and 'donor-darling' case of cooperation. The practitioners partially attributed cooperation in the Nile Basin to the character of the hydro-hegemon. As the basin downstream and very dry hegemon, Egypt is in a sense the most vulnerable riparian. At the same time, Egypt is also the primary user of the Nile waters, and has a keen interest in preserving the 'lion's share' of the flows it has grown accustomed to. It was considered clear that Egypt's water security would best be achieved through cooperation. Egypt's interests in cooperation were therefore identified primarily as pragmatic, namely: a) to keep vigilance and control over the development of upstream water infrastructure, even providing Egypt with veto power at the Nile Basin Initiative (NBI); b) to divert attention from the controversial 'water-sharing' paradigm towards a more consensual 'benefit-sharing' paradigm, and delay potential renegotiations of existing water agreements (most notably the 1959 Egypt–Sudan agreement); and c) to be able to use a greater share of Nile flows, once these have been made available through 'cooperative' river engineering projects in Sudan and

Ethiopia. Although participants from the other Nile riparians were not represented in the group of practitioners, it is very likely that they would not understand 'cooperation' in the same manner, however. The Nile upstream riparians have different perceptions on how to achieve effective cooperation, and consider that any sustainable cooperative arrangement in the Basin must necessarily be based on a new legal agreement founded on principles of equitable utilization of the water resources (Cascão, 2009b).

Lesson 4: The critical hydropolitics approach needs further refinement

The participants raised a number of issues they found with the approach, all of which serve to refine the theory. First, the approach appears to be less effective in less politicized or securitized conflicts. As previously discussed, water is not as politicized in North Africa as it is in the Middle East, where conflicts or lack of cooperation over water issues may be better explained through issues other than water, such as border disputes or regional rivalry.

Second, it was felt that examining the four pillars of power in separate boxes was too limiting. In reality, the fields of power are interrelated, and cannot be separated. Bargaining or ideational power have less traction without material power, for example. There was also a strong feeling that geographical power can indeed be a very strong source of bargaining power, when combined with the other dimensions. A suggestion proposed to improve the analysis of power asymmetry was to take into account how the strengths and weaknesses in one dimension mutually reinforce or weaken the others.

A further contribution to the theory was suggested by considering an additional dimension of power. There was a sense that the approach 'missed' a further political dimension that should 'encircle' or provide a background element of the fields of power. The 'missing' field of power was considered to derive from political and physical *contextual* elements which were enumerated as, for example:

1 changes in the regional power relations through war (e.g. the 2003 US/UK invasion of Iraq changed the balance of power in the Tigris–Euphrates);
2 changes in political circumstances, e.g. through a peace treaty (a potential treaty between Israel and Syria would have impacts in the water negotiations in the region);
3 changes in the domestic political scene (e.g. change of government, new political alliances, new negotiators);
4 political timing for negotiations and decisions;
5 support of new external partners (e.g. China, Gulf countries);
6 extended drought periods or unanticipated consequences of changes in climate.

Conclusion

This chapter has demonstrated the merit in viewing transboundary contexts through a lens of critical hydropolitics. The testing of the theory by mid-level water managers has confirmed its utility, and helped to substantiate it. While hydropolitics acknowledges the political nature of TWM, critical hydropolitics allows study of under-considered aspects. This chapter's focus has been on the role of power and power asymmetry in the purely political question of 'who gets how much water, when, where and why'.

The review of developments in hydropolitical theory has discussed the recognition that conflict and cooperation coexist, and that not all forms of 'cooperation' are good for all parties. Updates to hydro-hegemony theory were provided to consider four fields of power. These include material power (economic or military might) and riparian position, as a visible, or overt, form of power. The more hidden, 'covert', form of power included the fields of bargaining power (during the negotiations process) and ideational power (the ability to frame perceptions and establish mindsets). With the measures of power assessed relative to each other in a basin (considered visually as 'pillars'), the established form of hydro-hegemony becomes quite clear.

Brief consideration of the hydro-hegemonic configurations in the Jordan, Nile, and Tigris and Euphrates River Basins revealed that control by the hydro-hegemon in each case was carried out by a combination of the various fields of power. The theory emphasized, however, that configurations of power asymmetry are not static, and current configurations are changing or are likely to respond to changing political or physical circumstances. Power asymmetry can also provide opportunities for the collective good. And the so-called weaker riparian states have their own reserves of power that may be marshalled to challenge less equitable hydro-hegemonic configurations. The case of Ethiopia in 2009 negotiations of the NBI was taken as a case in point (see Nile case study, this volume Chapter 13).

Put to the test by practising water managers throughout the MENA region, the approach of critical hydropolitics was shown to stand up even as key improvements were suggested. Generally, the practitioners agreed about the relevance of power and power asymmetry in determining their operating contexts. Four key points for improvement were raised. First, there is a need for clarity and for distinguishing between 'transboundary' waters and 'shared' waters. It was felt that the common conflation of the terms in academic journals, policy reports or media served in some cases to veil the reality that many transboundary waters are indeed not shared equitably. Second, the approach should begin to refine the relative weight of the different fields and dimensions of power. It was pointed out, for instance, that riparian position was an influential source of power primarily when there was sufficient bargaining power to build upon it (as in the case of Turkey on the Tigris and Euphrates). The material field of power, furthermore, was shown to be much less significant in less asymmetrical contexts – as in the case of the North African states. Perhaps most importantly, bargaining power was identified as the primary field of utility for the non-hegemonic states. Each state has a measure of bargaining power that was not always fully recognized or exploited (e.g. Syria, Sudan and Ethiopia, and – though with less opportunity – Iraq and Palestine).

Building on the notion of coexisting conflict and cooperation, the third contribution to the theory was that some forms of water 'cooperation' are not always as beneficial as intended. There are instances of never-ending cooperative meetings whereby the hegemon shows inflexibility in retaining data or its preferred negotiating position rather than sharing or negotiating based on interests. Examples provided were from Egypt in the Nile Basin Initiative. The fourth manner in which the theory has been substantiated is through a number of suggested refinements. These include distinguishing the utility of the approach to non-securitized and securitized contexts; combination of the four fields of power rather than having them in artificially isolated 'pillars'; and most importantly the lack of consideration of the determining physical and political context. The approach would be strengthened by incorporating, for example, the timing of political events such as elections, treaties or coups, and physical events such as extended drought periods.

It is hoped that the readers will consider the approach of critical hydropolitics when considering the remaining contributions in the volume. The role of power and power asymmetry is also key and ever-present in legal processes, in attempts at institution-building and education, and in innovative attempts such as benefit-sharing to resolve water conflicts.

Notes

[1] Mirumachi has since conceived of an improved analytical method that works with the dual nature of conflict and cooperation in forming relations between riparians – through the plotting of cooperation and conflict continuums against each other, in the two-dimensional transboundary water interaction nexus (Mirumachi and Allan, 2007).

[2] The Sida-funded International Training Programme on Transboundary Water Management for the MENA Region is organized by Ramboll Natura AB and Stockholm International Water Institute (SIWI).

[3] The participants consisted of hydrologists, engineers, lawyers and political scientists holding positions in ministries, research centres, NGOs or

regional organizations. They came from Morocco, Algeria, Tunisia, Egypt, Jordan, Palestine, Lebanon, Syria, Turkey and Iraq.

References

Al Attili, Dr Shaddad (2008) head of the Palestinian Water Authority, personal communication, July

Allan, J. A. (2001) *The Middle East Water Question: Hydropolitics and the Global Economy*, I. B. Tauris, London

Bachrach, P. and Baratz, M. (1962) 'The two faces of power', *American Political Science Review*, vol 56, no 4, pp947–952

Brochmann, M. and Hensel, P. R. (2008) 'Peaceful management of international river claims', paper prepared for the *49th Annual Conference of the International Studies Association*, 26–29 March 2008, San Francisco, CA

Bullock, J. and Darwish, A. (1993) *Water Wars: Coming Conflicts in the Middle East*, St Dedmundsbury Press, London

Cascão, A. E. (2008a) 'Ethiopia – Challenges to Egyptian hegemony in the Nile Basin', *Water Policy*, vol 10 (S2), pp13–28

Cascão, A. E. (2008b) 'Ambiguity as strategy in transboundary river negotiations: the case of the Nile River Basin', in *Proceedings of II Nile Basin Development Forum*, Nile Basin Initiative/Sudanese Ministry of Irrigation and Water Resources, Khartoum, Sudan, 17–19 November 2008, pp569–579

Cascão, A. E. (2008c) (unpublished) 'Notes on learning experience in the International Training Programme on *Transboundary Water Management for the Middle East and North Africa (MENA) Region 2008*', Amman, Jordan, 2–3 November 2008

Cascão, A. E. (2009a) 'Changing power relations in the Nile river basin: unilateralism vs. cooperation?', *Water Alternatives*, vol 2, no 2, pp245–268

Cascão, A. E. (2009b) *Institutional Analysis of the Nile Basin Initiative: What Worked, What did Not Work and What are the Emerging Options?*, report prepared for the International Water Management Institute, Nile Basin Focal Project, Work Package 4

Daoudy, M. (2008) 'Hydro-hegemony and international water law: laying claims to water rights', *Water Policy*, vol 10 (S2), pp89–102

Daoudy, M. (2009) 'Asymmetric power: negotiating water in the Euphrates and Tigris', *International Negotiations*, vol 14, no 2, pp361–391

de Silva, L. and van der Zaag, P. (2009) 'Educational strategies: an integrative approach to water relations', presented to the transboundary water session *Cooperation as Conflict? Towards Effective Transboundary Water Interaction*, World Water Week August 2009, Stockholm International Water Institute, Stockholm, Sweden

Delli-Priscoli, J. (1996) *Conflict Resolution, Collaboration and Management in International Water Resource Issues*, Alternative Dispute Resolution Series, Institute for Water Resources (IWR) Working Paper 96-ADR-WP-6, Alexandria, USA, US Army Corps of Engineers, Alexandria, VA

Elhance, A. (1999) *Hydropolitics in the Third World: Conflict and Cooperation in International River Basins*, US Institute of Peace Press, Washington, DC

Falkenmark, M. (1989) 'Middle East hydro politics: water scarcity and conflicts in the Middle East', *Ambio*, vol 18, no 6, pp350–352

Frey, F. W. and Naff, T. (1985) 'Water: an emerging issue in the Middle East?', *Annals of the American Academy of Political and Social Science*, vol 482 (November), pp65–84

Furlong, K. (2008) 'Hidden theories, troubled waters: response to critics', *Political Geography*, vol 27, pp811–814

Gleditsch, N. P., Furlong, K., Hegre, H., Lacina, B. and Owen, T. (2006) 'Conflicts over shared rivers: resource scarcity or fuzzy boundaries?', *Political Geography*, vol 25, pp361–382

Hajer, M. A. (1997) *The Politics of Environmental Discourse: Ecological Modernization and the Policy Process*, Oxford University Press, Oxford

Hensel, P. R. and Brochmann, M. (2007) 'Armed conflict over international rivers: the onset and militarization of river claims', *Annual Meeting of the International Studies Association*, Chicago, IL, March 2007

The Independent (2006) 'Water wars – John Reid warns climate change may spark conflict between nations – and says British armed forces must be ready to tackle the violence', *The Independent*, London, 28 February, p1

Lasswell, H. D. (1936) *Politics: Who Gets What, When, How*, McGraw-Hill, New York, NY

Lewis, L. (2007) 'Water shortages are likely to be trigger for wars, says UN chief Ban Ki Moon', *Times Online*, 4 December

Lukes, S. (2004) *Power: A Radical View*, 2nd ed., Palgrave Macmillan, Basingstoke, UK

Lustick, I. S. (2002) 'Hegemony and the riddle of nationalism: the dialectics of nationalism and religion in the Middle East', *Logos*, vol 1, no 3, pp18–44

Markovitz, I. L. (1987) *Studies in Power and Class in Africa*, Oxford University Press, Oxford

Mehta, L. (2001) 'The manufacture of popular perceptions of scarcity: dams and water-related narratives in Gujarat, India', *World Development*, vol 29, no 12, pp2025–2041

Mirumachi, N. and Allan, J. A. (2007) 'Revisiting transboundary water governance: power, conflict, cooperation and the political economy', paper presented at the *International Conference on Adaptive and Integrated Water Management*, Basel, Switzerland, 12–15 November 2007

NATO (1999) *Environment and Security in an International Context – Final Report March 1999*, Committee on the Challenges of Modern Society Report No 232, North Atlantic Treaty Organization, Berlin, Germany

Schulz, M. (1995) 'Turkey, Syria and Iraq: a hydropolitical security complex', in: L. Ohlsson, (ed.) *Hydropolitics: Conflicts over Water as a Development Constraint*, Zed Books, London

Selby, J. (2003) 'Dressing up domination as "co-operation": the case of Israeli-Palestinian water relations', *Review of International Studies*, vol 29, no 1, pp121–138

Selby, J. (2007) 'Beyond hydro-hegemony: transnational hegemonic structures and national hegemonic projects', presentation given at the *Third International Workshop on Hydro-Hegemony*, London Water Research Group, London School of Economics, London, 12–13 May 2007

Tagar, Z., Keinan, T. and Bromberg, G. (2004) *A Seeping Time Bomb: Pollution of the Mountain Aquifer by Sewage*, Friends of the Earth–Middle East, Tel Aviv, Israel

Thomson, M. (2005) 'Ex-UN chief warns of water wars', *BBC News Online*, 2 February

Toset, H. P. W., Gleditsch, N. P. and Hegre, H. (2000) 'Shared rivers and interstate conflict', *Political Geography*, vol 19, pp971–996

Turton, A. (1999) *Water Scarcity and Social Adaptive Capacity: Towards an Understanding of the Social Dynamics of Water Demand Management in Developing Countries*, Occasional Paper 9, SOAS Water Issues Study Group, School of Oriental and African Studies/King's College London

Turton, A. (2002) 'Hydropolitics: the concept and its limitations', in A. Turton and R. Henwood (eds) *Hydropolitics in the Developing World: A Southern African Perspective*, African Water Issues Research Unit, Pretoria, South Africa

UNDP, UNEP, World Bank and World Water Resources Institute (2000) *World Resources 2000-2001: People and Ecosystems: The Fraying Web of Life*, World Water Resources Institute, Washington, DC

UNESCO (2008) *Groundwater Resources of the World: Transboundary Aquifer Systems*, (map), Worldwide Hydrogeological Mapping and Assessment Programme, UNESCO and IHP, Hannover, Germany, and Paris, France

Warner, J. (2008) 'Contested hydrohegemony: hydraulic control and security in Turkey', *Water Alternatives*, vol 1, no 2, pp271–288

Waterbury, J. (1979) *Hydropolitics of the Nile Valley*, Syracuse University Press, Syracuse, NY

Wegerich, K. (2008) 'Hydro-hegemony in the Amu Darya Basin', *Water Policy*, vol 10 (S2), pp71–88

Wolf, A. T. (1998) 'Conflict and cooperation along international waterways', *Water Policy*, vol 1, no 2, pp251–265

Wolf, A. T. (ed.) (2002) *Conflict Prevention and Resolution in Water Systems*, The Management of Water series, Edward Elgar Publishing, Cheltenham, UK

Wolf, A. T., Kramer, A., Carius, A. and Dabelko, G. D. (2005) 'Managing water conflict and cooperation', *State of the World 2005: Redefining Global Security*, Worldwatch Institute, Washington, DC

World Bank (2009) *West Bank and Gaza: Assessment of Restrictions on Palestinian Water Sector Development Sector Note April 2009*, Middle East and North Africa Region – Sustainable Development Report No 47657-GZ, The International Bank for Reconstruction and Development, Washington, DC

Zeitoun, M. (2006) *Power and the Palestinian-Israeli Water Conflict: Towards an Analytical Framework of Hydro-Hegemony*, PhD thesis, King's College London, University of London, UK

Zeitoun, M. and Warner, J. (2006) 'Hydro-hegemony: a framework for analysis of transboundary water conflicts', *Water Policy*, vol 8, no 5, pp435–460

4

Getting Beyond the Environment–Conflict Trap: Benefit Sharing in International River Basins

Marwa Daoudy

- Benefit sharing between basin states may induce cooperation over transboundary water management and assist in reducing conflicts.
- The need to identify the benefits is crucial and allows the development of positive-sum outcomes.
- Benefits can be of a monetary and non-monetary nature.
- To have useful effects, the process of benefit sharing needs to be more geographically inclusive, and needs to focus on the threat perceptions of the parties concerned.
- Benefit sharing should be sought in parallel with, and not versus, equitable allocations of water resources.

Introduction: Dilemmas of Conflict and Cooperation

Inequitable access to water can trigger conflict, especially when water is embedded in larger intra-state political conflicts or where limited economic diversification restricts the range of policy options open to governments. Potential solutions tend to involve agreements on volumetric water allocations. Sub-national conflicts are also of growing importance, and their potential is enhanced, as ever-increasing pressure is placed on the availability of water. Transboundary waters are of exceptional importance in this regard, especially in developing nations and among the poorest communities, which rely most heavily on natural resources. Conca (2006, pp73–91) notes that more than 50 per cent of the continental surface area lies in international river basins. Three dilemmas underlie the debate on conflict and cooperation in international river basins.

One dilemma in international river basins is that optimal water-usage solutions may not be congruent with the principle of equitable utilization (Grey, 2001). With spiralling water shortages, optimal solutions should be sought but this usually compromises the desire to achieve usage equity at the same time. Threat perceptions based on historic circumstances occurring outside the water sector then often push the issue of water-related equity in the direction of a purely technical point of view. It may be more efficient to locate dams in basin headwaters, but against this no guarantee may exist for the equitable distribution of benefits such dams may deliver. Unless specific, feasible, and enforceable guarantees can be constructed and unless the nature of regional politics supports such arrangements, this would render optimal technical solutions politically unsuitable.

Another dilemma relates to global public goods and ecological sustainability. It is well established that unregulated access to common pool resources results in unsustainable use, to the final disadvantage of all (Hardin, 1968). The inevitable consequence is the overexploitation of the resource, damaging the ecosystems and the services they provide. This is a particular problem with transboundary waters because upstream parties may overuse the resource and downstream states may be unable to counter these practices or unable to extract appropriate compensation.

Increasing water stress, the notion of an emerging threat to national resources and the inequitable distribution of water emphasize two dimensions of the theoretical debate: the link between environmental problems and the emergence of conflict, therefore highlighting the nature of water as a security issue – a third dilemma. Central to this is the role played by threat perception as an interceding variable, because it is through this mechanism that environmental issues such as water scarcity become perceived as national security threats, therefore driving securitization dynamics (Buzan et al, 1998; Turton, 2003b). Actors are often not merely concerned over the enforcement of an optimal distribution of water, but ultimately seek to achieve enforceable security arrangements. Security can be defined as a protection from threat (whether real or perceived), and the absence of fear (Brauch, 2003, p53). Security studies originally addressed concerns from the perspective of military and economic threats to a state's territory, its autonomy and its borders. This concept is at the core of traditional security studies (Buzan et al, 1998, p203; Brauch, 2003, p54). Critical or non-traditional security studies have introduced a new perspective on threats to national security, broadening the agenda from traditional

threats (military, economic) to new security threats which, for example, can be linked to the environment or to energy (Romm, 1993, p6; Krause and Williams, 1997; Spector and Wolf, 2000, p410). Other threats relate to national development policies, and establish a relationship between water and territory (Dalby, 1997, p3). For example, water scarcity and food-related scarcity often appear to constitute serious threats to the national security of developing countries (Korany et al, 1993, pp15–17).

The concept of *environmental security* has made its mark on the pre-existing debate on resource-oriented conflicts (Westing, 1986; Mathews, 1989, p162). Environmental scarcity is linked to very high risks of violent conflict because of acute change or stress in resources (such as water scarcity and extreme pollution), often accompanied by high population growth and a socially inequitable distribution of resources (Homer-Dixon, 1994, pp6–8). Environmental security can also be interpreted as the need to preserve local biospheres because of their established impact on a whole range of human activities, initiating a process of the securitization of environmental problems (Buzan et al, 1998, p76). Perceived as the protection of national resources, environmental security can be divided into categories, one of these involving transnational environmental problems, such as pollution, global climate change greenhouse effect, ozone depletion, depletion of forests and biodiversity, and resource-based problems (land degradation, forest and soil, threat of desertification) which traditionally affect territorial integrity and political stability (Romm, 1993, p15). Probabilities for increased military conflict seem to increase when rivers cross borders rather than forming borders, as this creates an upstream/downstream dynamic (Toset et al, 2000, p971). The predictions of water shortage and water stress were so alarming – particularly in some semi-arid regions – that by the mid-1990s, the idea of 'water wars' reached academic and policy-making circles, as well as the media.

Water has thus become a security issue, creating an 'environment–conflict trap'.[1] On the one hand, the link between environmental 'change' or 'degradation' and intra/interstate conflict is established, either as a cause of political conflict, social stress and ethnic tensions (Brauch, 2003, pp65–68) or as a contribution to armed conflict (Homer-Dixon, 1994), of which causes and intensities have been thoroughly quantified and their limits identified (Gleditsch, 1998). For others, in order to mitigate actual and potential conflict actually or ostensibly related to water, a 'power and security dilemma' needs to be overcome to deliver economic development (Buzan, 1991; Daoudy, 2009). Four categories of environmental change result in potential transboundary problems: degradation (pollution), scarcity (shortage), maldistribution (inequitable access) and disaster (Spector, 2002, p202). On the other hand, some would warn about the establishment of a paradigm in which all global threats are systematically linked to national security issues, making environmental security a loose concept (Ayoob, 1997; Deudney, 2004); others call for greater rigour in the use of security as a concept (Brauch, 2003), or object to any direct causal linkage between environmental change and violent conflict (Carius and Imbusch, 1999). Others consider the link between environmental degradation and/or change in water resource allocation as a threat to state security, without establishing this as a prime reason for armed conflicts (Lowi and Shaw, 2000, p149). The role of securitizing agents is also emphasized, when they seek to draw attention to the perceived plight of natural resource management by couching their communications in an alarmist manner (Buzan et al, 1998). This raises the question of how threats are perceived and used, rather than whether they are real. The securitization of threats helps to legitimize exceptional measures in the guise of defending national survival by taking the threat out of the normal domain of politics, and placing it in the security domain where it receives greater attention and means, but is also alienated from the general public.

Over the last decade, views have matured and begun emphasizing options for cooperation. This process has been cited as the 'desecuritization of water resource management' (Turton, 2003a, p96), and opens the way to negotiated agreements between states and the consequent *sharing of*

benefits. Recognition of the cooperative side of shared water resources became a mantra in the subsequent debate (Wolf, 1998; Conley and van Niekerk, 2000; Van der Zaag and Savenije, 2000; Kliot et al, 2001; Sadoff and Grey, 2002). Rather than military conflicts, weak water management and governance contributes to emerging 'social' conflicts over water, by affecting poor and rural communities (Conca, 2006). Shiva (2002) suggests that the privatization and pollution of water resources effectively constitute a 'war against the poor'. Boesen and Helle (2003) consider that social unrest is the more likely problem, and cautioned that without improved handling of international watersheds we may see a move 'from water wars to water riots'. The key issue remains the utilization of transboundary waters in a sustainable manner, while at the same time reducing poverty and ensuring that violent conflicts do not occur.

Characterized as a possible driver of regional integration, water resource management is also perceived as playing a major role in developing the foundation on which the future economic growth and prosperity of an entire region can be based. The best examples are found in the Southern African Development Community (SADC), the South Asian Association for Regional Cooperation (SAARC) and the Association of Southeast Asian Nations (ASEAN). A drive towards river basin management was initiated by ASEAN in 2002, with no significant contribution yet (Qaddumi, 2008, p12). When South Africa joined SADC in 1994, the very first Protocol that was signed in terms of the SADC Founding Charter was that on Shared Watercourse Systems, which became the foundation for regional economic integration (Ramoeli, 2002, p105). Even when military conflict was being waged in specific river basins, cooperation still occurred between water resource managers (Turton, 2003b).

Pursued as a policy tool especially at international conferences and workshops, where it appears to be more frequently cited than within academia, the concept of benefit sharing through international watershed management may help address these dilemmas. The establishment of river basin authorities with clear legal and organizational structures is commonly recommended, geographically covering the entire river basin, and functionally dealing with the multiple uses of water (Kliot et al, 2001, p252; Swain, 2004, p172). In recent years, options for cooperation and the amicable sharing of benefits resulting from professionally managed watersheds were identified (Sadoff and Grey, 2002; Phillips et al, 2006). The latter opens the way to negotiated agreements between states, the potential for conflict prevention, if not resolution. The present chapter seeks principally to analyse two important issues in the debate on conflict and cooperation in international river basins – whether cooperation on transboundary water management (TWM) can assist in reducing conflicts; and what role the sharing of benefits between co-riparians may play in this process.

The Rationale: Beyond the Environment–Conflict Trap, the Identification of Mutual Benefits

The fact that 'water wars' have not occurred does not imply that cooperation exists, or that water resources are shared equitably and reasonably by co-riparians. Lack of cooperation does carry security implications and can result in suboptimal water management, with adverse consequences for development and the environment. In this situation, some states prefer a non-cooperative situation to one where dominance and hegemony take over. Some have argued that, so far, many conflicts on competing claims for transboundary water resources have been solved by negotiations (so-called hydro-diplomacy) and through international bilateral and multilateral agreements (Barandat and Kaplan, 1998; Schiffler, 1998). Some note, however, that international agreements on water were commonly either abrogated (Swain, 2004), not implemented in full or replicated the asymmetrical distribution of power amongst riparians (Daoudy and Kistin, 2008). The following sections analyse conflict prevention in international river basins and the opportunities for cooperation introduced by the sharing of more broadly conceived potential mutual benefits.

In the peace and conflict literature, ecological well-being contributes to building 'positive peace', in contrast with 'negative peace' achieved through – among other things – diplomacy, negotiations and conflict resolution (Barash and Webel, 2002, pp460–484). More specifically, environmental conflict prevention seeks to reduce the impact of and the vulnerability to disasters, while addressing causes that can potentially contribute to environmentally induced tensions. The objective is to successfully and peacefully manage potential crises, and prevent their escalation into violent conflict. Examples of preventive negotiation are found in the Aral Sea Basin (1991) and the Zambezi River System Action Plan (1987), where agreements signed among concerned riparian states include conflict-prevention mechanisms (Spector, 2002, p211; Qaddumi, 2008).[2] By challenging the inevitability of the 'Tragedy of the Commons', Nobel Prize winner Elinor Ostrom also shows that common pool resources (CPR) are not always degraded, as individuals and societies adapt by devising rules and enforcement mechanisms. To foster the governance of natural resources, rules need to evolve over time, conflict resolution measures should be available, and the duties in preserving the common resource and the *benefits* from exploiting it should be balanced (Ostrom, 1990).

The need to identify the benefits is crucial. Benefit sharing is defined as 'any action designed to change the allocation of costs and benefits associated with cooperation' (Sadoff and Grey, 2005, cited in White et al, 2008, p7). Sadoff and Grey (2002, 2005) divide results from cooperation over a shared river basin into political, economic, environmental and catalytic categories, based on four general categories of benefits:

1   to the river (quantity, quality, regulation, soil conservation, other);
2   from the river (hydropower, agriculture, flood–drought management, other);
3   because of the river (cooperation versus conflict, economic development, food security, other);
4   beyond the river (markets and trade, regional stability, other).

Based on this framework, Phillips et al (2006) argue that benefit sharing is embedded in a larger set of bigger external issues, and therefore becomes the outcome of a process of issue-linkage. One of the complexities associated with the cause–effect linkage arises from the fact that water is fugitive in nature, moving through the landscape and biosphere over space and time. While many think of water as a stock, it is in fact a flow (Shaw, 2005, p5), otherwise known as a flux. And by virtue of crossing national boundaries, water forces riparian states into a situation of interdependence. The consideration of this phenomenon has been attempted in some of the literature on benefit sharing. For example, Klaphake (2005) looks narrowly at the benefits arising from specific engineering-related projects, adhering to those elements that can be easily measured. A bolder effort is found in the work by Claassen (2005), where the notion of benefits is linked to the overlap between ecosystems (as defined by watercourses) and economies (as defined by provincial borders). The Transboundary Water Opportunity (TWO) model later developed by Granit and Claassen (2009) provides a useful framework to identify the several opportunities offered within basins and their cumulative socioeconomic and environmental impacts for national economic development. The effect of this approach in terms of a benefit-sharing model is profound. Once specific issue-linkages are established, the model allows for a basket of possible trade-offs to be developed. The issue is shifted from the specific watershed, into the strategic domain of the 'problemshed' (Earle, 2003). In turn, this facilitates a dialogue of benefit sharing to take root and become viable.

Benefits can be of monetary and non-monetary nature. Phillips et al (2006) identify three broad sets of monetary and non-monetary benefits, which are identified as key motivating factors for decision makers: 1) security; 2) economic; and 3) environment-related benefits. In line with this, other models foster the 'optimization of values (economic, social, cultural, political and environmental) generated from water in its different uses and equitably distributing these benefits amongst water users and suppliers' (White et al,

2008, p4). From there, levels of conflict or cooperation are largely determined by the incentives co-riparians face and the need to identify joint watershed management benefits. On the one hand, security is a fundamental issue with a number of scales, in so far as water resource management can reduce uncertainty, and increase the assurance of supply needed to unleash the multiplier effects of good human health, stable economic employment and a belief in some future prosperity that motivates most human beings. In the security arena, regional benefits can accrue from the reduction in tensions that are multi-causal in nature, but amplified through specific water resource management paradigms and perspectives. Indicators relate to military expenditure per capita and percentage of GDP, water availability and dependency ratio, the history of water-related agreements, and levels of institutional cooperation and regional integration within the basin. Economic development, on the other hand, is a driver of human cooperation, and underpins any notion of security that informs the larger setting in which people find themselves. In the economic realm, issue-linkage can take place to the extent that a common currency is developed, through which benefit sharing can start to emerge. Human development can inform the negotiating positions of respective riparian states, developing a core logic of nested hierarchies arising from individual household security, through group security, societal security, and eventually up to national security levels. The framing of the logic in the language of economic development also means that common currency can be found via trade-offs that lead to potential benefit sharing. Similarly, if wealth is spread in society in a reasonably equitable fashion, then the incentive for conflict is reduced. Relevant indicators address GDP per capita, population below poverty line, life expectancy, literacy rate, energy use, and the contribution of agriculture/energy to the country's GDP.

Both of the above elements are nested in the environment. In the environmental sphere, a strong case can be made for the maintenance of some form of ecosystem integrity, either to sustain essential ecological goods and services on which human livelihoods are dependent, or to enable those ecosystems to function as sinks in a sustainable manner. Selected environmental indicators, such as the importance of flow regimes, the water quality index, sustainability of water use and biodiversity integrity, pave the way for potential benefit sharing in the environment. By providing security of access to environmental resources, threat perceptions are also allayed and desecuritization processes can start to gain acceptance. Similarly, humans are part of the environment, sustaining livelihood from it, but also impacting on it. Thus, the environment as a sink translates into thresholds of sustainability, and protection of the environment becomes a specific management objective that in and of itself can start to drive the type of cooperative spirit that underpins any form of benefit sharing. Similarly, the notion of the environment as a future renewable resource can be considered to underpin concepts of sustainable development. Finally, environmental health translates to good human health, specifically for poor and marginalized communities that are highly dependent on ecosystems for their livelihoods and basic food security. All of these become high-order outcomes that are benefits in their own right, easy to quantify and therefore possible to broadcast as components of a larger basket of benefits accruing from a mutually beneficial approach to the problem. In the environmental sphere, water is a foundation for all sustainable economic activities, with strong contributing factors to social stability and human well-being.

The way such benefits may best be shared is associated with trade-offs and scenarios for optimal use, while also been evaluated in relation to the cost of no agreement (or the cost of no cooperation). Grey et al note that an important driver towards cooperation is the 'management of water-related risks', such as flood control (2009, p16). Generally, flood control, drought mitigation, basin-wide yields of water, hydropower and optimal environmental management are perceived as positive outcomes of basin-wide cooperation (Qaddumi, 2008, p4). Consensus on the scope of benefits is enhanced through collaborative processes for the identification and quantification of benefits, by addressing, for example, negative

externalities generated by agriculture, such as pollution, soil erosion and reservoir siltation (White et al, 2008, p11). Poverty alleviation is also an important objective (White et al, 2008). Several mechanisms for the allocation of benefits include direct compensation for costs, side payments, revenue sharing, the establishment of development funds, direct payment for water, hydropower trade, preferential electricity rates, equity sharing with project-affected stakeholders or full ownership and financing procedures (Egré et al, 2002; Qaddumi, 2008). Possible trade between water resources and electrical power can be extended into many other areas, such as the exchange of water resources for traded food products deriving from these. Water allocations should however be agreed by the co-riparians if a basis is to be generated either for the sharing of benefits, or the calculation of compensation, which may be due from one co-riparian to another.

Van der Zaag and Savenije have also partially operationalized the concept, noting that one option for sharing the resources in a basin would be to identify development strategies that can thrive under an equitable division of water and other resources (2000, p14). They also address benefit sharing more concretely, referring to it as 'exploiting interdependencies', and making a reference to the gradual growth of a political situation such as that in the European Union, where countries dare to trust that benefits and costs will balance themselves out in the long term. Phillips et al (2006) note that, while clear, this perspective overlooks 'past or existing utilization' and 'existing and potential uses' (which also have legal standing, in the 1997 United Nations Convention), and fails to deal adequately with the need to avoid significant harm to the interests of other co-riparians. More importantly, the relative power usually determines the rationale for states to agree to such allocations. It therefore presumes the full acceptance by all stakeholders of the equitable use concept. This is clearly not always the case, as actors are sometimes reluctant to go beyond the water-security dilemma. In situations of dam construction in national river basins, the fair distribution of benefits from dams should reach populations affected by the project in the form of transfer of the revenues from hydropower projects to municipalities or regional entities, which acquire 'entitlements to parts of the ownership of the economic rent generated by the dam' (Egré et al, 2002, p8; Egré, 2007). The next section discusses possible ways to overcome these hurdles.

Resolving the 'Water-Security Dilemma': Acting on Perceptions, Asymmetries and Institutions

The securitization of water-related issues (linking water to national security concerns) is of a dual nature, with threat perception as a key variable because of its capacity to link issues of national security with perceptions of growing water scarcity (Turton, 2003a). In such scenarios, regional instability is generally increased, but short-term cooperation over water may in fact be promoted. There is also a risk that the use of a benefit-sharing approach may allow regional hegemons to exploit common resources under a formula that they themselves have defined, solely for their own benefit (Daoudy, 2006). Drawing on neo-realist, neo-liberal and historical structuralist perspectives, hegemony can carry different faces.[3] On one end of the spectrum, the benevolent hegemon is concerned with the absolute gains of states, while on the other end the exploitative hegemon is more concerned with relative gains and uses coercion to enforce compliance. It is therefore also crucial to address the asymmetry inherent to most water conflicts, and also to understand the way in which issue-linkage occurs through the mechanism of threat perception. The distribution of benefits, whether equitable or not, also depends on the power distribution within the basin (Phillips et al, 2006; Zeitoun and Jägerskog, 2009, p12). To have useful effect, the process of desecuritization and benefit sharing therefore needs to be more geographically inclusive, and needs to focus on threat perceptions. Perceptions are central, as 'states must believe the greater economic benefits will be gained and distributed equitably', and they cooperate when they perceive the benefits from cooperation as more important than the benefits of

not cooperating (Grey et al, 2009, pp1 and 15). The very process of engagement between key stakeholders changes their own perception of reality. This fosters learning, and serves to shift the dominant mode of thinking away from securitization to desecuritization, which is based on buy-in to a belief that cooperation *per se* generates mutually beneficial outcomes by providing accelerator processes through which benefits can be leveraged further. This requires a process-oriented approach, by enhancing communication and building information-sharing in what is sometimes described as 'informal information-based dialogue' (Grey et al, 2009, pp19–20). The Nile Basin Initiative (NBI) is referred to as a good illustration of this process (Sadoff and Grey, 2002; Grey et al, 2009). While some progress has been made concerning the sharing of benefits in the power sector, there has been little of real substance developed in relation to other types of benefits, as yet.

The debate has, also, been swaying between equitable allocations (and underlying water rights) versus benefits, or optimal allocation versus equity (Phillips et al, 2006; Qaddumi, 2008, p5). Phillips et al (2006) argue that the demand under customary international law for the equitable allocation of water resources, and the apparently competing approach for sharing benefits, are in fact 'two sides of the same coin'. One way forward would be to address water rights, but the prognosis for the success of this approach is limited if the role of threat perception as a mechanism of issue-linkage is ignored. The attainment of water rights (i.e. the equitable and reasonable utilization of the common water resource within a basin), has constituted the heart of the gradual process of codification of customary international water law, over many decades. This process led to the adoption of the United Nations Convention on the Law of the Non-Navigational Uses of International Watercourses (United Nations, 1997). The objective of the General Assembly of the United Nations in this instance was to promote harmonious practices of water management between upstream and downstream co-riparians, with a view to preventing unilateral abuses and the eruption of conflicts. This Convention has not as yet been ratified by a sufficient number of countries to enter into force as international law. It has, however, crystallized customary rules with relation to state behaviour, allowing riparians to appeal to this body of rules as enforceable international water law. It is argued here that the 'water rights approach' is critical to some transboundary basins, as the securitization dynamics do not allow anything other than agreement on specific volumetric allocations, which can be easily monitored and verified. In other basins, the partial desecuritization dynamic allows a much broader consideration of volumetric allocations alongside benefit-sharing scenarios. Solutions must therefore be sought on a case-by-case basis.

Within this context, robust regional and international institutions are important in providing much-needed infrastructure for the promotion and coordination of benefit sharing. Benefit-sharing is sometimes portrayed as 'an institutional arrangement for managing water' (White et al, 2008, p3). Strong institutions can also act in a major role in promoting economic growth and stability. By focusing on institutional strengthening, the dynamics of desecuritization are fostered. This results in institutional learning, as problems are redefined and issues are linked over time. By developing redundancy, the role of individual gatekeepers is factored out of the overall equation of interstate relations. By fostering coordination and cooperation, wider ranges of possible solutions are developed, making for a 'bigger basket' of potential benefits to be shared. The process of desecuritization can hence be supported by emerging legal rules and norms of state behaviour, which can be enforced both internally and externally.

Some authors have also questioned whether the sharing of benefits through some form of market allocation would in fact satisfy demands for equitable and reasonable allocations of the water resources, either theoretically or practically (Beyene and Wadley, 2004). For example, many states would encounter difficulties with employment in the agricultural sector if they were to agree to reduce operations in that sector in favour of benefit sharing with other co-riparians. While this could be offset by development in other sectors using a policy based on inter-sectoral allocative efficiency, the social implications would need to be clearly understood and addressed. Migration to urban centres would probably arise, along with other disruptive elements associated with a change in lifestyle from an agriculturally based society to an industrial nation. This is often seen by governments in developing countries as a threat to the foundations of their political power-base (Daoudy, 2005).

In concluding this part, it is worth noting that preventing or reducing water conflict should constitute a core objective of policy makers in the water community. In order to achieve this, we need to understand the dynamics of issue-linkage, of which threat perception is but one element in a complex array of factors. We now see a clear global trend towards the securitization of water-related issues, specifically where a prevailing threat perception allows such issues to be linked to fears of national survival, thereby unleashing the process of securitization which eventually leads to zero-sum outcomes which are not conducive to any possible benefit-sharing approach. In general terms, states will commonly tend to employ strategies to improve short-term gains, rather than engage in cooperative efforts towards the attainment of long-term solutions, and a comprehensive settlement is therefore not realized. This process of securitization occurs in stealth, and can lead to increased long-term instability between states, outweighing any immediate short-term benefits that individual parties may enjoy. Durable and peaceful relations between states require that benefits are shared, as only then can sustainable and equitable practices be realized. In this, the role of third parties is important, as they can mediate for an agreement on the benefits to be shared, provide technical assistance or inject investments within the basin (Qaddumi, 2008, p9).

Empirical Experiences of Benefit Sharing

Some national projects, such as the Shuikou dam in China which was commissioned in 1993–1996 for power generation and navigation, include the redistribution of project-related revenues or profits (Egré et al, 2002). It remains to be seen whether affected populations actually benefit from compensatory mechanisms. Egré et al (2002), Klaphake (2005), Egré (2007), Qaddumi (2008) and Grey et al (2009) analyse mechanisms that have been adopted in international river basins, mostly relating to dam construction for hydropower use and generation. Locally, compensation policies usually include access to irrigated land, generation of employment opportunities and market access (Egré, 2007, p2). The objective is to share the economic rent generated by the project with locally affected populations, in the form of revenue sharing, development funds, equity sharing, tax payment to local authorities and preferential water and electricity rates (Egré, 2007). In the Ganges–Brahmaputra–Meghna Basin (GBM), it has been argued that Bangladesh, Bhutan, China, India and Nepal would gain in considering ways to share benefits by jointly addressing future risks linked to increased monsoon intensity, high pollution, and burgeoning populations and industrial activities (Grey et al, 2009, pp16–17). Categories of benefits, linked to security, economics and the environment, have also been tested for potential conflict resolution in the Euphrates and Tigris Basin (Daoudy, 2007). Conclusions show the promising economic benefits which could be jointly reaped, and examples of mutual benefits can indeed be found in several international river basins. Power generation is shared by Brazil and Paraguay on the Itaipu (Egré et al, 2002). In the Senegal River Basin, development costs and benefits of joint water projects are shared between Senegal, Mali and Mauritania, on the basis of a 'burden-sharing formula' (Qaddumi, 2008, p5). In the Mahakali River, India and Nepal have agreed to share costs and trade power (Qaddumi, 2008). Two case studies, the Lesotho Highlands Project and the Columbia River Basin, however, provide illustrative examples of rent-sharing on the basis of mutual needs.

On the one hand, Lesotho and South Africa agreed, in 1986, on direct payment for water and power. The Treaty stipulated that a total of $70m^3$/second would be transferred in five phases from Lesotho to the water-scarce Gauteng province in South Africa, and a power plant constructed with a capacity of 72MW (Egré et al, 2002, p4). South Africa would pay royalties for water transfers and Lesotho would receive all of the hydroelectric power generated by the project. Phase 1A and Phase 1B were completed in 1998 and 2004 respectively, and a total of $38m^3$/second out of $70m^3$/second delivered to South Africa (Egré, 2007, pp86, 89). In 1999, the Lesotho Fund for Community Development was established to use royalty payments for poverty alleviation in Lesotho by supporting community-based activities. The flow of royalties has been substantial for Lesotho's revenues but experts note that the funds did not specifically reach the communities affected by the project. Considering the country's relatively small size, these represent an important part of the overall population (Egré, 2007, p92). Others note that, while contributing to joint institutions, data sharing and multilevel relationships within the basin, South Africa has also constrained the process of dialogue and strategic planning at the basin-wide level (Kistin, 2009).

On the other hand, the Columbia Treaty (1961) initiated an agreement for the construction, in British Columbia (Canada), of three large storage dams for flood control and the production of hydropower, which would be sold to users in the United States (Egré, 2007, pp46–47). In 1995, an institutional mechanism, similar to the Lesotho Fund for Community Development, was established: the Columbia Basin Trust Act (Egré, 2007, p50). Considering the profound socioeconomic, environmental and cultural impacts of the project, the purpose was to provide benefits to the project-affected residents of British Columbia, by addressing their social and environmental issues through rent distribution. Conditional on the strong involvement of community organizations, the delivery of benefits has indeed favoured sustainable development in the province of British Columbia (Egré, 2007).

In conclusion, successful benefit sharing can actually take place, as revealed by the empirical case studies. The extent of effectiveness is, however, limited by power asymmetry, as shown in the Lesotho Highlands Project, or conditional on the existence of a very active civil society, as witnessed on the Columbia River.

Conclusions

The present chapter addressed two issues – whether cooperation on TWM can assist in reducing conflicts; and what role the sharing of benefits between co-riparians may play in this process.

It may be concluded, at this stage, that the sharing of benefits cannot be considered a universal panacea, and does not provide a 'one-stop' alternative to the consideration of the equitable allocation of water resources. The issues faced in attempting to attain more equitable allocations of water resources or benefits (of various categories) must be addressed in a basin-specific fashion, with strategic trade-offs occurring at different levels of scale. Each basin presents distinct and unique characteristics, and no one pattern of future economic (or water-related) development will be successful throughout these – or indeed other – international watercourses. At best, benefit sharing will be highly complex to establish, and will not be implemented without the risk of failing to see the delivery of benefits reaching all stakeholders, at the state and sub-state level. Benefits arising from the generation of hydroelectric power cannot be divorced from those connected to other uses of water resources.

There is a clear need for further development of the concept as a whole, and this should involve much greater specificity as to why co-riparians of transboundary watercourses may either wish to or agree to share benefits, rather than simply dividing the water resources among themselves. They need to perceive net benefits as more advantageous than the duties and costs associated with mutual cooperation. In these cases, any successful benefit-sharing scheme will require the generation of a 'broad basket' of possible benefits to act as an inducement to each co-riparian to become involved. This emphasizes the compelling role of third-party intervention. It also implies a broad range of forms of trade, some including water resources or flows *per se*, and others involving benefits which relate to the water resources and their utilization, but are of a secondary nature. Finally, it is noted that the sharing of benefits is of a multi-directional nature, and will be possible only if the upstream–downstream dynamic which dominates many transboundary watercourses can be transcended by the relevant co-riparians, with a view to alleviating poverty, preventing conflict and promoting economic and social development.

Notes

1   This term is inspired from the 'poverty–conflict trap' (Collier et al, 2003). The contribution of development cooperation to peace-building has been emphasized on the basis of a 'conflict–development' or 'security–development' nexus (Paffenholz, 2005, p275).

2   In the case of the Aral Sea Basin, the 1991 agreement integrated a principle of equality among the five riparians (Kazakhstan, Kyrgistan, Tajikistan, Turkmenistan and Uzbekistan). In the Zambezi Basin, riparian states (Angola, Namibia, Zimbabwe, Botswana, Congo, Malawi and Mozambique) agreed to a set of procedures and mechanisms to deal with future conflicts over the shared waters.

3   The following sources represent, among others, the different schools of thought on hegemony: Organski, 1968; Gramsci, 1971; Wallerstein, 1983; Keohane, 1984; Gilpin, 1988.

References

Ayoob, M. (1997) 'Defining security: a subaltern realist perspective', in Krause and Williams (eds) *Critical Security Studies*, University of Minnesota Press, Minneapolis, MN, pp121–148

Barandat, J. and Kaplan, A. (1998) 'International water law: regulations for cooperation and the discussion of the international water convention', in Scheumann and Schiffler (eds) *Water in the Middle East, Potentials for Conflict and Prospects for Cooperation*, Springer, Berlin/Heidelberg, Germany, pp11–30

Barash, D. P. and Webel, C. (2002) *Peace and Conflict Studies*, Sage Publications, London

Beyene, Z. and Wadley, I. L. (2004) 'Common goods and the common good: transboundary natural resources, principled cooperation, and the Nile Basin Initiative', Presented at the *Breslauer Symposium on Natural Resource Issues in Africa*, University of California, Berkeley, CA

Boesen, J. and Helle, M. R. (2003) 'From water wars to water riots: lessons from transboundary water management', DIIS Working Paper No 2004/6, DIIS, Copenhagen, Denmark

Brauch, H. G. (2003) 'Security and environment linkages on the Mediterranean space: three phases of research on human and environmental security and peace', in H. G. Brauch, P. H. Liotta, A. Marquina, P. Rogers and S. M. El-Sayed (eds) *Security and Environment in the Mediterranean*, Springer, Berlin/Heidelberg, pp35–144

Buzan, B. (1991) *People, States and Fear: An Agenda for International Security Studies in the Post-Cold War Era*, Harvester Wheatsheaf, London

Buzan, B., Waever, O. and de Wilde, J. (1998) *Security, A New Framework for Analysis*, Lynne Rienner Publishers, Boulder, CO

Carius, A. and Imbusch, K. (1999) 'Environment and security in international politics – an introduction', in A. Carius and K. Lietzmann (eds) *Environmental Change and Security: A European Perspective*, Springer, Berlin/Heidelberg, pp7–30

Claassen, M. (2005) *Elands Catchment Reserve Assessment Study, Mpumalanga Province: The Ecological Reserve and the Economic Value of the Aquatic Ecosystem in the Elands River*, Department of Water Affairs and Forestry, Pretoria, South Africa

Collier, P., Lance, E., Havard, H., Hoeffler, A., Reynal-Querol, M. and Sambanis, N. (2003) *Breaking the Conflict Trap: Civil War and Development Policy*, Oxford University Press, Oxford

Conca, K. (2006) 'The new face of water conflict', *Navigating Peace*, Woodrow Wilson International Center for Scholars, Environmental Change and Security Program, Washington, DC, no 3, November 2006, pp1–3

Conley, A. H. and van Niekerk, P. H. (2000) 'Sustainable management of international waters: the Orange River case', *Water Policy*, vol 2, pp131–149

Dalby, S. (1997) 'Contesting an essential concept: reading the dilemmas in contemporary security discourse', in K. Krause and M. C. Williams (eds) *Critical Security Studies: Concepts and Cases*, University of Minnesota Press, Minneapolis, MN, pp3–32

Daoudy, M. (2005) *Le Partage des Eaux entre la Syrie, la Turquie et l'Irak: Negociation, Securite et Asymetrie des Pouvoirs*, CNRS Editions, Paris

Daoudy, M. (2006) 'The Euphrates and Tigris waters, a mixed process of water negotiation', Presentation at the Swedish International Water Institute, Stockholm, Sweden, 2 February 2006

Daoudy, M. (2007) 'Benefit sharing as a tool for conflict transformation: applying the Inter-SEDE model to the Euphrates and Tigris', *The Economics of Peace and Security Journal, Special Edition on Water, Economics and Conflict*, vol 2, no 2, pp26–32

Daoudy, M. (2009) 'Asymmetric power: negotiating water in the Euphrates and Tigris', *International Negotiation*, vol 14, pp361–391

Daoudy, M. and Kistin, E. J. (2008) 'Beyond water conflict: evaluating the effects of international water cooperation', paper presented at the *International Studies Association (ISA) Annual Meeting*, 26–29 March 2008

Deudney, D. (2004) 'The case against linking environmental degradation and national security', in K. Conca and G. Dabrelko (eds) *Green Planet Blues, Environmental Politics from Stockholm to Johannesburg*, 3rd ed., pp303–313

Earle, A. (2003) 'Watersheds and problemsheds: a strategic perspective on the water/food/trade nexus in Southern Africa', in A. R. Turton, P. Ashton and T. E. Cloete (eds) *Transboundary Rivers, Sovereignty and Development: Hydropolitical Drivers in the Okavango River Basin*, AWIRU, Pretoria and Green Cross International, Geneva, pp229–249

Egré, D. (2007) *Compendium on Relevant Practices – 2nd Stage, Revised Final Report, Benefit-Sharing Issue*, United Nations Environment Programme (UNEP), Dams and Development Project, 10 February 2007

Egré, D., Roquet, V. and Durocher, C. (2002) *Benefit Sharing from Dam Projects, Phase 1: Desk Study, Final Report*, The World Bank Group, Washington, DC, 15 November 2002

Gilpin, R. (1988) 'The theory of hegemonic war', *Journal of Interdisciplinary History*, vol 18, no 4, pp591–613

Gleditsch, N. P. (1998) 'Armed conflict and the environment: a critique of the literature', *Journal of Peace Research*, vol 35, no 3, pp381–400

Gramsci, A. (1971) *Selections from the Prison Notebooks*, Lawrence and Wishart, London

Granit, J. and Claassen, M. (2009) 'A path towards realizing tangible benefits in transboundary river basins', in A. Jagerskög and M. Zeitoun *Getting Transboundary Water Right: Theory and Practice for Effective Cooperation*, Report no 25, SIWI, Stockholm, Sweden, pp21–26

Grey, D. (2001) 'Sharing benefits of transboundary waters through cooperation', Presentation at the *International Conference on Freshwater*, Bonn, Germany

Grey, D., Sadoff, C. and Connors, G. (2009) 'Effective cooperation on transboundary waters: a practical perspective', in A. Jagerskög and M. Zeitoun *Getting Transboundary Water Right: Theory and Practice for Effective Cooperation*, Report no 25, SIWI, Stockholm, pp15–20

Hardin, G. (1968) 'The tragedy of the commons', *Science*, no 162, pp1243–1248

Homer-Dixon, T. (1994) 'Environmental scarcities and violent conflict: evidence from cases', *International Security*, vol 19, no 1, pp5–40

Keohane, R. O. (1984) *After Hegemony: Cooperation and Discord in the World Political Economy*, Princeton University Press, Princeton, NJ

Kistin, E. J. (2009) 'The dynamic effects of riparian interactions over transboundary water management in the Orange-Senqu Basin', paper prepared for LWRG/UPTW/UNESCO session on *Cooperation as Conflict? Towards Effective Transboundary Water Interaction*, World Water Week, Stockholm, Sweden, 16 August 2009

Klaphake, A. (2005) 'Kooperation an internationalen Flüssen aus ökonomischer Perspektive: Das Konzept der Benefit Sharing', Discussion Paper 6/2005, Deutches Institut für Entwisklungspolitik, Bonn, Germany

Kliot, N., Shmueli, D. and Shamir, U. (2001) 'Institutions for management of transboundary water resources: their nature, characteristics and shortcomings', *Water Policy*, vol 3, pp229–255

Korany, B., Brynen, R. and Noble, P. (1993) 'The analysis of national security in the Arab context: restating the state of the art', in B. Korany, R. Brynen and P. Noble, (eds) *The Many Faces of National Security in the Arab World*, St Martin's Press, New York, NY, pp1–23

Krause, K. and Williams, M. C. (eds) (1997) *Critical Security Studies: Concepts and Cases*, University of Minnesota Press, Minneapolis, MN

Lowi, M. and Shaw, B. (eds) (2000) *Environment and Security, Discourses and Practices*, Macmillan Press, London

Mathews, J. T. (1989) 'Redefining security', *Foreign Affairs*, vol 68, no 2, pp162–177

Organski, A. F. K., (1968) *World Politics*, Knopf, New York, NY

Ostrom, E. (1990) G*overning the Commons: The Evolution of Institutions for Collective Action*, Cambridge University Press, Cambridge

Paffenholz, T. (2009) 'Understanding the conflict-development nexus and the contribution of development cooperation to peacebuilding', in D. J. D. Sandole, S. Byrne, I. Sandole-Staroste and J. Senehi, J. (eds) *Handbook of Conflict Analysis and Resolution*, Routledge, London, pp272–285

Phillips, D., Daoudy, M., Öjendal, J., Turton, A. and McCaffrey, S. (2006) *Transboundary Water Cooperation as a Tool for Conflict Prevention and Broader Benefit-Sharing*, Global Development Studies, no 4, Ministry of Foreign Affairs, Stockholm, Sweden

Qaddumi, H. (2008) *Practical Approaches to Transboundary Water Benefit Sharing*, Working Paper 292, Overseas Development Institute, London

Ramoeli, P. (2002) 'SADC protocol on shared watercourses: its history and current status', in A. R. Turton, and R. Henwood (eds) *Hydropolitics in the Developing World: A Southern African Perspective*, African Water Issues Research Unit, Pretoria, South Africa, pp105–111

Romm, J. J. (1993) *Defining National Security: The Non-Military Aspects*, Council on Foreign Relations, PEW Project on America's Task in a Changed World, Council on Foreign Relations Press, New York, NY

Sadoff, C. W. and Grey, D. (2002) 'Beyond the river: the benefits of cooperation on international rivers', *Water Policy*, vol 4, pp389–403

Sadoff, C. W. and Grey, D. (2005) 'Cooperation on international rivers: a continuum for securing and sharing benefits', *Water International*, vol 30, no 4, pp420–427

Schiffler, M. (1998) 'International water agreements: a comparative view', in W. Scheumann and M. Schiffle (eds) *Water in the Middle East, Potential for Conflicts and Prospects for Cooperation*, Springer, Berlin/Heidelberg, pp31–45

Shaw, W. D. (2005) *Water Resource Economics and Policy: An Introduction*, Edward Elgar Publishing, Cheltenham, UK

Shiva, V. (2002) *Water Wars – Privatization, Pollution, and Profit*, Pluto Press, London

Spector, B. (2002) 'Transboundary disputes: keeping backyards clean', in A. Wolf, (ed.) *Conflict Prevention and Resolution in Water Systems: Procedures and Mechanisms*, Edward Elgar Publishing, Northampton, UK, pp200–221

Spector, B. and Wolf, A. (2000) Negotiating security: new goals, changed process', *International Negotiation*, vol 5, no 3, pp409–426

Swain, A. (2004) *Managing Water Conflict – Asia, Africa and the Middle East*, Routledge, London

Toset, H. P. W., Gleditsch, N. P. and Hegre, H. (2000) 'Shared Rivers and Interstate Conflict', *Political Geography*, vol 19, pp971–996

Turton, A. R. (2003a) 'The political aspects of institutional development in the water sector: South Africa and its international river basins', DPhil thesis, Department of Political Science, University of Pretoria, South Africa

Turton, A. R. (2003b) 'Hydropolitics: the concept and its limitations', in A. R. Turton and R. Henwood (eds) *Hydropolitics in the Developing World: A Southern African Perspective*, African Water Issues Research Unit, Pretoria, South Africa, pp13–19

United Nations (1997) *United Nations Convention on the Law of the Non-navigational Uses of International Watercourses,* United Nations General Assembly, A/RES/51/869, 21 May

Van der Zaag, P. and Savenije, H. H. J. (2000) 'Towards improved management of shared river basins: lessons from the Maseru Conference', *Water Policy*, vol 2, pp47–63

Wallerstein, I. (1983) 'The three instances of hegemony in the history of the capitalist world economy', *International Journal of Comparative Sociology*, vol 24, pp100 108

Westing, A. (ed.) (1986) *Global Resources and International Conflict: Environmental Factors in Strategic Policy and Action*, Oxford University Press, New York, NY

White, D., Wester, F., Huber-Lee, A., Hoanh, C. T. and Gichuki, F. (2008) *Water Benefits Sharing for Poverty Alleviation and Conflict Management: Topic 3 Synthesis Paper*, Challenge Program on Water and Food, Colombo, Sri Lanka

Wolf, A. T. (1998) 'Conflict and Cooperation along International Waterways', *Water Policy*, vol 1, no 2, pp251–265

Zeitoun, M. and Jagerskög, A. (2009) 'Confronting power: strategies to support less powerful states', in A. Jagerskög and M. Zeitoun, (eds) *Getting Transboundary Water Right: Theory and Practice for Effective Cooperation*, Report no 25, SIWI, Stockholm, Sweden, pp9–14

PART 2

TRANSBOUNDARY WATER MANAGEMENT POLITY AND PRACTICE

5

International Water Law: Concepts, Evolution and Development

Owen McIntyre

- The theoretical approaches that states adopt to the concept of sovereignty, and territorial sovereignty in particular, largely determine the legal principles which govern shared water resources and must, therefore, be reconciled.
- Though the governing principles of international water resources law are now almost universally accepted and understood, depending on their particular interests states may still disagree fundamentally on what such principles actually mean and on how they ought to be applied in practice.
- The practical implementation of substantive principles of international water resources law may be almost entirely dependent on the existence of appropriate interstate institutional machinery.
- International water resources law is hampered by the same shortcomings in relation to compliance and enforceability which hinder the application of international law more generally.

Introduction

International freshwaters have provided a source of conflict between states throughout history. Such conflict stems both from the absolute dependence of peoples upon water and the resulting interdependence of co-basin states, and from a measure of uncertainty historically as to the applicable principles of international law in this area. This interdependence stems from the fact that states only exercise effective control temporarily or partially over such waters as they flow through or along their territories and so each co-basin state's utilization or development of a shared water resource necessarily affects the quantity and quality of water available to lower or neighbouring basin states, or may have other detrimental results (Bruhacs, 1993, p42). This legal uncertainty has resulted primarily from the irreconcilability of various principles, invoked under general international law, which may be applied to the apportionment of quantum share or allocation of uses of shared water resources as between co-basin states. In this area, as in any other, legal uncertainty renders conflict avoidance and resolution problematic.

Of course, it must be borne in mind that the rules and principles of international law merely provide a backdrop for interstate cooperation and the resolution of interstate disputes where states consent to be bound by such rules and principles or agree to submit a water resources dispute to a tribunal which can apply them. The rules of international water resources law suffer from the same deficiencies in terms of enforceability as does international law generally. This lack of enforceability, as well as a lack of normative clarity, can function in some instances to permit determined strong states to disregard the restrictions imposed by international law and to promote their own interests inequitably to the detriment of co-riparian states (McIntyre, 2007).

Background

The use, development and apportionment of freshwater resources have been central concerns for societies throughout the ages (Teclaff, 1991, pp60–61). The potential for conflict is ever-increasing as the demands to utilize the water resources of all river systems increase. This increase in demand can be attributed to a host of factors including world population growth (Utton, 1996, pp153–4, increasing industrialization and the use of modern farming techniques. Water resources are being utilized at an ever-increasing rate, for agricultural irrigation, for sewage and sanitation, as a medium for the disposal of industrial waste, as a raw material in industrial processes, as a (carbon-free) source of energy, as a source of food and for recreational purposes. The problems caused by increasing demand are exacerbated by the deteriorating quality of the available supply due to pollution. Despite the closed nature of the hydrological system, water use and waste water production continue to rise at alarming rates, thus increasing competition among users (Tolba et al, 1993, p85). Increasing pressure on water resources and increasing technical capability add to the likelihood of development of river systems through the construction of dams, reservoirs and other projects which maximize water utilization (WCD, 2000, pp8–10). This increases the potential for conflict among co-basin states as development may prejudice the rights of lower basin states. Of course, the possible effects of global or regional climate change on the availability and distribution of freshwater resources are as yet unknown but are likely to further exacerbate interstate competition (see Falkenmark and Jägerskog, this volume Chapter 11).

General international law relating to the utilization of international freshwaters has traditionally suffered from uncertainty due, primarily, to the existence of a number of conflicting theoretical approaches or principles that have been invoked to identify and justify the respective rights of co-basin states regarding their exact quantum share of water, or the allocation and permissibility of uses of shared water resources. In addition, until relatively recently there had been little proven substantive general international law on the subject, with the community of nations being more concerned historically with navigational questions. Also, such international law as had existed tended to be dominated by bilateral and regional treaties, as opposed to multilateral or more widely applicable agreements. However, recent attempts at codifying this area of law have been of very considerable assistance.

Key Theoretical Background

Four seemingly incompatible theories and approaches have traditionally been put forward concerning the rights of co-basin states to utilize shared international freshwaters (Lipper, 1967, p18; Lammers, 1984, p556; Birnie and Boyle, 1992, p218), though this conceptual framework has also been categorized as including three key theories (Fitzmaurice and Elias, 2004, pp11–15). These include the principles of 'absolute territorial sovereignty', 'absolute territorial integrity', 'limited territorial sovereignty' and the 'common management' approach. Each of these theories attempts to reconcile a particular position in relation to state utilization of international watercourses with the concept of territorial sovereignty as recognized and protected under general international law.

Absolute territorial sovereignty

The first principle, that of 'absolute territorial sovereignty', provides that a co-basin state may freely utilize waters within its territory without having any regard to the rights of downstream or contiguous states. Having absolute sovereignty over water resources while they are within its territory, a state may extract or alter the quality of these waters to an unlimited extent but has no right to demand continued flow from an upper co-basin state or to assert its rights against a contiguous state. This approach is closely associated with the so-called 'Harmon Doctrine', named after the United States Attorney-General who first articulated the principle in 1895. In asserting the absolute right of the United States to divert the Rio Grande, Attorney-General Judson Harmon stated that:

> *The fact that the Rio Grande lacks sufficient water to permit its use by the inhabitants of both countries does not entitle Mexico to impose restrictions on the United States which would hamper the development of the latter's territory or deprive its inhabitants of an advantage with which nature had endowed it and which is situated entirely within its territory. To admit such a principle would be completely contrary to the principle that the United States exercises full sovereignty over its natural territory* (cited in Birnie and Boyle, 1992, p218; Fitzmaurice and Elias, 2004, p12).

Harmon goes on to conclude that the question of whether the United States should 'take any action from considerations of comity' is a question which 'should be decided as one of policy only, because, in my opinion, the rules, principles and precedents of international law impose no liability or obligation upon the United States' (cited in Bruhacs, 1993, p43). However, the United States did not assert this position long, concluding bilateral treaties with Mexico in 1906 and Canada in 1909, which are more consistent with the principle of equitable utilization. In a detailed analysis of the Harmon Doctrine, McCaffrey concludes either that this position was taken by the United States in the course of the dispute concerned as a matter of

advocacy which might prove useful in future negotiations, or that 'Mr Harmon's attitude seems to have been merely the caution of the ordinary lawyer who is determined not to concede unnecessarily a single point to the other side' (McCaffrey, 2001, pp93–4 and 111).

Indeed, having examined US practice in relation to a wide range of subsequent disputes over shared freshwater resources, McCaffrey can conclude that 'Harmon's opinion stands out as an unfortunate anomaly' (McCaffrey, 2001, pp102–111).

The doctrine has little support among commentators or in state practice (Birnie and Boyle, 1992, p218) though it has been referred to by the United States, Austria, Chile, the Federal Republic of Germany, Ethiopia and India (Bruhacs, 1993, p44). However, an examination of actual state practice reveals that even these states which have purported to rely on the principle of absolute territorial sovereignty in diplomatic exchanges have not acted accordingly. For example, India had asserted its 'full freedom' with regard to the utilization of the waters of the Indus but subsequently concluded the 1960 Indus Waters Treaty which provided for the equitable apportionment of these waters (Baxter, 1967, p453; Birnie and Boyle, 1992, p219). At one point during the dispute, India asserted that 'both [countries] have full and exclusive jurisdiction over the management, control and utilization of natural waters available in their territories' (Baxter, 1967, p456), a position which Pakistan characterized as striking at 'the very root of Pakistan's right to historic, legal and equitable share in the common rivers' (McCaffrey, 2001, p117). Similarly, in relation to the dispute over the construction by India of a barrage on the Ganges River at Farakka, 11 miles upstream from the Bangladesh border, India eventually stated its general position to the Special Political Committee of the UN General Assembly so as to expressly deny reliance on the Harmon Doctrine and to confirm that 'India, for its part, had always subscribed to the view that each riparian state was entitled to a reasonable and equitable share of the waters of an international river' (McCaffrey, 2001, p118). Indeed, it appeared that the Harmon

Doctrine had 'become a potent weapon in the hands of a downstream state accusing an upstream state of acting unreasonably' (McCaffrey, 2001, p118).

Likewise, though Austria had stated in the period immediately after World War II that '[I]n accordance with the law of territorial sovereignty, waterways are at the entire disposal of each country throughout the whole stretch within its territory', it consistently agreed to give notice of any plans to develop successive watercourses and to consider objections to those plans on 'legal, technical or economic grounds' (Fitzmaurice and Elias, 2004, p12). McCaffrey cites a senior Austrian official dealing with international watercourses who stated in 1958 that, though it is a predominantly upper-riparian state, Austria did not follow the Harmon Doctrine, which he considered to be 'dead' (McCaffrey, 2001, p119).

Though Chile would appear to have asserted an absolute territorial sovereignty approach in a 1921 dispute with Bolivia over the Rio Mauri, in both that dispute and a later one with Bolivia involving the Rio Lauca, the Chilean authorities took steps to ensure that the works contemplated would not cause significant harm to Bolivia or in any way prejudice its interests as a lower riparian (McCaffrey, 2001, p121). Also, Chile entered into the 1971 Act of Santiago concerning Hydrologic Basins with Argentina which, while 'expressly recognizing general rules of international law ... governing the utilization of the waters common to the two countries', provided, *inter alia*, that '[T]he waters of rivers and lakes shall always be utilized in a fair and reasonable manner'.

Finally, though Ethiopia would appear to continue to abide by the Harmon Doctrine, asserting as recently as 1978 'all the rights to exploit her natural resources' (Okidi, 1980, p440), McCaffrey cautions that these statements have been made in the heat of exchanges between Ethiopia and Egypt, against an historical background of Egypt and Sudan purporting to allocate all Nile waters between themselves exclusively, and so may have been in part designed to elicit Egyptian cooperation. He also refers to the fact that Ethiopia and Egypt, along with eight other Nile Basin countries, are currently engaged in the process of forming a cooperative framework agreement (McCaffrey, 2001, pp122–23).

This approach has little or no support in judicial or arbitral practice. For example, it would appear to have been roundly criticized by the Italian Court of Cassation in its 1938 decision in *Société énergie électrique du littoral méditerranéen v. Compagnie imprese elettriche liguri*, where it purported to apply principles of international law in the settlement of a dispute concerning the use of transboundary water resources.

As for support among publicists for the absolute territorial sovereignty approach, those authors who might be argued to have endorsed the principle mostly voiced their respective positions before non-navigational uses of international watercourses had taken on much significance, which 'may have led them not to appreciate fully the serious ecological, economic and other harm ... that could result from large-scale diversions or pollution' (Berber, 1959, pp15–19; McCaffrey, 2001, p126). Also, all the commentators referred to came from only four countries, Austria, Germany, Canada and the United States, each of which are upstream states, at least in relation to some of their watercourses and neighbouring co-riparians (McCaffrey, 2001, p124). At any rate, it is possible to conclude that no recent authoritative work supports this approach while very many, including even relatively early ones, completely eschew it (McIntyre, 2007). Indeed, Smith referred to the absolute territorial sovereignty approach generally as an 'intolerable' doctrine which was 'radically unsound' (Smith, 1931, p8; Bruhacs, 1993, p46). Likewise McCaffrey, having conducted a thorough and up-to-date study of purported support for this approach, can unequivocally conclude that '[I]t is at best an anachronism that has no place in today's interdependent, water-scarce world' (McCaffrey, 2001, p114).

Absolute territorial integrity

The second principle, 'absolute territorial integrity', confers a right on a lower co-basin state to demand the continuation of the full flow of waters of natural quality from an upper co-basin state but confers no right to restrict or impair the natural flow of waters from its territory into that of a still lower co-basin state. This principle is the antithesis of the Harmon Doctrine and would effectively grant a right of veto upon a downstream or contiguous state, as its prior consent would be required for any change in the regime of the international watercourse (McIntyre, 2007). As might be expected, this approach has generally tended to find favour among lower riparian states as it would be likely to operate to retard upstream states in the development of their water resources. Like the Harmon Doctrine, the principle is based upon sovereignty but it would also appear to have a basis in the equality of states and so may be seen as being more compatible with the principle of sovereign equality as enshrined in Article 2(1) of the Charter of the United Nations (Bruhacs, 1993, p47). It can also be seen as being somewhat more in step with the general principles of international law relating to environmental protection and with general obligations to cooperate.

This approach has its origins in riparian rights doctrines which had traditionally existed in national legal systems. The courts of the United Kingdom and the United States consistently held that the use of the waters of a stream was only permitted on riparian land, i.e. land contiguous to the stream, and that any riparian was permitted to use as much water as was needed for domestic purposes. He could also use the waters for other, extraordinary, purposes only if the use pertained to the riparian land and did not appreciably diminish the natural flow. Also, the use had to be one which the courts would consider reasonable (Teclaff, 1985, pp6–20; Teclaff, 1991, pp63–64).

The doctrine would appear to have been recognized, though in a moderate formulation, as established general international law by the International Law Institute in its 1911 Madrid Declaration. The Declaration recognized as fact that '[R]iparian states with a common stream are in a position of permanent physical dependence on each other', and identified two basic rules by which riparians were bound (Teclaff, 1991, p67). Firstly, that:

when a stream forms the frontier of two states … neither state may, on its own territory, utilize or allow the utilization of the water in such a way as to seriously interfere with the utilization by the other state or by individuals, corporations, … thereof

and secondly, that:

when a stream traverses successively the territories of two or more states … no establishment … may take so much water that the constitution [of the stream], otherwise called the utilizable or essential character of the stream shall, when it reaches the territory downstream, be seriously modified.

However, it is clear from the deliberations surrounding negotiation of the 1923 Geneva Convention relating to the Development of Hydraulic Power Affecting More than One State that states have always had serious reservations about this approach.

Regarding state practice, Argentina, Egypt, Spain, Bangladesh, and a number of Arab states have invoked the principle of absolute territorial integrity and the principle still has some state support (Bruhacs, 1993, p44). However, Bruhacs also points out that its application was unequivocally rejected in the 1957 case of *Lac Lanoux* (*Spain v. France*).

In the course of a thorough review of the purported support for this principle in state practice and academic writing, McCaffrey argues that reliance on these contrasting extreme positions demonstrates above all else that they were invoked as 'tools of advocacy' rather than as legal principles considered likely to assist in the resolution of concrete disputes (McCaffrey, 2001, pp129–130). In relation to Egypt, one of the most consistent and vociferous exponents of this approach, he notes that as recently as 1981 it has expressed the view in international forums that:

Each riparian country has the full right to maintain the status quo of the rivers flowing through its territory … it results from this principle that no country has the right to undertake any positive or negative measure that could have an impact on the river's flow in other countries … a river's upper reaches should not be touched lest this should affect the flow of quantity of its water ... in general any works at a river's upper reaches that may affect the countries at the lower reaches are banned unless negotiations have taken place.[1] (McCaffrey, 2001, p130)

It must also be noted, however, that in practice Egypt has agreed to projects in upper Nile Basin countries, such as the Owen Falls Dam in Uganda, where Egypt even agreed to pay substantial compensation for the loss of hydroelectric power resulting from measures taken to protect irrigation interests in Egypt, and also that Egypt is an active participant in efforts to develop a cooperative framework for the sustainable development of the Nile (Garretson, 1967, p272; McCaffrey, 2001, p131). Similarly, though Pakistan, the lower riparian on the Indus, would initially appear to have adopted a position consistent with absolute territorial integrity in its early diplomatic communications with India, it soon afterwards proposed a conference for the purpose of agreeing upon an 'equitable apportionment' of all waters shared by the two countries (Baxter, 1967, pp451 and 454). Also, though Bolivia has been cited as an example of a state relying on the principle of absolute territorial integrity in the course of disputes over the Rio Mauri and the Rio Lauca with Chile, its upstream neighbour, its arguments were based on the 1933 Declaration of Montevideo, rather than on general international law. Its extreme position is best understood in the context of the overall climate of tension between the two countries going back to the 1830s (McCaffrey, 2001, pp132–133). Thus, the political context remains very relevant in any discourse over the application of rules of international water law.

Of the few commentators who can be cited in support of the doctrine of absolute territorial integrity, even fewer stand up to serious scrutiny (Berber, 1959, pp19–22). Also, as in the case of absolute territorial sovereignty, most of the authors cited in support of absolute territorial integrity wrote before non-navigational uses had assumed much significance and thus before state practice concerning non-navigational uses had substantially developed. Ultimately, Birnie and Boyle are not convinced that this doctrine is estab-

lished or useful and state that it 'appears devoid of more than limited support in state practice, jurisprudence, or the writings of commentators' (Birnie and Boyle, 1992, p219). Similarly, commenting on both the opposing doctrines of absolute territorial sovereignty and absolute territorial integrity, McCaffrey concludes that:

Both doctrines are, in essence, factually myopic and legally 'anarchic': they ignore other states' need for and reliance on the waters of an international watercourse, and they deny that sovereignty entails duties as well as rights (McCaffrey, 2001, p135).

Limited territorial sovereignty/equitable utilization

The third approach, that of 'limited territorial sovereignty', which is usually articulated in the context of international watercourses as the principle of 'equitable utilization', entitles each co-basin state to an equitable and reasonable use of waters flowing through its territory. This principle may be understood as a compromise between the principles of absolute territorial sovereignty and absolute territorial integrity because the sovereignty of the upper basin state and the integrity of the lower basin state are restricted by recognition of the equal and correlative rights of the other state. The principle is based on the notion that international watercourses are shared resources in which there exists a community of interest among all co-basin states. The existence of a community of interest requires a 'reasonable and equitable' balancing of state interests which accommodates the needs and uses of each state. To permit flexibility, the concept of 'reasonable and equitable' is deliberately vague and can only be determined in each individual case in the light of all relevant factors, including of course considerations of human need and environmental protection (McIntyre, 2007). This principle, which is clearly the prevailing theory of international watercourse rights and obligations today, has its doctrinal origins in the sovereign equality of states, whereby all states sharing an international watercourse have equivalent rights to the use of its waters. However, it is not possible to apply this principle definitively to the resolution of abstract disputes.

Its application in the resolution of actual disputes will depend on the particular circumstances of each dispute and watercourse. It is worth noting that the principle of equitable utilization plays a pivotal role in the development of customary international law with regard to the environmental protection of international watercourses. It provides for the establishment of factors which are to be taken into account in determining water utilization and allocation of shares in the resource, and thus provides a theoretical structure or framework within which consideration of ecological conservation and environmental protection can easily be conducted.

This approach would appear to have its origins in widespread state practice, international treaty law, decisions of municipal courts, federal supreme courts and international courts, and, in judicial decisions and treaty law on the delimitation of other shared resources. An early example of relevant state practice is provided by a communication of 1856 from the Dutch government in relation to Belgian diversion of the River Meuse which asserted, *inter alia*, that 'both parties are entitled to make the natural use of the stream, but at the same time, following general principles of law, each is bound to abstain from any action which might cause damage to the other' (Smith, 1931, p217; McCaffrey, 2001, p139). Similarly, in the course of their dispute over use of the waters of the Paraná River, which primarily concerned questions of consultation, Argentina and Brazil agreed, in September 1972, to cooperate in the field of the environment by providing technical data regarding works to be undertaken within their jurisdiction in order to prevent any appreciable harm to the human environment of neighbouring areas (Lammers, 1984, p295; McCaffrey, 2001, p141). Indeed, in June 1971, both Argentina and Brazil had signed the Act of Asunción, containing the Declaration of Asunción on the Use of International Rivers. Paragraph 2 provides that in the case of 'successive international rivers, where there is no dual sovereignty, each state may use the waters in accordance with its needs provided that it causes no appreciable harm to any other state of the [La Plata] Basin'. Therefore, they would appear to have accepted a

form of limited territorial sovereignty whereby the sovereignty of each state over its territory is limited by the obligation not to use that territory in such a way as to cause appreciable harm to the other. Likewise, in the course of the dispute between Chile and Bolivia over use of the Rio Lauca, Chile, the upstream state, recognized that Bolivia had rights in the waters and stated that the 1933 Montivedeo Declaration 'may be considered as a codification of the generally accepted legal principles on this matter' (Lipper, 1967, pp27–8). Bolivia also invoked the Montivedeo Declaration, which in its opinion 'embodied international law', though it disagreed with Chile over its correct interpretation (Lammers, 1984, p289). The principle of equitable utilization has also received some limited support among the state practice of the several Middle Eastern states sharing the Jordan River and its tributaries (Lammers, 1984, p306; McCaffrey, 2001, p142).

In 1907, during negotiation of the 1909 Boundary Waters Treaty with the US, Canada articulated a number of principles, which it 'believed in general to be existing law' and on the basis of which all existing and future disputes between the parties could be resolved by an international tribunal. These principles included, *inter alia*, that '[N]either country could make diversions or obstructions which might cause injury in the other without the latter's consent' and that '[E]ach country would be entitled to an 'equitable' share of water for irrigation' (McCaffrey, 2001, p144). Indeed, the US, which is both an upper and lower riparian in respect of Canada and predominantly an upper riparian in respect of Mexico, has expressly accepted in a 1958 State Department memorandum that '[R]iparians are entitled to share in the use and benefits of a system of international waters on a just and reasonable basis' (McCaffrey, 2001, p143).

The principle has received consistent support in the case law of international tribunals, with McCaffrey observing that 'no known international decision supports a contrary rule' (McCaffrey, 2001, p145). In the *Lac Lanoux* Arbitration, the pleadings of both parties would appear to support particular formulations or interpretations of the doctrine of limited territorial sovereignty and the tribunal itself stated categorically that 'there is a rule prohibiting the upper riparian state from altering the waters of a river in circumstances calculated to do serious injury to the lower riparian state'. More recently, in the 1997 *Gabčíkovo-Nagymaros* case, the International Court of Justice left little doubt that equitable utilization comprises the governing principle in the field of international watercourses. Also, a long line of decisions of the US Federal Supreme Court, adjudicating in disputes between states over the utilization of shared water resources, apply the principle of 'equitable apportionment' of the waters. Decisions of other federal and national courts have also tended generally to support the doctrine of limited territorial sovereignty (Lipper, 1967, pp30–32).

The principle of equitable utilization has also been recognized as an established principle of customary international law in all recent significant codifications of this area, including the Institute of International Law's 1961 Salzburg Resolution; the International Law Association's (ILA) 1966 Helsinki Rules on the Uses of Waters of International Rivers; the United Nations Environment Programme's 1978 Principles on Shared Natural Resources; and the successive Draft Articles on the Law of Non-Navigational Uses of International Watercourses adopted by the International Law Commission (ILC) and recommended by the Commission to the United Nations General Assembly as the basis for the 1997 UN Convention on the topic. The principle has recently been emphatically confirmed by the ILA with the adoption of the Association's 2004 Berlin Rules on Water Resources Law. The commentary to Article 12 of the Berlin Rules, which articulates the principle of equitable utilization, states categorically that '[T]oday the principle of equitable utilization is universally accepted as basic to the management of the waters of an international drainage basin'. The principle of equitable utilization is now absolutely central to the purported regime for international watercourses created under the 1997 UN Convention on the Law of the Non-navigational Uses of International Watercourses.

By 2009, the 1997 UN Convention had only received 18 of the 35 ratifications required for its entry into force. However, because the Convention is the product of over 20 years of expert deliberation by the ILC, including detailed consultation with states, it is likely to be considered highly persuasive in identifying and interpreting relevant rules of general and customary international law. Indeed, in its 1997 decision in the *Gabčíkovo-Nagymaros* case, the International Court of Justice referred to the UN Convention as authority, despite the fact that it was only recently adopted by the UN General Assembly and nowhere near to entry into force. Further, the principle of equitable utilization is also recognized as the cardinal rule applying to shared groundwater resources and forms the core principle of the ILC's 2008 Draft Articles on the Law of Transboundary Aquifers (Eckstein, 2007).

The doctrine of limited territorial sovereignty is today supported by the overwhelming majority of commentators, who have grounded it on several different theoretical bases (Andrassy, 1952, pp116–118; Berber, 1959, pp25–40; Lipper, 1967, pp30–32; McCaffrey, 2001, pp147–148). On the basis of this broad acceptance alone it is hardly surprising that the principle of equitable utilization is widely regarded by commentators as the primary rule of customary international law governing the use and allocation of the waters of international watercourses (Dickstein, 1973, p492; Birnie and Boyle, 1992, p127). Customary international law is a primary source of international law and consists of rules of law derived from the widespread and consistent practice of states acting out of the belief that the law required them to act in that manner. Such rules are binding upon all states which have not actively rejected such obligations and apply in the absence of applicable conventional rules. However, the complexity of the process of balancing diverse interests and weighing up relevant factors, coupled with the uncertainty in application of the principle due to a lack of judicial elaboration, means that widespread agreement as to what the principle might actually mean in practice is somewhat less forthcoming. It is therefore suggested that the use of common management systems with developed institutional structures, including, in particular, permanent international river commissions or river basin organizations (RBOs) (see also Granit, this volume Chapter 10), might provide frameworks within which agreed formulae of the principle of equitable utilization might be adopted, and subsequently applied within particular basin regimes. This model, if widely applied, would create a wealth of state practice and regional treaty law which would serve to make more certain the application of the principle in general international law. The principle is therefore in no way incompatible with the fourth theoretical approach, that of common management. Indeed, common management might be regarded as conducive, if not essential, to the ongoing development and elaboration of the principle of equitable utilization.

Common management/community of interests

Under the 'common management' approach the drainage basin is regarded as an integrated whole and is managed as an economic unit, with the waters either vested in the community or divided among co-basin states by agreement, accompanied by the establishment of international machinery to formulate and implement common policies for the management and development of the basin. The institutional structure and purposes of common management regimes vary from basin to basin with most having a clear role in environmental management. Common management is an approach to managing water problems rather than a normative principle of international law, and as such it has been widely endorsed by the international community (Birnie and Boyle, 1992, p223), and adopted by international codification bodies, including the Institute of International Law and the International Law Commission. Recommendation 51 of the *Action Plan for the Human Environment* adopted at the 1972 Stockholm Conference called for the 'creation of river basin commissions or other appropriate machinery for cooperation between interested states for water resources common to more than one jurisdiction' and set down a number of basic principles by which such commissions should be guided.

Examples of common management institutional structures for water resources include the Danube Commission; the US–Canadian International Joint Commission; the Lake Chad Basin Commission; the River Niger Commission; the Permanent Joint Technical Commission for Nile Waters; the Zambezi Intergovernmental Monitoring and Co-ordinating Committee; the Intergovernmental Co-ordinating Committee of the River Plate Basin; and the Amazonian Cooperation Council. Indeed a 1979 survey conducted by the United Nations identified 90 common management institutions concerned with non-navigational uses, distributed throughout every region of the world, and recent estimates suggest that 'well over 100 international river commissions have been established by states' (McCaffrey, 2001, p159).

The idea that a community of interests exists in international watercourses, and the related idea that those interests can be identified and safeguarded on the basis of equity, has received some support in the deliberations of international judicial tribunals. In the 1929 *Territorial Jurisdiction of the International Commission of the River Oder* case, though concerned with rights of navigation, the Permanent Court of International Justice (PCIJ) referred to 'principles governing international fluvial law in general' and concluded that:

> [T]his community of interest in a navigable river becomes the basis of a common legal right, the essential features of which are the perfect equality of all riparian states in the use of the whole course of the river and the exclusion of any preferential privilege of any one riparian state in relation to the others.

Indeed, in the same passage, the PCIJ refers to 'the possibility of fulfilling the requirements of justice and the considerations of utility', suggesting that the Court anticipated a role for considerations of equity in giving effective protection to the rights of states. In the recent *Gabčíkovo-Nagymaros* case, the International Court of Justice quoted from the above passage from the *River Oder* case and stated that:

> [M]odern development of international law has strengthened this principle for non-navigational uses of international watercourses as well, as evidenced by

the adoption of the Convention of 21 May 1997 on the Law of the Non-Navigational Uses of International Watercourses by the United Nations General Assembly.

On the basis of this principle, the Court concluded that:

Czechoslovakia, by unilaterally assuming control of a shared resource, and thereby depriving Hungary of its right to an equitable and reasonable share of the natural resources of the Danube ... failed to respect the proportionality which is required by international law.

McCaffrey highlights this statement of the Court to illustrate that 'the concept of community of interest can function not only as a theoretical basis of the law of international watercourses but also as a principle that informs concrete obligations of riparian states, such as that of equitable utilization' (McCaffrey, 2001, p152).

In terms of state practice, the concept of community of interest is commonly traced back to a French decree of 1792 dealing with the opening of the Scheldt River to navigation. The position expressed in this decree was quickly adopted in a number of instruments concerned primarily with rights of navigation in international rivers (Vitányi, 1979, pp34–37), but also in some early agreements not restricted to navigational uses. For example, Article 4 of the 1905 Treaty of Karlstad between Sweden and Norway provides that '[T]he lakes and watercourses which form the frontier between the two states or which are situated in the territory of both or which flow into the said lakes and watercourses shall be considered as common' (Berber, 1959, p24). In terms of modern treaty practice, the original 1995 Protocol on Shared Watercourse Systems adopted by the Southern African Development Community (SADC) provided in Article 2 that the member states are to 'respect and abide by the principles of community of interests in the equitable utilization of [shared watercourse] systems and related resources'. However, the 2000 Revised SADC Protocol on Shared Watercourses, which supersedes the 1995 Protocol, does not contain any corresponding

provision but rather follows the approach taken under the 1997 UN Convention. Similarly, Article 1(2) of the 1992 Agreement between Namibia and South Africa on the Establishment of a Permanent Water Commission provides that the Commission's objective is, *inter alia*, 'to act as technical adviser to the Parties on matters relating to the development and utilization of *water resources of common interest to the Parties*'. McCaffrey observes that it is more usual for modern treaties 'to treat international watercourses as being of common interest than to *refer* to them expressly as common rivers or property'. He provides examples of, *inter alia*, agreements which entail the use of the territory of one riparian state by another for purposes such as storage, and agreements which relate to the production and division of hydroelectric power in a manner which entails an equitable division of the benefits of the shared waters (McCaffrey, 2001, p158).

Numerous commentators have advocated the principle of a community of interests in international watercourses and use of the associated common management approach, though few would contend that such an approach has evolved, or is likely soon to evolve, into a requirement of general or customary international law. For example, Godana observes that 'the idea has yet to develop into a principle of international law governing international water relations in the absence of treaties' (Godana, 1985, p49), while Kaya similarly concludes that 'there is not enough support for the theory of common management from customary international law' (Kaya, 2003, p205). Caflisch considers the merits of 'denationalizing' international watercourses and transferring their management from individual states to a joint organization, and concludes that 'while it is clear that a condominium could be established by treaty, one cannot maintain that, by virtue of the rules of customary law, the whole of an international watercourse, including its resources, forms a condominium' (Caflisch, 1992, pp59–61, cited in McCaffrey, 2001, pp163–164). Despite this lack of legal compulsion, 'the management of international watercourses by northern European states, however, clearly follows the theory of common management', involving 'joint management bodies – frontier river commissions – which regulate all uses of international watercourse in the region' (Fitzmaurice and Elias, 2004, p14).

Common management regimes must, therefore, necessarily be voluntary arrangements, established by treaty between basin states. The rules of general international law will not impose a positive obligation and compel basin states to create such regimes as 'international law limits only the state's freedom of unilateral action but does not require joint utilization' (Olmstead, 1967, p9). However, it is apparent that the accumulated practice of states in participating in such arrangements should serve to bolster the normative status, in customary or general international law, of the various rules comprising the duty to cooperate. This could, in turn, inform the substantive content of such procedural rules by making it clear that *bona fide* participation in common management institutions would go some way towards satisfying the obligations inherent therein. Interestingly, Article 9 of the 1992 United Nations Economic Commission for Europe (UNECE) Convention on the Protection and Use of Transboundary Watercourses and International Lakes actually requires parties to 'enter into bilateral or multilateral agreements or other arrangements' which 'shall provide for the establishment of joint bodies' having a wide range of environmental tasks. Furthermore, it seems reasonable to assume that a common management approach becomes more acceptable as recognition of the physical unity of the drainage basin gains ground in international law. Indeed, the ongoing evolution and development of the so-called 'ecosystems approach' to the environmental protection of international watercourses is likely to considerably enhance legal recognition of the physical unity of drainage basins and so to highlight the need for common management institutions (McIntyre, 2004, p1; Kaya, 2003, p189).

The 1997 UN Convention on the Law of the Non-navigational Uses of International Watercourses actively encourages watercourse states to enter into common management arrangements. The principle of 'equitable participation' set out under Article 5(2), which is closely linked to implementation of the rule of equitable utilization, suggests the nature and scope of the role potentially to be played by joint mechanisms. The ILC commentary to the 1994 Draft Articles, which preceded the Convention, explains that Article 5(2) involves 'not only the right to utilize an international watercourse, but also the duty to cooperate actively with other watercourse states in the protection and development of the watercourse', and Tanzi and Arcari argue that the provision 'not only requires coordination but also more significant forms of cooperation'. Indeed, they contend that a state's failure to participate actively in the procedural requirements inherent in equitable participation 'will make it difficult for that state to claim that its planned or actual use is … equitable under Article 5 of the Convention' (Tanzi and Arcari, 2001, p109). Therefore, any invitation to join or participate in a regional water body or river commission is likely to be considered carefully by riparian states. Also, in the context of the general obligation on states under Article 8 of the Convention to cooperate 'in order to attain optimal utilization and adequate protection of an international watercourse', Article 8(2) expressly proposes that states 'consider *the establishment of joint mechanisms or commissions*'. It is interesting to note that this reference was not included in the 1994 Draft Articles but inserted later, perhaps signalling growing acceptance of the common management approach and growing awareness of its merits. It is to be assumed that such arrangements would also generally be regarded as effective, if not essential, in facilitating the regular exchange of technical data and information required under Article 9. It is apparent that regular exchange of such information, facilitated by common management institutions, could have a significant role to play in determining an equitable regime for the utilization of an international watercourse in line with the principle of equitable utilization as elaborated under Articles 5 and 6 of the Convention. In addition, Article 21 provides, in relation to the 'prevention, reduction and control of pollution' that 'watercourse states shall, individually and, where appropriate, *jointly*, prevent, reduce and control the pollution of an international watercourse that may cause significant harm …' and that 'watercourse states shall take steps to harmonize their policics in this connection'. As the 'mutually agreeable measures and methods' envisaged under Article 21 for this purpose include, *inter alia*, 'setting joint water quality objectives and criteria', the potential role for common management machinery is obvious. Further, Article 24, which deals with the 'management' of international watercourses, provides that 'watercourse states shall, at the request of any of them, enter into consultations concerning the management of an international watercourse, *which may include the establishment of a joint management mechanism*'. Finally, the Convention envisages a role for common management mechanisms in relation to the settlement of disputes concerning the interpretation or application of the Convention, providing that:

> *If the parties concerned cannot reach agreement by negotiation … they may jointly seek the good offices of, or request mediation or conciliation by, a third party, or make use, as appropriate, of any joint watercourse institution that may have been established by them …*

In relation to its merits, most commentators would agree that 'the notion that all riparian states have a community of interests in an international watercourse reinforces the doctrine of limited territorial sovereignty, rather than in any way contradicting that doctrine' and put forward several advantages of such an approach where it is adopted, including that it 'expresses more accurately the normative consequences of the physical fact that a watercourse is, after all, a unity' and that 'it implies collective, or joint action' and 'evokes shared governance' (McCaffrey, 2001, pp168–169). Commentators have for some time expressed concern that, in the absence of common management arrangements, the traditional substantive rules of international watercourses law, including the obligation of states to prevent transboundary harm and the principle of equitable utilization, may be of limited avail in handling problems of water scarcity and quality (Utton, 1974, p182; Tanzi and Arcari, 2001, p18).

Conclusion

It is quite clear, therefore, that there exists general agreement that the principle of equitable utilization is now the pre-eminent legal rule of international law applying to the use and protection of international freshwater resources. However, the principle is necessarily somewhat flexible and normatively indeterminate as it must be capable of applying to the utilization of any international watercourse, despite the diversity of such watercourses in terms of their physical geography, demographics or the socioeconomic development of the riparian states. This drawback may be ameliorated to some extent through the establishment and operation of common management approaches involving permanent and technically competent institutional machinery.

Notes

1   Para. 3 of 'Country report, Egypt', a paper presented at the *Interregional Meeting of International River Organisations*, Dakar, Senegal, 5–14 May 1981, quoted in Godana (1985, p39).

References

Andrassy, J. (1952) 'Les relations internationales de voisinage' *Recueil des Cours*, vol 79, (1951-II)

Baxter, R. R. (1967) 'The Indus Basin', in A. H. Garretson, R. D. Hayton and C. J. Olmstead (eds) *The Law of International Drainage Basins*, Oceana Publications, Dobbs Ferry, New York, NY

Berber, F. J. (1959) *Rivers in International Law*, Stevens and Sons, London

Birnie, P. and Boyle A. E. (1992) *International Law and the Environment*, Clarendon Press, Oxford

Bruhacs, J. (1993) *The Law of Non-navigational Uses of International Watercourses*, Martinus Nijhoff, Dordrecht, The Netherlands

Caflisch, L. (1992) 'Règles générales du droit des cours d'eau internationaux', *Recueil des Cours*, vol 219 (1989-VII)

Dickstein, H. L. (1973) 'International lake and river pollution control: questions of method', *Columbia Journal of Transnational Law*, vol 12, p487

Eckstein, G. (2007) 'Commentary on the UN International Law Commission's draft articles on the law of transboundary aquifers', *Colorado Journal of International Environmental Law and Policy*, vol 18, pp537–610

Fitzmaurice, M. and Elias, O. (2004) *Watercourse Co-operation in Northern Europe – A Model for the Future*, TMC Asser Press, The Hague, The Netherlands

Garretson, A. H. (1967) 'The Nile Basin' in A. H. Garretson, R. D. Hayton and C. J. Olmstead (eds) *The Law of International Drainage Basins*, Oceana Publications, Dobbs Ferry, New York, NY

Godana, B. A. (1985) *Africa's Shared Water Resources: Legal and Institutional Aspects of the Nile, Niger and Senegal River Systems*, Frances Pinter, London.

Kaya, I. (2003) *Equitable Utilization: The Law of the Non-navigational Uses of International Watercourses*, Ashgate, Aldershot, UK

Lammers, J. G. (1984) *Pollution of International Watercourses*, Martinus Nijhoff, The Hague, The Netherlands

Lipper, J. (1967) 'Equitable utilization' in A. H. Garretson, R. D. Hayton and C. J. Olmstead (eds) *The Law of International Drainage Basins*, Oceana Publications, Dobbs Ferry, New York, NY

McCaffrey, S. (2001) *The Law of International Watercourses*, Oxford University Press, Oxford

McIntyre, O. (2004) 'The emergence of an "ecosystem approach" to the protection of international watercourses under international law', *Review of European Community and International Environmental Law*, vol 13, no 1, p1

McIntyre, O. (2007) *Environmental Protection of International Watercourses Under International Law*, Ashgate, Aldershot, UK

Okidi, C. O. (1980) 'Legal and policy regime of Lake Victoria and Nile Basins', *Indian Journal of International Law*, vol 20, p395

Olmstead, C. J. (1967) 'Introduction', in A. H. Garretson, R. D. Hayton and C. J. Olmstead (eds) *The Law of International Drainage Basins*, Oceana Publications, Dobbs Ferry, New York, NY

Smith, H. A. (1931) *The Economic Uses of International Rivers*, King and Son, London

Tanzi, A. and Arcari, M. (2001) *The United Nations Convention on the Law of International Watercourses*, Kluwer Law International, The Hague, The Netherlands

Teclaff, L. A. (1985) *Water Law in Historical Perspective*, Hein, Buffalo, NY

Teclaff, L. A. (1991) 'Fiat or custom: the checkered development of international water law', *Natural Resources Journal*, vol 31, pp45–73

Tolba, M. K., El-Kholy, O. A., et al (1993) *The World Environment 1972–1992: Two Decades of Challenge*, UNEP/Chapman & Hall

Utton, A. E. (1974) 'International water quality law', in L. Teclaff and A. E. Utton (eds) *International Environmental Law*, Praeger, Santa Barbara, CA, p154

Utton, A. E. (1996) 'Regional cooperation: the example of international water systems in the twentieth century', *Natural Resources Journal*, vol 36, p151

Vitányi, B. (1979) *The International Regime of River Navigation*, Sijthoff and Noordhoff, Alphen aan den Rijn, The Netherlands

WCD (World Commission on Dams) (2000) *Dams and Development: A New Framework for Decision-Making*, www.unep.org/dams/WCD/report.asp

6

Aquifer Resources in a Transboundary Context: A Hidden Resource? – Enabling the Practitioner to 'See It and Bank It' for Good Use

Shammy (Shaminder) Puri and Wilhelm Struckmeier

- Transboundary aquifer systems can be found in all parts of the world; the volume of water in them is many times greater than in analogous surface systems.
- In some arid and hyper-arid regions there are huge transboundary aquifers that need very careful cooperative stewarding as they were replenished in the last pluvial period (7000–10,000 years ago).
- Aquifers can 'bank' water in storage space that is free of cost and the water can be utilized in a manner analogous to a well-managed joint financial bank account.
- Transboundary aquifer–river connected water banking systems provide opportunities for sharing countries to adopt conjunctive use practices to enhance the resilience of their aquatic habitats in the face of the increasing amplitude of hydrological variability.

Introduction

This chapter aims to provide transboundary water practitioners, operating in countries sharing aquifers, with essential tools to enable them to 'see' the hidden resource. It will provide some guidance on how to 'bank' water in aquifers for their good use, much in the way that responsible individuals use their financial bank accounts. Why? Because unlike surface water in transboundary rivers, aquifers come already 'fitted' with storage. This means that in times of adversity the transboundary practitioners can *overdraw* on it, *and* just as an individual would then replenish the overdrawn financial account with extra deposits, an aquifer needs to be 'over-replenished' in times of above-average replenishment, requiring transboundary cooperation.

The substance of this chapter does not address the sociopolitics of transboundary aquifers; instead it will aim to equip the practitioner with an understanding of the dynamics of aquifer concepts, particularly as increasing climate variability kicks in with increasing intensity in the first quarter of this century (2025). The practitioners themselves may not need to be qualified hydrogeologists, but they do need to appreciate several of the ideas for sound negotiation with their counterparts in neighbouring countries, who may also draw on the same aquifer's resources.

Staying with the analogy of the bank account,[1] the guidance in the chapter will aim to use a comparison with a formal household partnership, to explain that if the cooperative partners draw on the same bank account, there should be no surprises if the account goes overdrawn – but if two unrelated and uncommunicative persons draw on one common account, there could be unacceptable surprises when the joint account goes empty at a time of adversity!

Background to the Hidden Transboundary Resources

How to make visible the 'hidden' resources in aquifers

As a first step towards providing guidance for the intended audience to this volume (see Chapter 1 in this volume), we set the stage for aquifer resources and their basic principles. This aspect of water resources, nationally and in the transboundary context, has been much neglected, leading to their designation as 'hidden resources', or 'hydroschizophrenia' (Jarvis et al, 2005). While there is limited global experience in the practical governance of transboundary aquifers, towards the end of this chapter some of the key issues are discussed with regard to good policy and practice to achieve forms of sound international transboundary aquifer agreements in the face of socioeconomic and environmental change, as discussed by Falkenmark and Jägerskold (this volume Chapter 11).

Basic principles of aquifer systems

Aquifers are geological formations capable of storing and transmitting water in significant quantities, such that the water can be extracted by manual or mechanical means. Their dimensions can vary greatly, from a few hundreds of hectares in surface area to thousands of square kilometres, while the depth can range from several metres to hundreds or thousands of metres (Puri, 2009) (Figure 6.1). The term 'aquifer' denotes the geological rock formation and the water within the rock formation, both in hard rocks and in uncon-

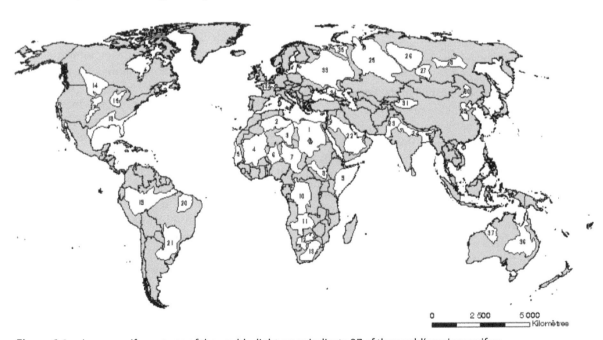

Figure 6.1   Large aquifer systems of the world – light areas indicate 37 of the world's major aquifers

solidated formations. The term 'groundwater' is only used in this chapter when referring to the *water* alone, once it has been extracted from the host, that is the rocks. The ultimate source of the water in aquifers is recharge from rainfall. Aquifers are constantly renewed through natural recharge which passes through aquifers very slowly.

Large aquifer systems (approximately 1 million square kilometres) rarely coincide with political boundaries and may extend beyond national borders (Figure 6.1). Globally about 2 billion people and innumerable riverine aquatic habitats[2] rely almost completely on water in aquifers, with many of them being transboundary in nature. The groundwater stored in aquifers may be compared to the dependent populations holding financial assets in diversified bonds and stocks. However, just as in the case of consistent financial asset management, the hidden resources of aquifers have first to be recognized by countries as their natural capital (*Le Temps*, 2007) and second to be managed coherently (Jarvis et al, 2005), to do justice to all the 'stockholders' who may have a share in the sustainability of an aquifer system.

Aquifers as reservoirs of transboundary water resources

The global volume of water held in aquifers is many times greater than the volume of water in rivers and lakes, as demonstrated graphically in Figure 6.2. In principle the water in aquifers is accessible to all the inhabitants of the land surface that overlies an aquifer system, unlike water in streams which is only directly accessible to those living close to streams. Shallow aquifers, if present, provide the most practical source of water that an individual family or farming unit can access at the least unit cost; however, it requires drilling and well construction. This is one of its advantages over surface water, which requires large upfront capital investment for diversion works, and often burdensome administrative regulation that can become much amplified in a transboundary context. Generally aquifers contain water that is a local resource par excellence for rapid deployment, with significantly lower capital investment costs.[3]

In the same way that you would conduct a thorough investigation for a dam site for surface water storage, an investigation is required of the subsurface aquifer in order to identify the heterogeneities that are prevalent in the underground. It is well known to groundwater practitioners that surface basin boundaries and the underlying aquifer system boundaries do not necessarily coincide, as illustrated by the conditions found on the African continent (Figure 6.3). This poses a particular challenge for transforming the concept of integrated water resources management (IWRM) into practice, because these natural conditions need to be adequately taken into account.

Evidence that the 'hidden resources' are in significant use around the world

Withdrawal of water from aquifers and its utilization

Of all the raw materials extracted from the underground, the greatest tonnage by far is that of water

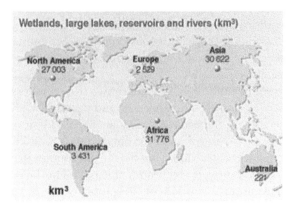

Figure 6.2   A graphical representation of the volumes of water in aquifers and in river–lake basins

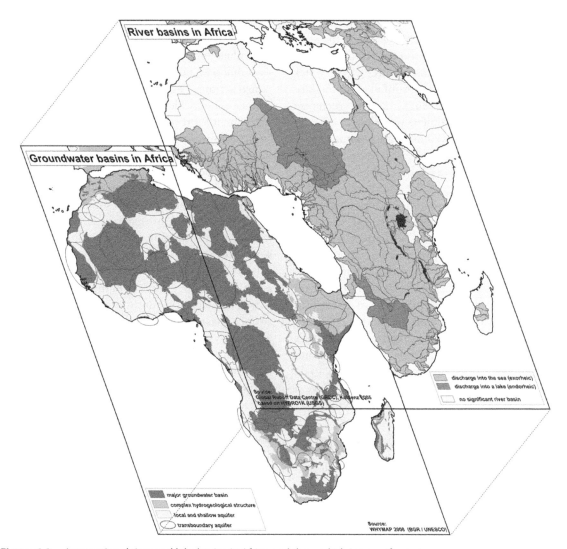

Figure 6.3   International river and lake basins in Africa and the underlying aquifer systems

from aquifers. Table 6.1 shows the global withdrawal of water from aquifers by continent, illustrating that contiguous countries exploiting their national water resource may impact on their neighbours.

The utilization of water drawn from aquifers is a worldwide phenomenon and practically all sectors of the economy may access one, if it is present. Most commonly groundwater is used for domestic water supply, but the use of groundwater for irrigation is steeply on the rise. The large consumptive portion of irrigation water is of concern for the sustainability of groundwater irrigation, because the number of aquifers being

Table 6.1   Withdrawal of water from aquifers by continent

Groundwater extraction	km³/year
Africa	35
North and Central America	150
South America	25
Asia	500
Australasia and Oceania	10
Europe	80

Source: Margat, 2008, figures rounded

Table 6.2   Utilization of water drawn from aquifers in selected countries and regions

Water for potable supply from groundwater (%)		Water for irrigation needs from aquifers (%)	
Austria, Denmark	100	Saudi Arabia, Libya	90
Pakistan	100	India	89
Italy, Hungary	90	Greece	87
Switzerland	84	Bangladesh	86
Russia	79	Tunisia	85
Germany	74	South Africa	84
EU (average)	70	Argentina	79
Poland	66	Spain	71
India	64	USA	70
Belgium, France, The Netherlands	>60	Turkey	60
		Mexico	64
		Italy, Australia	57
		China	54

Source: Margat, 2008, figures rounded

depleted by irrigation is increasing worldwide (Table 6.2).

Critical value of aquifers in arid zones of the world

While aquifers play an important role all over the world, it is in the arid regions where their value is critical because of unreliable surface water resources. One such aquifer is the transboundary Nubian Aquifer System (UNESCO, 2000a), shared by Egypt, Sudan, Chad and Libya, which has resources stored in it for use potentially over many tens of future decades,[4] but, much like a bank account, it will be overdrawn if the sharing partners (in this case Egypt, Sudan, Chad and Libya) do not adopt some form of legally binding domestic partnership over its shared use and management (Mechlem, 2003). Even though the Nubian Aquifer System may not be 'typical' there are important observations that can be drawn from it for other transboundary regions of high natural aridity. These are mentioned later in this chapter.

Freely Available Storage as a Shared Benefit

Aquifers are pre-fitted with storage properties

All aquifer systems have three properties as part of their broad characteristics – *inflow*, that normally results from recharge at the aquifer outcrop; *throughflow*, that is the volume of the inter-annual average outflow from the aquifer; and *storage*, the volume of water that essentially remains in the body of the aquifer system. Often the interconnectedness of aquifer properties is forgotten by practitioners when dealing with the benefits of transboundary aquifers, since the recharge may be located in one country, while the remainder of the aquifer system is in others. This property of aquifers to store water provides unique opportunities for ensuring sustainability during climatic extremes of both the built environment and natural ecosystems but requires multifaceted cooperation between transboundary water institutions. As elaborated by Jarvis and Wolf (this volume Chapter 9), perceived incompatibilities over the size and functions of a transboundary water resource are a greater source of grievance than real water scarcity, and this grievance is often exploited for political gain.

Limited inflow, throughflow and 'enormous' storage

A number of aquifer systems found in arid regions of North Africa (Figure 6.3) receive very limited natural inflow, although in past geological periods, known as the last Pluvial (7000 to over 10,000 years ago), significant recharge took place. In the current climatic period of the anthropocene the Nubian Sandstone Aquifer System (mentioned above), in common with others, receives limited contemporary inflow, has a limited throughflow and theoretically speaking an 'unlimited storage' in comparison to present-day withdrawal. The volume held in storage is equivalent to many thousand times the annual flow of the River Nile. The guidance to practitioners for the management of this type of aquifer system has been given in several authoritative reports (e.g. UNESCO, 2006b; Foster and Loucks, 2006) and is not repeated here, except to note that some aquifers have large potential to store water, for use in times of deep droughts, and that several such aquifers are found in a transboundary context.

How to share the benefit from the storage in aquifers

Deploying aquifer storage and manipulating it for improved resource reliability

The 'freely available' storage in aquifers suggests a similar idea to that of a dam on a national or transboundary river system, built for storage and release of water. In the case of an aquifer, such storage is naturally available below ground at no capital cost, and even if man-made measures are included to enhance the recharge, the unit cost is a fraction of the cost of dam construction. Experience has shown that 'aquifer storage' is one of the more difficult concepts to promote to many surface water experts and is doubly difficult to promote to decision makers, even though the principle is quite straightforward and requires just a basic understanding of physics. If there are also certain corresponding political tensions in the transboundary context, they simply exacerbate the issue, creating blind spots that practitioners find difficult to evade.

A schematic example of deploying aquifer storage and manipulating it for improved resource

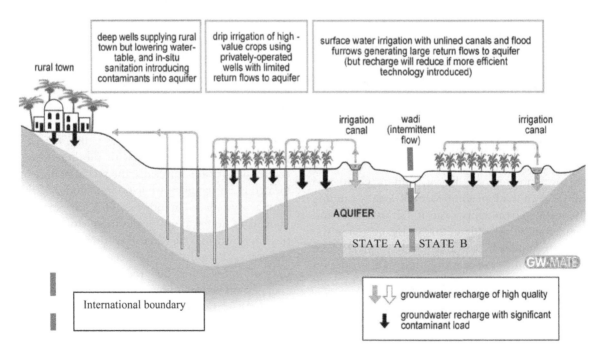

Figure 6.4   Use of the 'storage property' of aquifers to improve resource reliability – national and international context

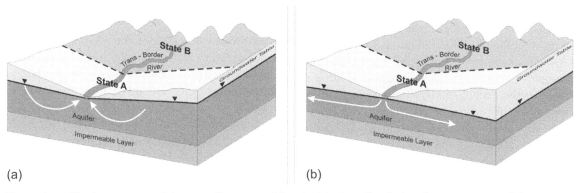

Figure 6.5   The interconnected river–aquifer system: (a) replenished aquifer discharging to a connected river; (b) river water infiltrating into aquifer storage

reliability in a transboundary river–aquifer system is shown in Figure 6.4. Water flowing in irrigation channels or in ephemeral streams replenishes an underlying aquifer, which in turn is used for municipal and irrigation needs. Since the storage in the aquifer is many times greater than the flow in the surface channels, the water stored in the aquifer can be drawn throughout the year, and this is replenished by the infiltration. While the schematic shown in Figure 6.4 may appear as a 'local configuration', the value of transboundary cooperation is apparent, where the aquifer on one side of the boundary (State A) which has significantly better storage, can support the other side (State B) at times of drought, given that other sociopolitical factors permit this to take place. Joint water management and water banking operations in aquifers may thus create good win–win situations.

Use and abuse of storage properties

While the storage in an aquifer system can be put to good use for the improvement of the reliability of resources, it can also be unwittingly abused. This is illustrated in Figure 6.5, which shows that poor management of the water infiltrating into the aquifers from irrigated fields where excessive agrochemicals are applied, or in rural settlements where untreated sewage infiltrates into aquifers, can result in the stored water in the aquifer becoming unusable. The analogy with a surface reservoir into which untreated sewage is discharged is obvious.

Analogies to and contrasts with storage in reservoirs

As mentioned above, the analogy to a surface reservoir is clear, although there are some contrasts that are worth mentioning. If poor quality water infiltrates into the rock medium, then water moving slowly through a porous medium will undergo a quality improvement provided the contaminants are biodegradable, since the natural biochemical processes that occur can reduce the level of contamination. The provisos are that the rate and the quality of the infiltrated water do not exceed the potential of the porous medium to improve the quality, as can be demonstrated by the principles of biogeochemical evolution of the quality.

Banking of Water in Aquifer Storage

Why 'bank' water in aquifers?

The 'current and deposit accounts' of water resources

The discussion in the previous section indicates that cost-free aquifer storage is available in all aquifers, and that by drawing on this property, countries sharing aquifers can help to improve the reliability of their mutual water resources. It is evident that it is possible to 'bank' water in aquifers for later use through a number of well-understood and tested techniques known as 'aquifer storage and recovery'. However, this technique is only usable if there is water available for 'banking'. Generally this water is only available as rain or as excess flows in rivers, although treated wastewater has been used in many locations. An excellent example of a practical legally binding example of transboundary aquifer cooperation is found in the French–Swiss Commission on the Geneva Aquifer established by Canton de Geneve and Prefecture du Haute Savois in 1996 (Mechlem, 2003).

The analogy of financial banking can be readily applied to rivers and aquifers: while rivers provide the *current* account of resources, the connected aquifer can be thought of as a *deposit* account – as represented by Figure 6.4 where States A and B may store their periodic wadi flows in times when there is rainfall. Since rivers, especially wadis, respond fast and cyclically to annual rainfall, they perform the role of an individual's current financial account, with frequent and uneven inputs and outputs of funds. A linked deposit account of the same individual can fulfil the function of either receiving a fixed and predetermined amount each month or, vice versa, of releasing an amount. An analogous conceptualization can be applied to linked aquifers and rivers. While the cyclic input from a river is easier to perceive, the same principle applies to direct rainfall that falls on the aquifer surface.

Banking water in suitable transboundary aquifers can have significant advantages over surface dams; these include the comparatively low capital cost, the possibility of incremental construction of the required infrastructure, no loss of land through flooding and no evaporative losses. These physical and geographical advantages are, however, rarely fully exploited because of a lack of awareness and knowledge in states sharing transboundary aquifers. As the developments can be progressive, political challenges can be ameliorated through building upon staged transboundary confidences among the main stakeholders and seeking sequentially increased cooperation, based on knowledge and education, as elaborated by Salamé and van der Zaag (this volume Chapter 12).

Overdrawing on the accounts: political challenges

Overdrawing on either of the two accounts – the surface water resource or the aquifer resource – would have the expected negative results. Poorly planned overdrafts can and do have dire impacts, as has been seen in many examples of transboundary water systems (for example the Aral Sea Basin and the Ogallala Limestone Aquifer lying across several US states). However, planned overdrafts, with a commitment to replenish the overdraft, changes the picture dramatically and has been carried out in several national examples cited in literature. The Genovese Aquifer (mentioned above), a transboundary resource, is a unique example of this. A planned environmentally sound overdraft immediately means that transboundary negotiating parties involved in the overdraft management are in good communication with each other and have considered, and taken account of, the financial, technical and legal issues, and may be following sound and common 'business models' discussed by Granit (this volume Chapter 10). International norms and values applicable to transboundary aquifers are evolving to support countries that engage in this type of dialogue (Stephan, 2006).

It may be worth noting that the political challenges of exploiting certain transboundary natural resources have been easily overcome in the oil and gas sectors, but appear insurmountable in the water sector. Countries appear to agree rapidly on letting concessions for oil exploration and exploitation even on disputed territory (such as on the Nigeria–Cameroon border),[5] sharing the benefits through the levying of taxes on the developer. This topic of developing transboundary aquifers through a concession to a commercial operator is beyond the scope of the present chapter and will not be discussed further, except to state that there are lessons that might be learnt from it. The UN International Law Commission (ILC)'s proposed consideration of oil and gas as a shared natural resource provides an example of this (Yamada, 2009).

Managing a shared account: cooperative versus adversarial approaches

Following on from the foregoing and staying with the analogy of financial bank accounts, the principles that apply in managing a shared account can apply equally to a shared water resource, whether an aquifer system or a river basin. (See also application of business models in transboundary resources, Granit, this volume Chapter 10.) If an analogy can be found in a formal household partnership, such as a conventional marriage, the partners would operate a joint account in concordance with each other. In the case of an aquifer, with elements as described in the previous sections, even if the actual magnitude and a definite quantitative analysis of the aquifer system is not available (or more probably there is a large uncertainly in the assessments), it would still be possible to jointly and cooperatively draw on the resources for mutual benefit as a planned, time-bound, joint action based on precautionary principles. On the other hand, in a non-cooperative and adversarial relationship, there would be a competition for the resource and it would lead to mutual disbenefits (Puri et al, 2007). In the real world, various users accessing common resources in aquifers frequently have different and possibly mutually exclusive interests, thus demonstrating that the application of the analogy of a bank account that goes into overdraft and then collapses, is apt. Aquifers developed on the basis of competition for the commons, generally result in lose–lose outcomes. Fortunately there are few if any transboundary aquifers that have reached such a condition of collapse, except the aquifers of the West Bank of Palestine, where the overriding political stresses that outweigh all others, are beyond the scope of this chapter.

Connecting the river and the aquifer

One resource – two systems

It is evident that 'water' is a single resource and that once it has been withdrawn from a river or an aquifer, its use, care, treatment and disposal is not differentiated. *Before* that water is extracted from a river or an aquifer, there is a difference that needs to be recognized by the practitioner (Puri, 2006).

In Figure 6.5a, the transboundary river system receives groundwater inflows from either side of the aquifer when the aquifer is in a high state of replenishment. In the same system the river water can infiltrate into the aquifer when the hydrostatic level of the water falls below the bed of the river (Figure 6.5b). Many of the world's major transboundary river systems (the Mekong, Ganges, Brahmaputra, Rhine and Danube to name a few) have features in many ways similar to those shown in Figure 6.5. Major populations live above such a connected river–aquifer transboundary system, and water both from the rivers (if irrigated lands are close to the channels) and aquifers (if the land is not commanded by an irrigation channel) is used by land owners. In a transboundary context, such as that of all the rivers mentioned previously, the sound management of both the rivers (using the principle of hydrology) and the aquifers (using the principles of hydrogeology) is essential.

Integrated surface and groundwater use or conjunctive use?

Experience has shown that in many of the world's largest river–aquifer systems practitioners have paid rather insignificant attention to underlying aquifers and significantly more to irrigation infrastructure, without integrating the aquifers as a store of water. Evidence of this 'hydroschizophrenia' (Jarvis et al, 2005) comes from the frequent waterlogging of lands (e.g. the alluvial plains of the Syr Darya and Amu Darya), salinization, desertification and general land degradation. The over-hyped phrase 'integrated surface and groundwater use' has fast simply become a cliché, since the integration has not been meaningful. This phrase may well have to be dropped for the better conceptualized expression of 'conjunctive use of river basin and aquifer resources'.

Conjunctive use[6] implies the science-based management approach of using surface and subsurface water resources jointly in a complementary manner. The science of hydrogeology has established all the required principles but their application has been limited. Figure 6.6 shows four different configurations in which conjunctive use of basin and aquifer resources could be implemented. Among the many advantages of sound

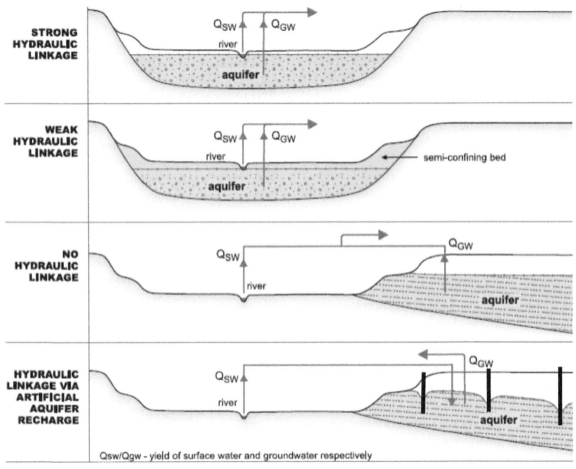

Figure 6.6   Various configurations for conjunctive use of river basin and aquifer resources

conjunctive use of river basin and aquifer resources is one that can include both river–aquifer connected systems as well as unconnected river–aquifer resources into the design of conjunctive use schemes. Conjunctive use of the water of the Syr Darya and the aquifers in the Fergana Valley shared by Kazakhstan, Kyrgyzstan, Uzbekistan and Tajikistan has great promise. However, the institutional tradition of the practitioners of the region has meant an 'out of sight out of mind' approach to this option (see the EU's EuropeAid programme in the Central Asian region, EU EuropeAid, 2009).

The lessons from the Fergana Valley of Central Asia suggest that the adoption of any of these alternatives faces several institutional hurdles at the outset, even in the national context. In the transboundary context, the hurdles can be further magnified, which the practitioner could resolve through cooperative and science-based interinstitutional communications.

Withdrawals and banking – generating the transboundary interest

Abstraction and replenishment in aquifers

While the incentives and the interest in withdrawal of water from aquifers is well understood and remains a priority for those seeking to use the water, the replenishment side is generally either ignored or simply set aside – the *do nothing* approach. These result in consequences that could be drastic. The North China Plain aquifers, which lie across several provinces, have been excessively overdrawn with no explicit plan to replenish them

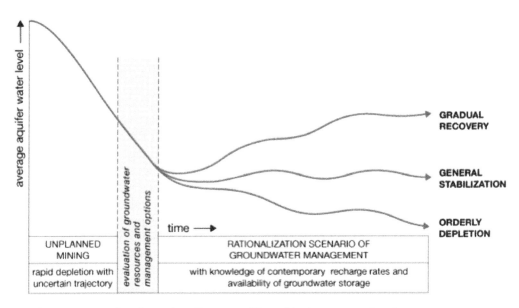

Figure 6.7   Consequences of unplanned aquifer use and 'do nothing' to recover after closure

(Wang et al, 2006), since diverted surface water from major rivers is expected to replace the lost resource. Ceasing pumping for many years might result in the aquifer's self-recovery, as illustrated in Figure 6.7. The 'do nothing' scenario is associated with excessive abstraction rates, which over time result in excessive water level drawdowns, high cost of power and the abandonment of pumps with the eventual closure of the scheme – after this, given much time, natural recovery might sometimes take place.

As a response to this 'do nothing' approach, which is more or less widespread in many water resource development schemes, the initiative known as 'managed aquifer recharge' has taken on a new urgency. Overdrawn aquifer resources in many parts of the world have become matters for urgent concern even though such techniques are already being deployed and show promise.

The question arises, 'why is there no interest in replenishing overdrawn aquifers?' The answer may be found in the well-rehearsed problem of lack of attention to the externalities in exploitation of natural resources, but in the case of aquifers, also the lack of recognition that replenishment is an essential element in their utilization.

The analogy with financial bank accounts is all the more apt, since withdrawal from deposits with no attention to the interest rate will result in bankruptcy. While such 'interest' or the natural replenishment rate, is ignored, measures can be taken to create incentives (or, financially, interest) for users. Unless the users are convinced that tangible returns arise from making investments in the replenishment of aquifers, there would be limited interest. Sound institutional strengthening, and capacity development of agencies charged with balancing abstraction and encouraging replenishment (Fried et al, 2007), could respond to this in the same way that other naturally occurring common pool resources are managed – through a system of incentives and fines (e.g. the manage-ment of fisheries by quotas and natural forests through encouraging conservation).

Dynamics of Transboundary Aquifers: The Coming Challenges

Climate uncertainty versus aquifer demand certainty

Increasing amplitude of climatic variability

Other than the political and social challenges that can be anticipated, this section considers the forth-coming physical challenges. While climate-related uncertainty can be expressed in terms of the variability and a change in the amplitude of hydrological extremes, the challenges in the dynamics of withdrawals and their replenishment can similarly be taken into consideration in planning for management of transboundary resources.

All climate change scenarios agree broadly that aquifer recharge will change. By combining this with scarcity, dependence on aquifers, general sensitivity and adaptive capacity, an overall sensitivity index (SI) has been developed by Döll (2009). Mapping these indicators at a global scale, and comparing with other relevant indicators, suggests some key conclusions on the anticipated challenges in the management of transboundary aquifers for future policy development.

The estimated SI is highest in larger regions such as India, Pakistan, Iran, Saudi Arabia, Jordan, Morocco and Tunisia, the majority of which are dependent to a large degree on their transboundary aquifers. The Sahel Zone in Africa, and the Arabian Peninsula, also show high SI values.

The challenges for practitioners in these regions can be expected to be experienced in multiple contexts, first in the altering replenishment and second in socioeconomic changes. Both of these contexts require early capacity development in the widest sense for joint and coordinated management as described, for example, by Fried et al (2007). The outcome of this training and capacity development would be in sound adaptive management.

Structuring the dynamics of aquifer replenishment under conditions of high variability

From the foregoing it may be noted that there is a need to equip practitioners with, among other things, guidance on the future dynamics of aquifers (especially in the regions mentioned above) that are expected to increase their reliance on shared transboundary aquifers. It would seem that three aspects of management and policy development would have priority:

1   response to anticipated climate variability;

2   Preparation for sound negotiation with equally affected neighbours; and

3   Evolving a cooperative and communicative atmosphere for negotiations.

Since other chapters in this volume address the two last issues with regard to transboundary water management (TWM) (see, for example, Falkenmark and Jägerskold, Chapter 11; Jarvis and Wolf, Chapter 9; Salamé and van der Zaag, Chapter 12), they will not be discussed further here. Instead the subsequent discussion will focus on guidance to practitioners on building and enhancing resilience in transboundary aquifers, considered to be the appropriate response to the anticipated climate variability.

Aquifer storage as a buffer to drought

Building resilience in national resources

The water held in storage in aquifers serves as a cost-free, effective buffer to drought. The regions of the world where these buffer resources could lose resilience to climate shocks from 2025 onwards can be identified with some degree of certainty, as discussed in the previous section. We have also indicated that many of these regions are underlain by transboundary aquifers (see Puri and Aureli, 2009). A simple representation of such an aquifer system is shown in Figure 6.8. Since the recharge in such a system takes place entirely in one of the two Aquifer System States,[7] efforts to build the resilience of all natural resources of State A will also benefit State B. Therefore, in structuring the response to climate change, the practitioner must consider land-use change policies in the 'recharge zone' of State A that will include strong incentives to land owners to include 'managed aquifer recharge' (MAR) in their land-use practices. Since MAR technology (sometimes also expressed as the '3Rs': recharge–retention–reuse of rainwater) is well described elsewhere[8] it will not discussed here, except to state that the incentives for implementation of MAR would have to be linked to broader national policies, and perhaps also relate to international relations with State B. Such land-use policies

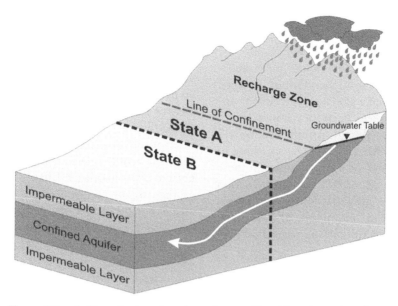

Figure 6.8   Setting of a deep transboundary aquifer with recharge in only one state

incorporating MAR or the 3Rs would build the resilience of the aquifers as a response to climate variability. The practitioner would have to decide how to package the policy measures in national and international terms, for example with food production and its import–export (or virtual water, discussed in Allan and Mirumachi, this volume Chapter 2), as well as other social development programmes, some of which may be associated with the globalization of the world economy (Villholth, 2005; Puri, 2006).

Enhancing resilience in transboundary resources: policy measures

In addition to policy measures that build resilience in natural resource systems such as transboundary aquifers, it is evident that they need to be enhanced. The conditions shown in Figure 6.8 demonstrate that any enhanced measures for the improved resilience of the aquifer to continue to provide reliable resources, and perhaps an increase in yield, would be of significant benefit to State B. It would thus be in the interest of State B to adopt measures in its own territory that help in enhancing the resilience of the aquifer system that is shared by the two states. Policy measures would include balancing the production from the aquifer,

to the additional resources occurring as a result of MAR and 3R concepts in State A, also known as sharing of benefits in the transboundary discourse (see Jarvis and Wolf, this volume Chapter 9).

Such policy measures can be adopted through centralized control based on water production licences, which is institutionally burdensome in some economies, or through water user actions based on technical advice from hydrogeological experts. This implies the need for self-restraint in the amount of water abstracted in State A as a function of the multi-year annual recharge taking place in State A. In a large aquifer system such as that of the Nubian Sandstone Aquifer System, these policy measures should be developed incrementally over a decade because the response of the aquifer is very slow and the resources are many tens of times greater than the amount abstracted – nevertheless such measures need to be contemplated and negotiated as soon as possible. In other aquifer systems which are of a smaller physical dimension and where production is already close to the annual recharge and storage is smaller, such policy measures need to be taken with the utmost urgency.

Setting aside the aspects of hydro-hegemony and power asymmetry which will drive the relations between states (see Cascão and Zeitoun, this volume Chapter 3), it is evident from the foregoing that new policy packages will be required. Since 'one size does not fit all' the policies should be a direct function of the configuration and hydrogeological operation of aquifers. It is for this reason that a global inventory of transboundary aquifers has been developed under the Internationally Shared Aquifer Resources Management (ISARM) programme, as a means to provide assistance to UN member states in these issues (Puri and Aureli, 2005). The next section briefly outlines the global inventory of transboundary aquifers (Puri and Aureli, 2009) that can be used by the practitioner in identifying an aquifer system.

Transboundary Aquifers Global Inventory – A Basis for Cooperation

Knowledge pyramid for a global inventory of transboundary aquifers

From country to bilateral and subregional cooperation

In 2000 the UNESCO International Hydrological Programme launched the initiative on Transboundary Aquifer Resource Management (TARM) based on the poor recognition of such resources by the International Association of Hydrogeologists (IAH) TARM Commission (Puri and Aureli, 2005). The resulting cooperative ISARM programme was soon joined by other UN and international bodies. The initial bilateral discussions were expanded to subregional levels through the involvement of such groupings as the Southern African Development Community (SADC), the Organization of American States (OAS) and the Economic and Social Commission for Western Asia (ESCWA). With support from financing agencies such as the Global Environment Facility (GEF), the ISARM programme rapidly achieved its target of compiling a global atlas (Puri and Aureli, 2009).

Harmonization of the key elements in transboundary aquifer management

In parallel with the preparation of inventories of transboundary aquifers, expert groups also focused on five key issues that relate to the sound management of transboundary aquifer resources – science, socioeconomics, legal frameworks, institutional set-ups and environmental issues – with a view to gaining a better understanding of the elements that need to be harmonized. Clearly, maps and technical data are the first elements to be harmonized to enable practitioners to grasp the basic information on the areal extent, distribution and the population utilizing these resources (Struckmeier et al, 2006, 2008). The legal frameworks in neighbouring countries have been found to be disparate, with some provision for transboundary cooperation, as found in the survey conducted by the Food and Agriculture Organization of the United States (FAO) and regional experts in the ISARM of the Americas. Experts found that there was a need to undertake revisions of national frameworks so that transboundary cooperation would be more fruitful. The institutional set-ups were often found to be sufficiently different to require a significant effort to harmonize their activities in relation to their common aquifers. Finally, the assessment of the socioeconomic importance and the means to conduct harmonization would prove to require a firm commitment from the political structures. This aspect of transboundary water resources management is complex and is the subject of discussion in other chapters and therefore will not be pursued further, except to state that effort on this issue is still ongoing under the ISARM programme.

The knowledge pyramid shown in Figure 6.9 represents the process that has been adopted in the course of the decade of work under ISARM. As indicated, the effort commenced with country-level cooperation between expert groups. This fruitful collaboration was extended into subregional groupings where incentives already exist for neighbouring countries to cooperate, due to broader concerns of social security and regional economic development. In parallel with the support of non-partisan bodies of the UN it has

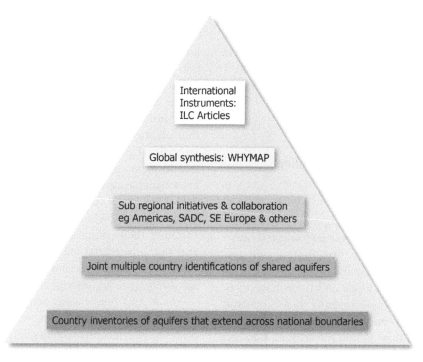

Figure 6.9   Building the knowledge base for the sound management of transboundary aquifers – from national inventories through to an international legal instrument

been possible to compile the global map showing the distribution of the 273 aquifers that are included in the ISARM Atlas. In addition, through the parallel efforts in the ILC (Yamada, 2005), guidance in the form of a potential international legal instrument has been created, as discussed in the next section.

Developing global policy instruments

Drafting articles on the law of transboundary aquifers

The regional and subregional analysis of present-day management of transboundary aquifers demonstrated the need to develop an instrument or guidance for countries to adopt some form of minimum legal agreement. The analysis also suggested that since transboundary aquifers are conceptually problematic for decision makers, sound science would be deployed to assist in formulating the principles, to allow them to gain a full understanding. In the period from 2003 to 2008, the ILC requested the hydrogeological community to provide expert support (Yamada,

2005). This was fortunately possible through the ISARM programme, under which science community from all regions of the world were able to provide their input. The drafting process for the Draft Articles and the full logic behind their formulation is described in a number of readily available publications (Yamada 2007, 2009) and also on the ILC website (www.un.org/law/ilc), and will not be repeated here.

As a related point, it may be noted that the UN Convention on International Watercourses took 27 years to negotiate and agree in the ILC (see McIntyre, this volume Chapter 5). In contrast, the Articles on transboundary aquifers were formulated in five years – demonstrating that the conflictive and (hydro) hegemonic relations (Cascão and Zeitoun, this volume Chapter 3) that arise in the case of the surface water resources appear to be significantly ameliorated in the case of the 'hidden' aquifer resources. A scientific comparative study of this contrast has not been conducted, but the preceding statement is the subjective view of the authors.

In December 2008, the UN General Assembly adopted a Resolution to take note of the Draft Articles on the Law of Transboundary Aquifers and agreed to review these Draft Articles in 2011 in order to decide whether they may be made into a UN Convention.

Developing Good Management Policy and Practice

As will be clear from the scope of this volume, management policies and practices in transboundary aquifer management are multifaceted and cannot be one-dimensional; science alone cannot provide the required tools, and these have to be complemented with solutions to socioeconomic, legal and political constraints. As referred to in the Introduction to this volume (Earle et al, Chapter 1), transboundary water resources face a range of difficulties that may be generically captured for transboundary aquifer management, such as:

- scale issues in aquifer management;
- innovative approaches for the upscaling from the single borehole to Aquifer System States;
- consideration of complexity in the recognition and management of aquifer systems, thereby taking into account stakeholder interests in aquifer systems;
- river/canal channel controls versus spatial withdrawals;
- other important governance issues such as fragmentation in incompatible sector laws, state hegemonies over surface water contrasting with aquifer resource management, as well as international trade and invisible influences on aquifers.

This concluding section briefly addresses the above-listed issues that are constraints to policy and practice in connection with transboundary aquifers. First, in transboundary aquifers in particular the *scale* issue (i.e. small aquifers versus large aquifers) has a very great significance for policy and practice. Although in scale the Nubian Sandstone Aquifer System is large by any measure (2.2 million km^2),

the development plans of two of the sharing countries (Egypt and Libya) have aroused much interest among involved and non-partisan stakeholders. Despite reiterated assessments made by experts that the resources contained in the aquifer are significantly larger than the current utilization, there has been some scepticism that planned abstractions are viable. While such scepticism by the uninformed is not totally unjustified in terms of the utilization of the water for unsustainable agriculture, all Aquifer System States (Chad, Egypt, Libya and Sudan) are willing to reach a consensus on the long-term future use and conservation of their common resource in a cooperative and non-adversarial manner, as discussed previously (see *Managing a shared account: cooperative versus adversarial approaches*, above).

Scaling up from a single borehole on a farmer's land, to the position where two Aquifer System States have joint responsibility, requires innovative approaches to the bottom-up approach. Some of the scale-related issues in aquifers transcend into the *fragmentation* of the governance of the water sector through sometimes incompatible sector laws and institutions not fully equipped to deal with them. While aquifer systems do suffer from the *complexity* of their physical features and the conceptual illusiveness for decision makers there is also a tendency for the hydrogeological community to sometimes complicate issues to the point of obscurity! Finally, stakeholder *interests* in aquifer systems are very entrenched – much more so than in surface water management, because groundwater does not occur in a well-defined channel, where its withdrawal can be controlled. In an aquifer the extraction may take place by stakeholders from their privately owned boreholes, making the governance of aquifers an issue of an order of magnitude more complex than in the case of the stakeholders who rely on river-irrigation canal systems. Monitoring systems are therefore required to control the groundwater regimes in aquifers. The concerns over state control for surface water management, in contrast to aquifer resource management, is an issue that is evolving. Further discussions drawing on this issue of stakeholders from borehole to basin levels can be expected in coming years.

Closing Remarks

It is not the purpose of this chapter to offer silver bullets that will automatically result in good management and practice for the sound use of transboundary aquifers. The body of the chapter has covered all the basic concepts that allude to aquifers and the hidden resources in them, to equip the practitioner to visualize issues and remove the mystery that seems to constrain policy formulation, adoption and implementation. The chapter has also identified several unique properties of aquifers that may assist the practitioner in adopting innovative and 'out-of-the-box' solutions which may not be as constrained by political expediency as surface water resources seem to be. The chapter also attempted to draw some analogies to financial management of personal funds, as this is most easily appreciated by every section of stakeholders.

Finally, the chapter has also addressed some forthcoming challenges in connection with climate variability, which is expected to be at its most intense by 2025. Some of the solutions are also examples of adaptive management that will need to be implemented. In the penultimate section of this chapter the work of the ISARM programme has been outlined. This took the bottom-up approach of developing the global inventory of transboundary aquifers and coupled it with the top-down development of an international instrument that has been adopted by a majority of the world's governments in a resolution of the UN's General Assembly. The lessons from the decade-long action has provided many reasons for optimism, but there is still work to be done, if only to translate the Draft Articles of the ILC into national legislative frameworks among countries that share transboundary aquifers.

Acknowledgements

The authors acknowledge the support of international agencies that have contributed to the work of the TARM Commission of the IAH, in particular UNESCO's IHP.

Notes

1 The 'bank account' analogy applied to sound aquifer management is being used as the backdrop to a 'game and role playing' educational tool for senior water managers in the Middle East and North Africa (MENA) region and will undergo initial trials towards the end of 2009.

2 The majority of the world's aquifers naturally discharge into rivers, providing base flows – a global estimate suggests that 30 per cent of global river flow is derived from aquifer discharge.

3 At the time of the pollution from a chemical spill at a petrochemical plant in Jilin Songhua River (Amur Basin, shared by Russia and China), the most rapid solution to replace polluted water supplies was from the nearby transboundary Heilongjiang aquifers.

4 The Nubian Sandstone Aquifer System extends 2.2 million km^2 and holds approximately 373,000 km^3 of freshwater. Further details can be found in UNESCO (2000b).

5 Oil and gas in the Nigeria–Cameroon maritime boundary dispute: an alternative legal approach. Annotated bibliography compiled by Okafor, C. B. (LL.M candidate, University of Calgary, Canada).

6 'Conjunctive use' is a collective term used by water resource managers to describe the many ways in which water from a river and the water from an aquifer can be used, e.g. use of river water when the flows are high in the wet season, and using aquifer water in the dry season.

7 'Aquifer System States': this terminology has been adopted in the Draft Articles on the Use of Transboundary Aquifers prepared by the ILC and noted by a Resolution of all member states of the UN in December 2008, which encourages member states to adopt and implement them.

8 Event held at Stockholm World Water Week in 2009, organized by the Rainwater Harvesting Implementation Network (RAIN) and ACACIA Water, together with Southern and Eastern Africa Rainwater Network (SEARNET) and the World Agroforestry Centre (ICRAF, www.iah.org/mar).

References

Döll P. (2009) 'Vulnerability to the impact of climate change on renewable groundwater resources: a global-scale assessment', *Environmental Research Letter*, vol 4, 035006

EU EuropeAid (2009) *Call for Project Proposal: Sustainable Use and Management of Transboundary Aquifers in Central Asia (SUMTACA)*, University of Calgary, Department of Law, Canada

Foster, S. and Loucks, D. P. (2006) *Non-renewable Groundwater Resources: A Guidebook on Socially Sustainable Groundwater Management for Water Policy Makers*, UNESCO IHP VI Series on Groundwater no 10, UNESCO Paris

Fried J. et al (2007) 'Education and training for trans-boundary groundwater management as an instrument of dialogue and communication', in C. J. G. Darnault (ed) *Overexploitation and Contamination of Shared Groundwater Resources: Management, (Bio)Technological, and Political Approaches to Avoid Conflicts*, Springer, Germany

Jarvis, T. et al (2005) 'International borders, ground-water flow, and hydroschizophrenia', *Groundwater*, vol 43, pp764–770

Margat, J. (2008) *Les Eaux Souterraines dans le Monde*, UNESCO/BRGM, Paris, France

Mechlem, K. (2003) 'International groundwater law: towards closing the gaps?' *Yearbook of International Environmental Law*, vol 14, pp47–80

Puri, S. (2006) 'Transboundary aquifers and their management in the context of globalization', *3rd International Symposium on Transboundary Waters Management*, Ciudad Real, Spain, 30 May–2 June 2006

Puri, S. (2009) 'Transboundary aquifer resources', in G. E. Likens (ed.) *Applied Aspects of Inland Aquatic Ecosystems. Encyclopedia of Inland Waters Volume 1*, Elsevier, Oxford, pp359–366

Puri, S. and Aureli, A. (2005) 'Transboundary aquifers: a global programme to assess, evaluate and develop policy', *Groundwater*, vol 43, pp661–668

Puri, S. and Aureli, A. (2009) *A Global Atlas of Trans-boundary Aquifers of the World*, UNESCO IHP, Paris

Puri, S., Aureli, A. and Stephan, R. (2007) 'Shared groundwater resources: global significance for social and environmental sustainability', in C. J. G. Darnault (ed) *Overexploitation and Contamination of Shared Groundwater Resources: Management, (Bio)Technological, and Political Approaches to Avoid Conflicts*, Springer, Germany

Stephan, R. (2006) 'Evolution of international norms and values for transboundary groundwater gover-nance', in A. R. Turton, D. R. M. Claassen and J. Hattingh (eds) *Governance as a Trialogue: Government–Society–Science in Transition*, Springer-Verlag, Berlin

Struckmeier, W. F., Gilbrich, W. H., Gun, J. v. d.,

Maurer, T., Richts, A., Winter, P. and Zaepk, M. (2006) *WHYMAP: World Map of Transboundary Aquifer Systems*, special edition for the *4th World Water Forum*, Mexico City, March 2006

Struckmeier, W. F., Gilbrich, W. H., Gun, J. v. d., Maurer, T., Richts, A., Winter, P. and Zaepk, M (2008) *Groundwater Resources of the World*, Worldwide Hydrogeological Mapping and Assessment Project (WHYMAP), UNESCO and German Geological Survey (BGR), www.whymap.org, accessed February 2010

Le Temps (2007) 'Les pays doivent inscrire l'eau dans leur capital nature', *Le Temps* (Geneva), 16 August

UNESCO (2000a) *Internationally Shared Aquifer Resource Management: A Framework Document*, IHP Series on Groundwater, no1

UNESCO (2000b) *CEDARE Report on the Nubian Aquifer System*, ISARM Framework Document, UNESCO, Paris, France

UNESCO (2006a) 'Sharing Water', *2nd UN World Water Development Report*, UNESCO–WWAP, Paris, pp372–397

UNESCO (2006b) 'Aquifer Characterisation Techniques', in *Non-Renewable Groundwater Resources*, UNESCO–IHP Series on Groundwater no 10, Paris

Villholth, K. G. (2005) 'Groundwater assessment and management: implications and opportunities of globalization', *Hydrogeology Journal*, vol 14, no 3

Wang, J. et al (2006) 'Privatization of tubewells in North China: determinants and impacts on irrigated area, productivity and the water table', *Hydrogeology Journal*, vol 14, pp275–285

Yamada, C. (2005) *Containing a Complete Set of 25 Draft Articles of a Draft Convention on the Law of Transboundary Aquifers, Together With an Addendum Setting Out Provisions of Legal Instruments Relevant to Some of the Draft Articles Proposed by the Special Rapporteur*, Third Report of the Special Rapporteur, Document no UN A/CN.4/551, International Law Commission

Yamada, C. (2007) *Initial Consideration of Oil and Natural Gas, Together With a Discussion of the Relationship Between the Work on Groundwaters and that on Oil and Gas*, Fourth report of the Special Rapporteur, Document no UN A/CN.4/580, International Law Commission

Yamada, C. (2009) *Special Rapporteur's Paper on Oil and Gas in the Context of Managing Shared Natural Resources*, Document no A/CN.4/608, International Law Commission

7

Governance in Transboundary Basins – the Roles of Stakeholders; Concepts and Approaches in International River Basins

Nicole Kranz and Erik Mostert

- Involvement of stakeholders at the transboundary scale is key in order to ensure adaptive water management (being able to respond to challenges before they become problems).
- Participative processes require careful planning and implementation and usually require more resources than planned.
- The institutional framework for participation needs to be legitimate and capacitated, allowing stakeholders to contribute meaningfully to management processes.
- There exists no blueprint for participatory processes, but certain key elements can be identified which aid the development of a planned and structured stakeholder engagement process, suited to a specific social, economic and political context.

Introduction

Stakeholder participation is widely considered a key prerequisite for adaptive and integrated water resources management (Pahl-Wostl, 2008). Organizing stakeholder participation is, however, not an easy task, especially not in transboundary basins. In moving beyond the boundaries of nation states, the influence of individual states is reduced (Holling and Meffe, 1995). While this situation could potentially lead to the creation of an 'open space' for participation of non-state stakeholders, it also leads to several additional challenges. The number of different stakeholders is usually very high. Moreover, their cultural backgrounds may differ and political systems in the basin can be quite diverse.

The following questions emerge concerning the set-up of participative processes in transboundary basins in particular. For instance, should participation be conducted as an add-on or as an integral part of planning processes (Ridder et al, 2005)? Which stakeholders should be involved and how to stimulate their involvement? At which level or levels should they be involved: at the river basin, the national or the local level? Who should initiate the participatory process: an international commission, a national institution or perhaps one of the stakeholders themselves? And which participation methods should be used?

This chapter discusses some of the approaches to participation in transboundary water governance developed over past years, and the experiences that have been gained therewith. Where possible, we also look at the effects of stakeholder participation and whether it actually supports international cooperation (Karkkainen, 2005; Fritsch and Newig, 2007). The first part of this contribution discusses the importance of non-state stakeholders in transboundary water management (TWM), the theory of public participation and setting up participatory processes. The second part presents four case studies of stakeholder participation in transboundary river basins. The conclusion wraps up the main findings of this chapter and points to linkage with other contributions in this book.

The Importance of Stakeholders

Traditionally, TWM is seen as an issue between sovereign 'states'. In practice, however, the 'state' is an abstraction (Nicol, 2003; Trottier, 2003; Merrey, 2009). The state is a legal concept and can constitute an important symbol and a source of identity for many people. However, when we look at the practice of international negotiations, we do not see 'states' interacting but different individuals who speak on behalf of their country but may, as the case may be, primarily represent specific organizations, sectors and social groups within their country. The question of legitimate representation of interests emerges prominently in this regard.

When we decompose the concept of 'state', we see a large number of different stakeholders. These include groups and individuals possessing legitimacy, formal authority and/or other important resources for developing or implementing international agreements, such as money, political influence, information and expertise – the influential stakeholders. They also include groups and

individuals who may be affected by water management but are unable to exert any significant influence – the affected stakeholders (Trottier, 2003). And, finally, they include groups and individuals who are both influential and affected.

Not involving influential stakeholders may severely reduce the effectiveness of TWM. International agreements may for instance be negotiated by representatives of the ministry for water resources but may not be supported by the agricultural sector. This can then result in political problems and the treaty may not be ratified in parliament. Or treaties may be ratified, but not implemented because the bodies responsible for implementation do not support the treaty and cannot be forced to implement it (e.g. Dieperink, 1998; Meijerink, 1999; Marty, 2001). States might be too weak or unwilling to manage an international river basin, due to political reasons or a lack of information on key management challenges. In those cases, non-state stakeholders can fill this 'governance gap' at least to some extent, their own financial capacity and expertise allowing.

Not involving the affected stakeholders affects transboundary water resources management in a different way (Nicol, 2003). Focusing exclusively on the interests and goals of 'states' means in practice focusing on the interests and goals of national governments and the social groups they represent. In a democratic setting, this will ideally involve all citizens and societal stakeholders. However, one needs to take into consideration conditions of limited democratic representation (Börzel et al, 2006; Börzel, 2007), which may result in too little attention for the interests and needs of underprivileged stakeholders. International cooperation on the Senegal River, for example, could be seen as a success: large dams have been built and irrigated agricultural land has increased. However, reportedly, this proved to be at the expense of flood-recession farming, fisheries, the environment and the health of the local population (Adams, 2000). This points to the relevance of stakeholders in implementing the much-promoted benefit-sharing approach (Sadoff and Grey, 2002), as only vibrant stakeholder networks will allow for the full consideration of all issues related to water management in international basins.

The stakeholders involved in transboundary river basin management may include different national ministries and agencies, regional and local governments, regional economic development organizations, such as the European Union, river basin organizations, local communities, individual farmers and farmer organizations, fishermen, water works, energy companies, water-using industries, environmental NGOs, international donors, men, women, etc. (Mostert, 2005). In this chapter we focus on the non-state stakeholders. They include individual citizens and farmers (the 'broad public' or 'general public'), as well as organized stakeholders, such as national and international NGOs, business associations and local governments.

The Theory of Stakeholder Participation

There is a lot of literature on stakeholder participation that is relevant for TWM as well. According to this literature, stakeholder participation may be organized for several reasons (Pateman, 1970; Delli Priscolli, 1978; Webler and Renn, 1995; CIS, 2002; Reed, 2008). For pragmatic reasons, water managers or other organizers can involve stakeholders who have relevant information or expertise, are creative, possess necessary funds, can veto plans and projects, have to implement the plans and projects, or have some leverage over stakeholders whose cooperation is needed. By involving these stakeholders, decision making can become better informed and implementation can be improved. Normative reasons for involving stakeholders include increased transparency of decision making, more responsive and democratic government, and empowerment of the public, even leading to self-management by the people themselves.

It has to be stressed that there is no generally accepted terminology concerning public and stakeholder participation. While some authors distinguish between public participation and stakeholder participation (for example, Pahl-Wostl, 2002; Jonsson, 2005), others do not. When a distinction is made, 'public participation' usually refers to participation by individuals, and 'stakeholder participation' to participation by organized groups, such as NGOs and associations, but the distinction is not always made in the same way. In this contribution we will use 'public participation' and 'stakeholder participation' as synonymous, covering participation by all types of stakeholders except state stakeholders.

Another issue is the meaning of 'participation'. Public participation can be defined as direct involvement in planning and decision-making processes. This excludes elections, which are a form of indirect involvement, and demonstrations, which are more a call for involvement, but it may include financial contributions to water resources management and contributions in-kind, self-management by the beneficiaries through for instance water users' organizations, and litigation (Meinzen-Dick, 1997; Ebbesson, 1998; REC, 1999).

Participation may take place at different intensities or levels. Already in 1969, Sheila Arnstein (1969) came up with the idea of a 'ladder of citizen participation' consisting of eight rungs. More recently, simpler ladders with three to six steps have become popular, such as: (1) information, (2) consultation, (3) co-thinking or 'discussion', (4) co-designing, (5) co-decision making (sharing responsibility), and (6) self control (Roberts, 1995; CIS, 2002; Mostert, 2006; Henriksen et al, 2009).

An important limitation of Arnstein's original ladder is its exclusive focus on issues of power and power sharing. While power issues are important, public participation is also about developing trust, learning from each other and eventually coping constructively with differences. In this respect the notion of social learning is of interest. Social learning can be summarized in one phrase as 'learning together to manage together' (Ridder et al, 2005; Muro, 2008). It first requires that water managers and the other stakeholders realize their dependence on each other for reaching their goals. Next, they need to start interacting, share their problem perceptions and develop different potential solutions. This in turn requires the development of mutual trust, recognition of diversity and critical self-reflection. Finally, the stakeholders need to take joint decisions and make arrangements for implementation.

There is some evidence that public participation can actually result in social learning. For example, the HarmoniCOP project (www.harmonicop.uos.de) involved participatory processes in ten national European river basins (Mostert et al, 2007). In several basins, relations between different stakeholders improved. In most cases, stakeholders obtained a better understanding of the management issues at stake and got to know and appreciate each others' perspectives, which opened up possibilities for win–win solutions, and in a few basins tangible improvements for the stakeholders and for the environment could be identified (see also Cuff, 2001; Huitema and van de Kerkhof, 2006). At the same time, different obstacles to effective public participation could be identified. In five of the ten cases, the status of the initiative in which the stakeholders could become involved was not clear and many stakeholders doubted that their input would make a difference. Quite often, the existing governance style was not participatory and it took a lot of convincing to move towards a more collaborative approach. In many cases, the authorities lacked experience with multi-party approaches, relied heavily on technical expertise, and feared losing power.

Less prominent in the HarmoniCOP cases, but known from others, is the issue of representativeness. Often, only the 'usual suspects' show up at participatory events: middle-aged highly educated men (Korsten and Kropman, 1976). Rather than empowering the public at large, public participation may reinforce existing power relations (Sabatier and Jenkins-Smith, 1993; Cooke and Kothari, 2001; Leach and Pelkey, 2001; Olsson et al, 2004; Videira, et al, 2006; Warner, 2006; Huitema et al, 2007).

In some parts of the world, such as Europe and the United Nations Economic Commission for Europe (UNECE) region, public participation in river basin management is required by law (e.g. Ebbesson, 1998; CIS, 2002), while in others it may be advisable for pragmatic and normative reasons. Yet, if it is not well organized, public participation may prove counter-effective and result in less mutual trust, worse relations and generally less effective management.

What implications do these concepts have for implementing participation on the ground? The following section provides for an introduction on the 'how to' of setting up participatory processes.

Setting up Participatory Processes

Especially in an international or transboundary context, the initiative for participatory processes can originate from diverse actors, resulting in different types of participatory processes. Participatory processes set up by national governments or water authorities often aim to improve the implementation of national or internationally agreed water policies, whereas those set up by non-state stakeholders, such as NGOs or local community groups often aim to improve their access to and influence on policy formation and water management processes. Yet, there are a number of issues that all organizers of participatory processes have to address before any participatory activity can take place (Ridder et al, 2005). These are discussed below.

Stakeholders and purpose

The first two issues concern the identification of the stakeholders to be involved and the purpose of the participatory process. Stakeholders may identify themselves, but many interesting participation methods allow for only a limited number of participants, and hence the participants have to be selected carefully. In addition, some stakeholder groups have to be targeted specifically to promote their participation and prevent a limited or unrepresentative response. A further important aspect to consider is the respective power vested with each stakeholder group.

The choice of which stakeholders to target depends on the purpose of the participatory process. If the purpose is to improve decision making, stakeholders with relevant information, skills and expertise should be targeted. If the purpose is to improve implementation, stakeholders with financial resources, legal competencies and political influence are more relevant. For improving democracy and empowerment, affected stakeholders without much influence should be targeted specifically. In these cases it might be necessary to engage in capacity-building activities for these stakeholder groups, which otherwise might not be able to participate (James, 1994).

Strategic considerations

Organizing a participatory process is not a neutral activity. As discussed above, participatory processes may have different purposes. In addition, manipulation may take place. For instance, organizers of participatory processes may ensure that the majority of the participants support their ideas by inviting or reimbursing the travel costs of supporters only. Such strategies may be effective, but they may also reduce the legitimacy of the participatory process and increase political tension.

Roles and rules

The next issue concerns the roles of the different stakeholders. These can range from merely listening, to answering questions, participating in the discussions, designing policy and actually taking decisions. For some public participation methods, specific rules may be needed, for instance with respect to speaking time, voting, and deadlines for submitting documents. The roles and rules should be clear at the start of the participatory process. For this reason it is advisable to discuss them explicitly at the start.

Process managers

A role of particular importance is that of process manager. This is the person or group of persons in charge of designing the participatory process, facilitating meetings and liaising with the various government bodies involved. Process managers should not only have the necessary communication and organizational skills, but they should also be seen as impartial, and the latter requirement may be especially difficult to fulfil in transboundary basins. Rather than appointing a government official from one of the basin countries as process manager, an external process manager could be hired. More recently, representatives from interna-

tional river basin commissions have been assigned as process managers. The managerial capacity of the manager is, however, an important issue, especially in recently established river basin organizations. Moreover, there is the issue of preferential allegiance to a specific country, which needs to be addressed with the utmost transparency (Kranz, 2008).

Process organization

The process manager should have sufficient administrative support for organizing meetings, sending notifications to the media, processing written comments, etc. This requires adequate funding. International donors may play a role in this, and in this way promote participatory processes to some extent (Mostert, 2005). To facilitate proper follow-up, it may be advisable to establish a steering committee with all relevant government bodies. International river basin commissions could perform this function.

Scope

A crucial issue that the organizers have to agree on is the scope of the process: what can be discussed and what cannot? This depends partly on the various legal and political constraints, including existing international obligations. The scope must be known in advance to the potential participants so that they can decide better whether or not to become involved and do not get frustrated later on in the process. It is advisable to be flexible with respect to the scope in order to accommodate the interests of the potential participants, who may otherwise decide not to participate, and to allow the development of broad package deals that can benefit all concerned.

Scale issue

A very pertinent issue in transboundary basins is the question at which scale or scales public participation should be organized. Many issues in transboundary basins have a national, regional or even local scale, such as the national implementation of international agreements and local water problems. Even if they have a basin-wide scale, there may be good reasons for organizing public participation at lower scales. If only one international event is organized, issues can be discussed at a general level only and few people can participate. Instead, several regional or local public participation meetings could be held. To be effective, the results of local or regional participatory events should be communicated in time to the scale where decisions are taken, and afterwards feedback has to be given at the local scale regarding the results. This poses huge organizational challenges and may be costly. For this reason, organizers may decide that participation should be at the transboundary river basin level. This allows for direct input into decision making at the transboundary level but for practical reasons only large stakeholders, such as international environmental NGOs, will be able to participate (CIS, 2002, pp19–21).

Timing

Public participation may be conducted during different phases. If conducted too late, much of the input provided by the public cannot be used without causing serious delays, but if conducted early, before important choices have been made, it may be difficult to generate public interest. A way out of this dilemma might be to target different stakeholders at different times: only (semi-) professional NGOs and large companies in the early phases, and all stakeholders in later phases. In following this approach, utmost transparency of the decision as well as the process must be upheld.

Methods

Last but not least, public participation methods have to be chosen. Many are available, such as a variety of techniques for informing the public (leaflets, information evenings, etc.), written consultations, interviews, public hearings, and many different types of small- and large-group meetings. The choice of method or methods depends on many factors, including:

- The level of participation foreseen. Higher levels usually require interaction in small groups.

- The number of participants.
- The type of participant. For instance, if there are one or two dominant participants, or if some participants are reluctant to speak in large groups, it may be advisable to use break-out sessions.
- Phase of the management process. In early phases methods with little structure such as 'open spaces' may be used, but when concrete proposals are tabled and decisions have to be taken, more structured meetings may be more appropriate.
- The cultural context. For instance, role plays may be appreciated in some countries but not in others. Furthermore, stakeholders might devise completely different (financial) means and (cultural) approaches towards participation. A language barrier might aggravate these differences.
- Resources of the organizers. Some methods require more time, money and skills than others. If the approach foreseen requires more resources than available, either more resources should be made available, or the approach should be modified.

Communication represents an important element of the participatory process and often is an integral part of participatory methods applied. Thus, communication tools also need to be carefully catered to the set-up and specifics of the participatory process as indicated above. Communication can be designed to support different aspects and stages of the participatory process as well as speak to different stakeholder groups, taking into consideration their varying degrees of knowledge with regard to the water management issue at hand. Effective communication is not only tailored to the respective recipient, but is also necessarily two- or multi-way, thus not only allowing for information but also for the exchange of positions and opinions. Challenges in this regard are evident in transboundary basins, featuring very diverse cultures and possibly even languages.

New visual communication tools are therefore gaining in significance; other examples are internet-based tools, such as the river awareness tool kits. Similarly, indicator systems, reflecting the situation in the river basin in easily accessible terms, can assist in communicating with a wide range of stakeholders.

Structural issues

There are a number of issues that form the background for all participatory efforts in transboundary basins. These include institutional structures at the international as well as the national level, including government but also the composition of the stakeholders. The degree to which participation is realized in the individual riparians also plays a very important role. Equally pertinent are financial resources available with different actors, such as national governments, but also other stakeholders involved. Participation can be costly and limited financial resources hamper effective participative approaches, lead to power imbalances and misbalanced representation. In some instances, donor organizations provide support for these processes and need to be considered among the stakeholders.

The previous steps provide guidance on possible approaches to participatory processes in transboundary basins, including issues to be considered as well as possible strategic avenues. Obviously each participatory process will be a result of or needs to be tailored to the specific situation in each basin. The following section presents three cases (the Okavango Basin, the Danube River Basin and the Dniester River Basin plus an additional case study in Chapter 13, Orange–Senqu River Basin) on the challenges encountered and the solutions found in setting up transboundary participatory processes.

Experiences in Selected International River Basins

The cases represent unique approaches to participation in very different transboundary settings. They were chosen in order to represent different cultural backgrounds, scopes and stages in the process outlined above.

Case 1: Okavango Basin, Southern African Region

The Okavango Basin, with its three riparian countries Angola, Botswana and Namibia, not only features the Okavango delta, an internationally recognized and unique ecosystem, but also constitutes the major source of water for drought-prone Botswana and Namibia (see Figure 7.1). The political situation in the region had been strained due to the protracted civil war in Angola, which only ended in 2002 (Ashton and Neal, 2003).

The creation of the international Okavango River Commission in 1994 established the foundation for what has been described as of the most advanced public participation programmes in a southern African transboundary basin (M. M. Montshiwa, pers comm, November 2006). While the commitment to public participation is already vested in the agreement establishing the Permanent Okavango Basin River Water Commission (OKACOM), evidencing the political will of the countries involved, it was the strength of community-based organizations and NGOs in the basin, and a strong perception of the linkage between the protection of the river and sustained livelihoods of several stakeholder groups, which

Figure 7.1   Map of the Okavango River Basin

inspired a unique networking activity and joint efforts at the local level (Ashton and Neal, 2005).

A series of issues pertinent to integrated water resources management (IWRM) in the basin, including biodiversity, but also economic sustainability, were taken up by NGOs and thus provided a basis for collaboration. At the same time, the basin had been endowed with a considerable number of donor-funded projects, providing support for setting up governance systems which promoted stakeholder involvement. Of instrumental relevance for advancing public participation in the basin was the 'Every River has its People' project, run by the Botswana-based Kalahari Conservation Society (KCS) and the Namibia Nature Foundation (NNF),[1] which provided for a link between affected communities on the ground and decision makers at the transboundary level, and established a connection between environmental conservation and the socioeconomic situation of local communities (KCS, 2009). The activities of 'Every River has its People' are explicitly targeted at raising awareness through enhancing communication and thus instilling an understanding of responsibility for the resource among local communities.

The work between communities, community-based organizations, NGOs and stakeholders culminated in the development of a Basin-Wide Forum, consisting of ten representatives from each basin country, to discuss the challenges facing the basin. In an initiative without precedent in the region, which gives testimony to the recognition and legitimacy of the Basin-Wide Forum, OKACOM has granted observer status to the Forum, allowing for a representative of the Basin-Wide Forum to attend the Commission's meetings. In addition, initiatives for a 'shadow commission' made up of grassroots organizations and supported by international donors are currently being developed in the basin.

The Okavango basin is a good example of parallel development of participatory processes at the international and at the local level, which are aligned through an intermediary project. The degree of involvement is quite high, due to the increasing awareness of TWM issues among local communities, which are translated into activities

on the ground and thought to link water issues with other livelihood aspects (e.g. health, economic development etc.).

The experiences with civil society participation in the Okavango River Basin have been very positive due to their enthusiastic take-up by local communities, as they proved pertinent to stakeholders' concerns at the local level. For this reason, the scheme has inspired other river basin commissions. Nevertheless, bringing in stakeholders from the Angolan side has proven to be difficult, as the country is still recovering from the consequences of the civil war (Porto and Clover, 2003). At the same time, the localized strategy, which is currently being replicated in the Angolan part, offers a well-suited strategy in order to provide for a truly integrative river basin management approach, which is of relevance to local communities and their livelihoods. Current activities seek to continue capacity-building for local stakeholders, for example with a view to biodiversity and nature conservation (e.g. the Biokavango project).

The Okavango case demonstrates the challenge of identifying and empowering those stakeholders who are the most affected by basin management decisions, but who might not possess the necessary resources to participate. This also points to the relevance of choosing methods which are suitable to the stakeholder groups addressed. In terms of scale and scope, activities are designed to cater to the local level, but the results should be considered at the transboundary level as well. In order to create these linkages across levels as well as issue areas, a strong process management, such as that provided by the 'Every River has its People' project, is absolutely crucial.

Case 2: Danube River Basin, Europe

The Danube, with a length of 2857km and a catchment area of 817,000km², has more than 19 riparian countries, covering parts of Western, Central and Eastern Europe, including EU member and non-member states, and thus easily qualifies as the most international river basin in the world (see Figure 7.2). This diversity brings with it a history of very different political systems in the region, which continue to have repercus-

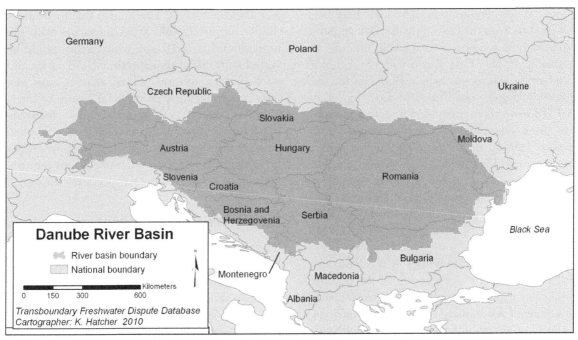

Figure 7.2   Map of the Danube River Basin

sions on the ways and means of public participation in the area today. In the former communist countries, for example, civil society organizations developed in protest against the old regime and consequently had a highly political character (Greenspan Bell and Jansky, 2005). This continues to complicate collaboration with government in participatory processes today. In the 'western' parts of the basin, civil society organizations have a different origin and character, complicating collaboration with civil society at the international level even further.

The public participation process at the international level gained momentum with the signing of the 1994 Danube River Protection Convention (DRPC) and the establishment of the International Commission for the Protection of the Danube River (ICPDR) in 1998. Efforts for public participation under the DPRC were driven and influenced by the following factors:

- acute pollution of the river, resulting from industrial effluents and malfunctioning municipal sewage plants, as well as hydromorphological alterations, affecting livelihoods in the immediate catchment as well as the Black Sea (ICPDR, 2005);
- activities of the Global Environment Facility (GEF) and the EU Phare Programme since 1991, intended to build institutions and a viable civil society as a partner in river basin management (ICPDR, 2004);
- the EU accession process for a large number of riparian countries and the entry into force in 2000 of the Water Framework Directive (WFD).

The ICPDR came to understand public participation as an integrative and ongoing effort, requiring strong civil society organizations with access to main planning and decision-making forums. The Danube Environmental Forum (DEF) was established in 1999 as an international civil society network, and observer status to all expert working groups of the ICPDR was granted to civil society groups (Danube Environmental Forum, 2009). In addition regional support projects continue to provide capacity-building and finance to the DEF and individual NGOs (ICPDR et al, 2006). In the canon of ICPDR's activities, public outreach and participation play major roles, and the commission has a dedicated staff position for addressing these issues. While preliminary activities had been undertaken from the beginning, the official public participation process was kicked off on the occasion of the First Basin-wide Stakeholder Conference, which was held in Budapest in 2005. This conference constituted the first step of the ICPDR towards participation in actual planning processes. These activities were continued in the process leading up to the Danube River Basin Management Plan, which was finalized in 2009, through informing stakeholders on the current planning stages, the elicitation of feedback, focus groups discussions and the organization of the Second Danube Stakeholder Forum (ICPDR, 2009).

The stakeholder activities at the basin level reflect the different target groups of the ICPDR's public participation approach. Stakeholder activities directed at organized stakeholder groups, such as farmers, energy utilities, environmental NGOs, those with navigation interests, and industries, are intended to provide a platform for informed discussions. Activities targeting all inhabitants of the basin are aimed at public information and awareness-raising. These activities are epitomized by the annual Danube Day, which is aimed at demonstrating the relevance of the Danube River for the livelihood of the catchment's inhabitants, and the journal *Danube Watch*, which documents the activities of the ICPDR and the riparians in an easily accessible way. Danube Day furthermore is a good example of how communication to a broad range of actors throughout the basin could be designed. Most recently, the ICPDR has developed a so-called Danube Box, an interactive learning tool for younger stakeholders in the Danube Basin.

Participation in Danube management has come a long way and thus all the actors involved have gone through many learning experiences. It shows, however, that guidance from the international level, such as provided by GEF and the EU, can eventually lead to a transformation of old patterns of civil society involvement. It also demonstrates that international river basin commissions can only do so much and establish a favourable atmosphere for public participation, but action needs to be nested in the riparian

countries themselves. This implies that the responsibility for successful participation lies with the individual riparian states, which again might be vested with differing institutional capacities and participatory cultures. These aspects, however, offer promising entry points for the work of international basin organizations in fostering participative efforts at the national level by helping build strong institutions for that purpose, while taking into consideration the needs and requirements framed by the respective cultural setting.

Case 3: Dniester River Basin, Eastern Europe

The Dniester River has a length of 1362km and a basin area of $72,000km^2$, shared between Ukraine and Moldova. Flowing from the Carpathians, through Moldova into the Black Sea, the Dniester features the rare specificity that Ukraine is both an upstream and a downstream country (see Figure 7.3). While the basin was managed as an entire catchment during Soviet times, a national approach has prevailed since the break-up of the Soviet Union. Currently, treaties exist regulating the use and allocation of water between the two states while ignoring ecological aspects of river basin management. The situation is aggravated due to the stalled conflict involving the enclave of Transdniestria, hampering cooperation between the two riparians in many aspects. As a consequence the Dniester carries a considerable pollution load, significantly contributing to the deterioration of the Black Sea region.

Major problems in the catchment originate from spills of municipal sewage due to management deficits in both riparians. The current institutional situation in the two riparian countries is not set up to address these problems, as existing institutions are mostly catering to issues of water use and economic development, such as the joint committee, set up by representatives from Apele Moldovei and the Ukrainian State Water Committee for the Dniester, and the joint commission on economic cooperation between Moldova and Ukraine (Trombitsky, 2007; Trombitsky et al, 2007).

In this problematic political, environmental and institutional situation, environmental NGOs

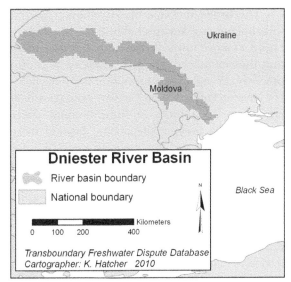

Figure 7.3   Map of the Dniester River Basin

Source: www.dniester.org

from both Moldova and Ukraine have played a major role in bringing environmental issues to the forefront, and mobilizing national and international dialogues on these topics. Over the last 15 years, they have engaged in numerous activities, such as expeditions along the river, newspaper articles discussing the unfavourable environmental conditions, and have also organized several international conferences. It was in this context that an international NGO, the International Association of the Dniester River Keepers (ECO-Tiras) was formed in 1999 and the first draft of an international convention on 'the Use of Water and Biological Resources and Conservation of Landscape and Biological Diversity' was brought forward by the NGO community and proposed to both governments (Trombitsky, 2002). Much importance is attributed to the adoption of the convention in terms of aiding with the establishment of a basin-wide coordinative action as well as with the attraction of donor funding.

In an effort to further strengthen the NGO capacity in the region, ECO-Tiras aided with the establishment of NGOs in the area, which were brought together in the Joint Commission for Environmental Cooperation in the Lower Dniester by 2000. Important contributions to these efforts were the Dniester projects, initiated

by the Organization for Security and Cooperation in Europe (OSCE) and UNECE and aiming at providing an institutional basis for transboundary cooperation in the basin. The projects commenced in 2004 and helped to strengthen the stakeholder community in the basin and also assisted in laying ground rules for stakeholder participation in the basin (UNECE and OSCE, 2005, 2007; ENVSEC et al, 2009). Currently, the Dniester Basin has a vibrant NGO community, which is currently taking on the role of custodian of the catchment.

The Dniester case study is an example of very active public participation activities, which emerged despite the lack of adequate institutional structures at the transboundary level (Kranz and Vorwerk, 2006). Citizen initiatives and groups, such as ECO-Tiras and others from Moldova, Ukraine and Transdniestria took on the role of forming an active civil society, filling the 'governance gap' created by the political situation in the region. A supportive element in these efforts proved to be the international process in relation to the UNECE convention, which provided the necessary legitimacy for the activities undertaken by the NGO community in addition to the much-needed motivation and support of those active in the Dniester basin.

Conclusions and Outlook

This chapter provided an overview of the theory behind stakeholder participation in transboundary river basins as well as an introduction to some of the main elements of successful participatory processes at the international level. The case studies investigated demonstrated the relevance of recognizing, capacitating and involving key stakeholders in well-designed processes with clearly defined targets, but also the challenges in implementing such processes.

Persons or entities mandated with or responsible for facilitating and implementing participative processes need to possess sufficient process management skills as well as political clout. At this point also the questions of legitimacy and power are raised. In most cases the task of facilitating and implementing participatory processes is assigned to international river basin organizations. Vested with the mandate and trust of the riparian governments they are in a good position to organize participatory processes, but they may lack process management skills and moreover often have no direct links to the local level. A completely different situation emerges in the Dniester River Basin, where no designated entity exists yet and the participative process is driven by stakeholder groups.

Similar issues emerge on the side of the stakeholders. Capacitated stakeholder groups are necessary in order to allow for a fair representation of all pertinent interests. Capacity in this regard refers to a number of characteristics, which include financial and organizational resources, but also touch on the issue of legitimate representation. Different stakeholder groups do not necessarily possess sufficient resources in this regard. How to account for this and address potential discrepancies is one of the main challenges in transboundary participative processes. Variation among stakeholder groups is often aggravated by differences due to national legislation and institutional structures.

Next, process or the 'how to' moves into the centre of attention. What are the issues that can actually be addressed at the transboundary level, and how can local-level issues gain in relevance at the transboundary level? When and in what order should issues be addressed? Participatory methods play an important role in providing answers. The development and implementation of effective communication techniques represent one side of the coin; educational aspects, such as training and capacity-building (see Salamé and Van der Zaag, this volume Chapter 12, for more detail) the other. River basin awareness kits as well as other innovative instruments, as for example the Stockholm International Water Institute (SIWI) TWO analysis (Phillips et al, 2008), the Global Water Partnership's IWRM toolbox, as well as a broad range of locally tested tools, are gaining in significance. These, however, must be carefully selected and possibly adapted in order to correspond with the capacities and needs of the stakeholders to be involved and the requirements of the process chosen.

The broad message of this chapter is that stakeholders are key in enabling transboundary basins to move towards sustainable development and thus enhanced adaptive capacity. It is also precisely through this contribution that stakeholders have an influence on the way countries may choose to collaborate on international water resources, by moving the debate beyond mere allocation issues, and thus creating a new basis for discussion. This, however, will be subject to future research.

The coming years will see the increasing emergence of stakeholders' positions and voices, and these need to be heard and effectively integrated into decision making for water resources management at the transboundary level. This chapter provides insights into possible approaches and avenues to be taken by governments as well as stakeholders.

Notes

1   The initiative was supported by the Swedish development agency, Sida.

References

Adams, A. (2000) *Social Impacts of an African Dam: Equity and Distributional Issues in the Senegal River Valley*, Working Paper, World Commission on Dams, Thematic Reviews, Social Issues, www.ideas.repec.org/p/ess/wpaper/id514.html

Arnstein, S. (1969) 'A ladder of citizen participation in the USA', *Journal of the American Institute of Planners*, vol 35, no 4, pp216–224

Ashton, P. J. and Neal, M. (2003) 'An overview of the strategic issues in the Okavango Basin', in A. R. Turton, P. J. Ashton and T. J. Cloete (eds) *Transboundary Rivers, Sovereignty and Development: Hydropolitical Drivers in the Okavango Basin*, Green Cross International, Geneva

Ashton, P. and Neal, M. (2005) 'Public involvement in water resource management within the Okavango River Basin', in C. Bruch, L. Jansky, M. Nakayama, and K. Salewicz (eds) *Public Participation in the Governance of International Freshwater Resources*, United Nations University Press, Tokyo, pp169–195

Börzel, T. A. (2007) 'State capacity and the emergence of new modes of governance', Paper within the European Union NewGov New Modes of Governance Project

Börzel, T. A., Buzogany, A. and Guttenbrunner, S. (2006) 'New modes of governance in the new member states: when soft modes meet hard constraints', CONNEX Workshop on *Democracy, Rule of Law and Soft Modes of Governance*, Roskilde, Denmark

CIS (2002) *Guidance on Public Participation in relation to the Water Framework Directive: Active Involvement, Consultation, and Public Access to Information. Prepared in the Framework of the Common Implementation Strategy of the European Commission and the EU Member States*, Office for Official Publications of the European Communities, Luxembourg

Cooke, B. and Kothari, U. (2001) *Participation: The New Tyranny?*, Zed Books, London

Cuff, J. (2001) 'Participatory processes: a tool to assist the wise use of catchments; a guide based on experience', *Wise Use of Floodplains Project Report*, www.floodplains.org.uk/pdf/technical_reports/Participatory%20Processes%20Report.pdf

Danube Environmental Forum (2009) 'Public participation process in the Danube River Basin', paper presented at the *Black Sea Regional NGO Workshop*, Kyiv, Ukraine

Delli Priscolli, J. (1978) 'Implementing public involvement programs in federal agencies', in S. Langton (ed.) *Citizen Participation in America: Essays on the State of the Art*, Lexington Books, Lexington, MA

Dieperink, C. (1998) 'From open sewer to salmon run: lessons from the Rhine water quality regime', *Water Policy*, vol 1, no 5, pp471–485

Ebbesson, J. (1998) 'The notion of public participation in international environmental law', *Yearbook of International Environmental Law*, vol 8, pp51–97

ENVSEC, UNECE, OSCE and UNEP (2009) *Report on the First Meeting of the Project 'Transboundary Cooperation and Sustainable Management in the Dniester River Basin: Phase III - Implementation of the Action Programme'*, Chisinau, Moldova

Fritsch, O., and Newig, J. (2007) 'Under which conditions does public participation really advance sustainability goals? Findings of a meta-analysis of stakeholder involvement in environmental decision-making', presented at the *Amsterdam Conference on the Human Dimensions of Global Environmental Change*, Vrije Universiteit Amsterdam, 24–26 May 2007

Greenspan Bell, R. and Jansky, L. (2005) 'Public participation in the management of the Danube River', in C. Bruch, L. Jansky, M. Nakayama and K. Salewicz (eds) *Public Participation in the Governance of International Freshwater Resources*, United Nations University Press, Tokyo, pp101–117

Henriksen, H. J., Refsgaard, J. C., Højberg, A. L., Ferrand, N., Gijsbers, P. and Scholten, H. (2009) 'Harmonised principles for public participation in quality assurance of integrated water resources modelling', *Water Resources Management*, vol 23, no 12, pp1–16

Holling, C. S. and Meffe, G. K. (1995). 'Command and control and the pathology of natural resource management', *Conservation Biology*, vol 10, pp328–337

Huitema, D. and van de Kerkhof, M. (2006) 'Public participation in water management: the outcomes of an experiment with two participatory methods under the Water Framework Directive in the Netherlands: analysis and prospects', in V. Grover (ed.) *Water: Global Common and Global Problems*, Science Publishers, Enfield, NH, US, pp269–296

Huitema, D., van de Kerkhof, M. and Pesch, U. (2007) 'The nature of the beast: are citizens' juries deliberative or pluralist?', *Policy Sciences*, vol 40, no 4, pp287–311

ICPDR (2004) *ICPDR Joint Action Programme 2001–2005: Implementation Report, Reporting Period 2001–2003*, ICPDR, Vienna

ICPDR (2005) *WFD Roof Report for the Danube River Basin, River Basin Characteristics, Impact of Human Activities and Economic Analysis required under Article 5, Annex II and Annex III, and Inventory of Protected Areas required under Article 6, Annex IV of the EU Water Framework Directive (2000/60/EC)*, ICPDR, Vienna

ICPDR (2009) *Draft Danube River Basin District Management Plan*, ICPDR, Vienna

ICPDR, UNDP/GEF and UNOPS (2006) *15 Years of Managing the Danube River Basin 1991–2006*, ICPDR, Vienna

James, R. (1994) *Strengthening the Capacity of Southern NGO Partners*, Occasional Papers Series Number 5, International NGO Training and Research Centre, Oxford

Jonsson, A. (2005) 'Public participation in water resources management: stakeholder voices on degree, scale, potential, and methods in future water management', *AMBIO: A Journal of the Human Environment*, vol 34, no 7, pp495–500

Karkkainen, B. C. (2005) 'Transboundary ecosystem governance: beyond sovereignty?', in C. Bruch, L. Jansky, M. Nakayama and K. Salewicz (eds) *Public Participation in the Governance of International Freshwater*

Resources, United Nations University Press, Tokyo, pp73–87

KCS (2009) *Every River has Its People Project Overview*, Kalahari Conservation Society, Gaborone, Botswana

Korsten, A. F. A. and Kropman, J. A. (1976) *Inspraak bij de ontwikkeling van het streekplan Veluwe: samenvatting van een onderzoek naar de inspraak bij de ontwikkeling van het streekplan Veluwe tijdens de programmafase, in vergelijking met de inspraak in Midden-Gelderland, en bij enkele andere plannen*, Instituut voor Toegepaste Sociologie, Nijmegen, The Netherlands

Kranz, N. (2008) 'Participation in water management – fostering adaptive capacity', presented at *ISA's 49th Annual Convention: Bridging Multiple Divides*, San Francisco, CA

Kranz, N. and Vorwerk, A. (2006) 'Strategies for public participation in integrated water resources management. Experience from the Rhine and contribution to the Dniester Basin', *Academician Leo Berg – 130: Collection of Scientific Articles*, Leo Berg Educational Foundation, Bendery, Moldova

Leach, W. D. and Pelkey, N. W. (2001) 'Making watershed partnerships work: a review of the empirical literature', *Journal of Water Resources Planning and Management*, vol 127, no 6, pp378–385

Marty, F. (2001) *Managing International Rivers: Problems, Politics and Institutions*, Peter Lang, Bern, New York

Meijerink, S. V. (1999) *Conflict and cooperation on the Scheldt River Basin: A Case Study of Decision Making on International Scheldt Issues between 1967 and 1997*, Kluwer, Dordrecht, The Netherlands

Meinzen-Dick, R. (1997) 'Farmer participation in irrigation; 20 years of experience and lessons for the future', *Irrigation and Drainage Systems*, vol 11, no 2, pp103–118

Merrey, D. J. (2009) 'African models for transnational river basin organisations in Africa: an unexplored dimension', *Water Alternatives*, vol 2, no 2, pp183–204

Mostert, E. (2005) *How Can International Donors Promote Transboundary Water Management?*, Deutsches Institut für Entwicklungspolitik, Bonn, Germany

Mostert, E. (2006) 'Participation for sustainable water management', in C. Giupponi, A. J. Jakeman, D. Karssenberg and M. P. Hare (eds) *Sustainable Management of Water Resources: An Integrated Approach*, Edward Elgar Publishing, Cheltenham, UK/Northampton, MA, pp153–176

Mostert, E., Pahl-Wostl, C., Rees, Y., Searle, B., Tàbara, D. and Tippett, J. (2007) 'Social learning in European river basin management; barriers and supportive mechanisms from 10 river basins', *Ecology and Society*, vol 12, no 1, art 19

Muro, M. (2008) 'A critical review of the theory and application of social learning in participatory natural resource management processes', *Journal of Environmental Planning and Management*, vol 51, no 3, pp325–344

Nicol, A. (2003) *The Nile: Moving Beyond Cooperation*, UNESCO-IHP, Paris

Olsson, P., Folke, K. and Berkes, F. (2004) 'Adaptive comanagement for building resilience in social–ecological systems', *Environmental Management*, vol 34, no 1, pp75–90

Pahl-Wostl, C. (2002) 'Participative and stakeholder-based policy design, evaluation and modelling processes', *Integrated Assessment*, vol 3, no 1, pp3–14

Pahl-Wostl, C. (2008) 'Requirements for adaptive water management', in C. Pahl-Wostl, P. Kabat and J. Möltgen (eds) *Adaptive and Integrated Water Management*, Springer, Berlin, Heidelberg, New York

Pateman, C. (1970) *Participation and Democratic Theory*, Cambridge University Press, Cambridge

Phillips, D., Allan, A., Claassen, M., Jägerskog, A., Kistin, E., Patrick, M. and Turton, A. (2008) *The TWO Analysis: Introducing a Methodology for the Transboundary Water Opportunity Analysis*, vol 23, SIWI, Stockholm

Porto, J. G. and Clover, J. (2003) 'The peace dividend in Angola: strategic implications for the Okavango Basin Cooperation', in A. R. Turton, P. J. Ashton and T. J. Cloete (eds) *Transboundary Rivers, Sovereignty and Development: Hydropolitical Drivers in the Okavango Basin*, Green Cross International, Geneva

REC (1999) *Healthy Decisions; Access to Information, Public Participation in Decision-making and Access to Justice in Environment and Health Matters*, Regional Environmental Center, Szentendre, Hungary

Reed, M. S. (2008) 'Stakeholder participation for environmental management: a literature review', *Biological Conservation*, vol 141, pp2417–2431

Ridder, D., Mostert, E. and Wolters, H. A. (eds) (2005) *Learning Together to Manage Together: Improving Participation in Water Management*, USF, University of Osnabrück, Osnabrück

Roberts, R. (1995) 'Public involvement: from consultation to participation', in F. Vanclay and D. A. Bronstein (eds) *Environmental and Social Impact Assessment*, J. Wiley, Chichester, UK/New York, NY

Sabatier, P. A. and Jenkins-Smith, H. C. (eds) (1993) *Policy Change and Learning: An Advocacy Coalition Approach*, Westview Press, Boulder, CO

Sadoff, C. and Grey, D. (2002) 'Beyond the river: the benefits of cooperation on international rivers',

Water Policy, vol 4, pp389–403

Trombitsky, I. (2002) 'The Role of civil society in conservation and sustainable management of the NIS transboundary watercourse, the Dniester River', *10th OSCE Economic Forum 2002*, Prague, Czech Republic

Trombitsky, I. (2007) 'Integrated management of transboundary Dniester river and consolidating role of NGOs in conditions of frozen conflict', paper presented at the *OSCE Economic and Environmental Forum 2007*, Zaragoza, Spain

Trombitsky, I., Slesarenok, S. and Ignatiev, I. (2007) 'Towards integrated management of transboundary Dniester River and role of public participation', *OSCE Economic and Environmental Forum – Part 2: Key Challenges to Ensure Environmental Security and Sustainable Development in the OSCE Area 2007*, Prague, Czech Republic

Trottier, J. (2003) 'The need for multiscalar analysis in the management of shared water resources', in F. A. Hassan, M. Reuss, J. Trottier, C. Bernhardt, A. T. Wolf, J. Mohamed-Katerere, and P. van der Zaag (eds) *History and Future of Shared Water Resources*, UNESCO-IHP, Paris, pp1–10

UNECE and OSCE (2005) *Transboundary Diagnostic Study for the Dniester River Basin*, UNECE and OSCE, Geneva

UNECE and OSCE (2007) *Action Programme to Improve Transboundary Cooperation and Sustainable Management of the Dniester Basin*, UNECE and OSCE, Kyiv, Ukraine

Videira, N., Antunes, P., Santos, R. and Lobo, G. (2006) 'Public and stakeholder participation in European water policy: a critical review of project evaluation processes', *European Environment*, vol 16, pp19–31

Warner, J. F. (2006) 'More sustainable participation? Multi-stakeholder platforms for integrated catchment management', *International Journal of Water Resources Development*, vol 22, no 1, pp15–35

Webler, T. and Renn, O. (1995) 'A brief primer on participation: philosophy and practice', in O. Renn and T. Webler (eds) *Fairness and Competence in Citizen Participation: Evaluating Models For Environmental Discourse*, Kluwer Academic, Dordrecht, The Netherlands/Boston, MA, pp17–33

8

Environmental Flows in Shared Watercourses: Review of Assessment Methods and Relevance in the Transboundary Setting

Cate Brown and Jackie King

- Ecological and subsistence issues should be factored into water-resource development plans automatically and in a structured manner.
- Environmental flows (EFs) should be done at the basin level at an early stage of water-resource planning.
- For transparent decision making to occur it is essential that trade-offs should be made and that objective scenarios capturing the full spectrum of outcomes, positive and negative, are generated.
- It is never too early to get started.
- Involvement of stakeholders is essential at every stage of the EF assessment process.
- Commitment to EFs is commitment to a long-term complex management process.

Introduction

This chapter is an overview of the concept of environmental flows (EFs) for ecosystem maintenance, how they can be assessed, and their perceived role in the development and management of transboundary river basins. We highlight some of the approaches to EF assessments (EFAs) as well as specific challenges in the transboundary water management (TWM) setting, and make recommendations on how to overcome some of these.

We introduce the belief that freshwater ecosystems can be managed to be at different levels of condition (health), from pristine, when they provide a range of natural services of benefit to humans; through various stages of change from pristine, when the original services disappear and others appear; to serious degradation, when virtually all natural services essentially disappear.

We outline the evolution of EF concepts from their early beginnings to present-day understanding of the importance of the variability in magnitude, timing, frequency and duration of water in determining the nature and condition of freshwater ecosystems. We also discuss the array of different methods used to assess EFs, their history and their applicability to different situations. Important advances in techniques and approaches are highlighted, in particular the move towards multidisciplinary, basin-wide, scenario-based assessments that allow for provision of information on the full suite of costs and benefits of water-resource development.

We use case studies to illustrate some of the ways in which EFAs have been used in a trans-boundary setting, and discuss the role of EFAs in planning and managing water-resource develop-ments in transboundary river basins. Then we discuss possible avenues for promoting the inclu-sion of ecological considerations into integrated water resource management (IWRM), particularly those pertaining to the basin-wide implications of flow and sediment changes brought about by water-resource developments. We draw attention to some important ecosystem considerations for water-resource developments, and explore the concept of 'water-resource development space' as a viable option for the provision of EFs as a genuine, and lasting, component of sustainable water-resource management.

Terminology

EF is just one term in a sequence that has devel-oped over the last four decades to define the water needed for maintenance of freshwater ecosystems. These include instream flow requirements (IFRs), an earlier, less comprehensive term, usually focused on flows for maintaining the fish habitat in rivers; minimum flow, a general term that origi-nated as a dry-season streamflow standard to constrain the abstraction of water and has been used in several ways to describe a flow to maintain some feature of a river ecosystem; and eFlows, a modern day shorthand term for EFs. All are designed to aid, in some way, maintenance of the condition (health) of freshwater ecosystems, in contrast to other uses of water, including offstream uses, which do not embrace the objective of managing ecosystem health (Brown and King 2003). Minimum flow is still widely used in engineering and planning circles, but is misleading

as it implies that there is a single number that represents the amount of water that will prevent degradation of the targeted ecosystem. This is not the case, for freshwater ecosystems increasingly degrade as their water is manipulated for human needs, and so the final allocation of water to them is a trade-off by society between what will be gained by development of their water resources and what will be lost in terms of the natural resources they supply.

EFAs traditionally focused on rivers but are increasingly addressing the water needed for maintenance of wetlands, estuaries and aquifers. In this chapter, for simplicity, we use the term EF in the wider sense of water needed for mainte-nance of all kinds of inland water ecosystems, while recognizing that 'environmental water' might be more appropriate than EF in some cases where flow *per se* is not important.

Freshwater Ecosystems and the Importance of Flow and Sediments

Freshwater ecosystems are the foundation of every country's social, cultural and economic well-being. Healthy freshwater ecosystems – rivers, lakes, floodplains, wetlands and estuaries – provide clean water, food, fibre, energy and many other benefits that support economies and livelihoods around the world. They are essential to human health and well-being (Brisbane Declaration, 2007).

Freshwater ecosystems have evolved in response to the natural geology, topography, climate and vegetation of their basins, and the volume and pattern of water draining them. Together, these create the habitats that support a wide variety of animals and plants, and provide a host of services to people. The suite of ecological services provided by inland water ecosystems is valued at about US$6 trillion per annum (Postel and Richter, 2003), and they have among the highest value per unit area of any of the world's ecosystems (Costanza et al, 1997, cited in Postel and Richter, 2003). The services they provide include:

- production services (edible plants and animals; water; raw materials – wood, rocks and sand for construction, firewood, genetic resources and medicines; ornamental products for handicrafts and decoration);
- regulatory services (groundwater recharge; flood attenuation; soil stabilization; water purification);
- cultural services (spiritual; religious; aesthetic appeal; inspiration for books, music, art and photography; national symbols; advertising);
- supporting services (nutrient cycling; soil formation; pollination; carbon sequestration);
- nursery services (breeding areas for marine fishes), and more (King, 2009).

Different flows are important for maintaining ecosystems (e.g. Carter et al, 1979; Poff and Ward, 1990; Richter et al, 1997; Bunn and Arthington, 2002). In rivers, floods and the sediments they carry shape the channel, creating a diverse array of habitats for animals and plants (Richter et al, 1997). Floods also inundate floodplains, trigger fish migrations and breeding, and maintain the vegetation on floodplains and riverbanks that, in turn, protect the riverbank from erosion and act as a buffer against chemicals and sediments flowing off the catchment (Tabacchi et al, 1998). The low flows maintain the basic ephemeral, seasonal or perennial nature of the river, and determine the animals and plants that can survive there (Brown and King, 2003). The different magnitudes of low flow in the dry and wet seasons create more or less wetted habitat and different hydraulic and chemical conditions, which directly influence the balance of species. The timing of these different flows is as critical as their size, as the reproductive and other behaviours of plants and animals are attuned to, and dependent on, the seasonal fluctuations of flow and temperature. Flow variability, on a daily, seasonal or annual basis, maintains biological diversity through increased heterogeneity of physical habitats. Variability also means that conditions are optimal for different species at different times, which ensures that no one species proliferates to 'pest' proportions (e.g. Zakhary, 1997). In wetlands and lakes the influence of flow is less apparent but no less important. The timing

and magnitude of seasonal fluctuations in inundated area and depth drive the chemical and thermal characteristics and set in motion the breeding and growing cycles of plants, fish and other animals (e.g. Welcomme, 1979; Karenge and Kolding, 1994). Seasonally flooded areas provide grazing for migratory species such as antelope (e.g. Mendelsohn and el Obied, 2004), breeding grounds for fish and birds (e.g. Welcomme, 2001; Kamweneshe and Beilfuss, 2002), and fertile land for recession agriculture (e.g. Heeg and Breen, 1982).

Similarly, freshwater inflow to estuaries is a fundamental part of estuarine chemistry, morphology and biology. The mixing of fresh- and salt-water creates and sustains a unique type of environment that is among the most productive of any on earth (e.g. Nixon et al, 2004). The amounts, duration and intensity of flow events influence estuarine geomorphology, water temperature, salinity, pH, turbidity, nutrient status, organic inputs, dissolved oxygen concentrations, olfactory cues, mouth status, tidal prism, habitat diversity, primary and secondary productivity, fish recruitment, food availability and competition (Whitfield and Wooldridge, 1994).

Because they are the drainage points of whole basins, these ecosystems are highly vulnerable to changes in their basins (e.g. Baron et al, 2002). Land-use changes, water-resource developments, point- and diffuse-source pollution, and other interventions all impact on the freshwater environment and the services they afford people, sometimes profoundly and possibly for hundreds of kilometres downstream. This is particularly true for changes in the flow of water and sediment, the influences of which tend to track farthest (and fastest) down the system. Nor are the impacts always unidirectional. The effects of developments such as dams can also be felt upstream if, for instance, in-channel dams act as barriers to fish migrating upstream to breeding and feeding grounds (King, 2009). Hence, for sustainable development to occur, these ecosystems need to be managed holistically within their whole drainage basin (King and Brown, 2009a).

Removal of water from freshwater ecosystems and/or manipulation of their flow regimes will always be a trade-off between loss of ecosystem function and resilience, on the one hand, and benefit derived from the use of that water elsewhere, on the other. The greater the divergence from a natural flow regime (in terms of both the volume and the timing of different magnitude flows), the more the ecosystem could be expected to change. Furthermore, if the water-resource infrastructure changes the natural sediment supply, these changes will be even more pronounced (Petts and Gurnell, 2005).

This presents particular challenges for sustainable management in transboundary river basins, where political borders often separate development pressures and the need for ecological protection (Fox and Sneddon, 2007). The solution lies in countries recognizing the critical importance of the hydrological and sediment regimes as the primary drivers of ecological processes in river–floodplain systems (Carter et al, 1979; Richter et al, 1997; Tharme, 2003) and engaging in cross-border collaboration to assess the impacts of water-resource developments on the integrity of downstream ecosystems, and on the suite of services offered to people by those ecosystems. Specifically, they need to reach agreement on future condition of those systems, on which of the valued services will be maintained and the volume and timing of the flows required to sustain them (Table 8.1) – the so-called EFs.

EFs refer to 'the quality, quantity, and timing of water set aside to maintain the components, functions, processes, and resilience of aquatic ecosystems that provide goods and services to people' (after Hirji and Davis, 2009). They arose as an explicit response to the need for sustainable development of the world's water resources (Gleick, 2002; United Nations, 2007) and as such do not focus on conservation or protection of nature and the environment, although they can be used as a tool for both. Rather, they reflect the recognition that water-resource development does and should occur for the betterment of people, but that this will affect the ecological condition of the targeted system. Aquifers, rivers, wetlands, lakes and estuaries can be managed to be at different levels of condition (health), from pristine, when they provide a range of natural services of benefit

to humans; through various stages of change from natural, when the original services disappear and others appear; to serious degradation, when many of the natural services are lost (King and Brown, 2009a). The finally agreed condition should be a societal choice based on the consideration of all options. For instance, in the case of a large dam development, the decision may be to release a portion of the inflows for maintenance of the downstream river, which would reduce the yield of the dam, but retain valued ecosystem attributes in the reaches downstream. In Lesotho, 10–14 per cent of the mean annual runoff (MAR) of the rivers is released from the Lesotho Highlands Water Project dams, which will maintain the rivers in a condition that is reduced from natural but that will still support some of the fish and woody vegetation that the riparian people use (LHDA, 2003).

The science of EFs has five main objectives (after Brown and King, 2006):

1   to understand the nature and functioning of freshwater ecosystems;
2   to be able to predict how these ecosystems will change with flow change;
3   to understand how people use the freshwater ecosystem resources;
4   to be able to predict how these uses will be affected by change in these ecosystems;
5   to combine the predictions into scenarios reflecting the costs and benefits of a range of development options.

EF assessments are thus part of a comprehensive approach to water-resource management that can guide more sustainable use of aquatic ecosystems. They provide hitherto absent information on the consequences of water-resource developments for downstream freshwater ecosystems and the people who depend on them, which can empower and inform decision making (Richter and Postel, 2004; Arthington et al, 2007; Brown and King, 2006).

EFs can be incorporated into water-resource developments at virtually any stage (Postel and Richter, 2003), but consideration of them should preferably be done early in the planning process of water development, together with an analysis of

Table 8.1   Examples of valued features of freshwater ecosystems that could be protected through EFs

Feature	Explanation of value	Examples of EFs required
Aquatic animals	Freshwater fish are a valuable source of protein for rural people. Other valued attributes include: angling fish, rare water birds, or the small aquatic life that forms the base of the food chain	• flushing flows to maintain the physical habitat • flows to maintain suitable water quality • flows to allow passage for migratory fish • small floods to trigger life-cycle cues such as spawning or egg-laying
Riparian vegetation	Stabilizes riverbanks, provides food and firewood for rural people and habitat for animals, and buffers the river against human activities in the catchment	• flows that maintain soil moisture levels in the banks; • high flows to deposit nutrients on the banks and distribute seeds.
River sand	Used for building	Flows to transport sand and to separate it from finer particles
Estuaries	Provide nursery areas for marine fish	Flows that maintain the required salt/freshwater balance and mouth conditions in the estuary
Aquifers and groundwater	Maintain the perennial nature of rivers through acting as sources of water during the dry season	Flows to recharge the aquifers
Floodplains	Support fisheries and flood-recession agriculture for rural people	Floods that inundate the floodplain at the appropriate time of the year
Aesthetics	The sound of water running over rocks, the smells and sights of a river with trees, birds and fish	Sufficient flow to maximize natural aesthetic features, including many of the flows mentioned above
Recreational and cultural features	For example, clean water and rapids for river rafting or clean pools for baptism ceremonies or bathing. Also features valued by anglers, birdwatchers, and photographers	Flows that flush sediments and algae, and that maintain the water quality – see also Aquatic animals
Ecosystem services	Maintain the capacity of aquatic ecosystems to regulate essential ecological processes, for instance: purify water, attenuate floods or control pests	Flows that maintain biodiversity and ecosystem functioning
Overall environmental protection	Minimize human impacts and conserve natural systems for future generations	Some or all of the above type of flows

Source: Brown and King, 2003

the economic benefits of the proposed scheme, so that the agreed trade-off between development benefits and natural resource degradation can guide project design and operation (Watson, 2006).

Providing both sides of the development picture in this way for discussion and negotiation by stakeholders contributes to the requirements of integrated water resources management (IWRM) (King and Brown, 2009a). IWRM is 'a process that promotes the coordinated development and management of water, land and related resources, in order to maximize the resultant economic and social welfare in an equitable manner without compromising the sustainability of vital ecosystems' (GWP, 2003). It is a concept that promotes sustainable use of water, encouraging people to move away from traditional project-driven ways of operating and towards a larger-scale basin or regional approach that takes into account the overall distribution and scarcity of water resources and the needs of other potential water users. In essence IWRM is a political procedure that aims for sustainability of use; a process of balancing all water demands and supplies including those for environmental maintenance; an iterative approach that recognizes the need for adaptive management; and a way of life (King and Brown, 2009a).

Development of the EF Concept and Assessment Methods

The concept of EFs first arose in the 1940s with pioneering work in the USA aimed at protecting river flows for valued angling fish, predominately trout and salmon (King et al, 1999; Tharme,

2003). Changes in legislation in that country in the early 1970s provided extra impetus for EF work and resulted in the development of numerous methods for setting such flows (Stalnaker, 1982; Tharme, 2003), such as the Tennant Method (Tennant, 1976) and IFIM (Stalnaker et al, 1994) (Table 8.2). It was not until the late 1980s, however, that EFs were more widely addressed, with Australia, England, New Zealand and South Africa bringing new thinking to the topic, Brazil, Czech Republic, Japan and Portugal becoming active (Tharme, 1996, 2003), and other countries in Africa, Asia, Europe and South America following suit later. As was the case in the USA, the increased interest in EFs was often either initiated or supported by changes in national legislation (e.g. DWAF, 1998; URT, 2002; DEWHA, 2007).

Since the conclusion of the World Commission on Dams (WCD, 2000), which strongly advocated the need for the inclusion of EFs in the planning and operation of major water-resource developments, some form of EF assessment has been completed for freshwater ecosystems in many countries in the world, and the inclusion of EFs in international and national water laws and policy has gained more momentum (Richter and Postel, 2004). For instance, the European Union adopted the Freshwater Directive, which aims to reinstate 'good' ecological status in the rivers, lakes and wetlands of all member states by 2015 (European Union, 2000). The Southern African Development Community (SADC) included EFs in the SADC Vision for Water and Life, and at present more than half of the member countries have made provision for EFs in their water laws and others are currently in the process of making relevant revisions to their laws (McIntyre, this volume Chapter 5).

This increased global attention to ecosystem health has been accompanied by a broadening of the objectives of EFs, with the focus moving away from maintaining suitable conditions for single species, to more comprehensive objectives that seek to describe the effects of flow manipulations on whole aquatic ecosystems, and the social and economic networks that depend on them (e.g. Metsi Consultants, 2000; Berkamp et al, 2000).

Tharme (2003) reported that there were more than 200 EF assessment methods in use worldwide, and it is likely that that number has increased significantly since then. Most of these methods are based on understandings of the importance of variability, pattern and volume in determining the nature and condition of freshwater ecosystems, and many share strikingly similar features, such as division of the flow regime into ecologically significant flow categories (e.g. Arthington et al 1992; Richter et al, 1997; King et al, 2003). The main differences between them lie in the extent to which they consider different components and aspects of freshwater ecosystems.

EF methods have been categorized in several different ways. For instance, Tharme (2003) identified six categories of assessment methods for rivers: hydrological; hydraulic rating; habitat simulation; holistic; combination-type; and 'other' methods (Table 8.2). Habitat-rating methods tend to be concentrated in the northern hemisphere and in developing countries aided by the US and Europe, whereas the holistic approaches are centered in the southern hemisphere, e.g. South Africa and Australia, and are the most rapidly growing set of methods globally at the present time (Tharme, 2003; King and Brown, 2006). Others have distinguished between methods that prescribe a single flow regime to maintain a prescribed ecosystem condition (prescriptive methods) and those that indicate the consequences for a range of flow manipulations (interactive or scenario-based methods, e.g. Brown and King, 2003) (Table 8.2).

Experience has shown that selection of an EF method to suit the aims and constraints of the assessment is crucial to its success, and countries may find that they need to implement a range of methods depending on the aims of an assessment, the available budget and information, the experience of the personnel and the severity of the pressure on the environment (Hirji and Davis, 2009). The choice of EF method is particularly important in developing countries where rural people rely on freshwater ecosystems to support their livelihoods, and hence there is a greater need to understand the implications of water-resource development for both them and the resources they

Table 8.2   Characterization of EF methods, with examples

Category	Description	Examples	Prescriptive/ interactive	Reference
Hydrological	Use fixed percentages or thresholds in the hydrological record to represents the flow intended to maintain some ecological feature(s) at some presumed level	Tennant (or Montana) Method	Prescriptive	Tennant (1976)
		Range of Variability Approach (RVA)	Prescriptive	Richter et al (1996, 1997)
		Texas Method	Prescriptive	Matthews and Bao (1991)
		Desktop Method	Prescriptive	Hughes (2001); Hughes and Hannart (2003)
		Ecological Limits of Hydrologic Alteration (ELOHA)	Prescriptive	Poff et al (2009)
Hydraulic rating	Single cross-section methodologies that use change in a hydraulic variable, such as wetted perimeter or maximum depth, to develop a relationship between habitat and flow recommendations	Wetted Perimeter (generic) Method	Prescriptive	e.g. Gippel and Stewardson (1998)
		R-2 Cross Method	Prescriptive	Anon (1974), cited in Stalnaker and Arnette (1976)
Habitat simulation	Physical microhabitat is modelled in terms of various hydraulic variables, and changes in the availability of simulated combinations of hydraulic habitat are combined with information on suitable and unsuitable microhabitat conditions for particular species to recommend optimum discharges	Instream Flow Incremental Methodology (IFIM)	Interactive	Stalnaker et al (1994)
		River Hydraulics and Habitat Simulation Program (RHYHABSIM)	Interactive	Jowett (1989)
Holistic	Considers the relationship between different parts of the flow regime and many components or attributes of the ecosystem to either recommend a flow regime to maintain some pre-agreed condition, or to provide insight into the implications of flow changes for the ecosystem	Building Block Methodology	Prescriptive	King and Louw (1998)
		Flow Restoration Methodology (FLOWRESM)	Prescriptive	Arthington (1998); Arthington et al (2000)
		Habitat-Flow Stressor Response (HFSR)	Interactive	O'Keeffe et al (2002)
		Downstream Response to Imposed Flow Transformations (DRIFT)	Interactive	King et al (2003)
		Murray-Darling Flow Assessment Tool (MFat)	Interactive	Young et al (2003)
Combination-type	Use a combination of above approaches	Managed Flood Release Approach	Mixed	Acreman et al (2000)
		Integrated Basin Flow Assessment (IBFA)	Mixed	King and Brown (2009a)
		The Water Resource Classification System	Mixed	Dollar et al (in press)

Source: after Tharme, 2003

use. Thus, holistic EF assessment methods (Table 8.2) are often more appropriate for developing countries.

An EF assessment is a multidisciplinary activity that uses experts from a wide range of disciplines. Local circumstances and requirements will affect almost every aspect of an assessment, including the disciplines required for the social and biophysical components of the study (Brown and King, 2006). Typically, however, disciplines include: hydrology; hydraulics; geomorphology; water quality; riparian and aquatic botany; aquatic invertebrates; fish, as well as sociology;

public health; animal health; and resource economics. Increasingly, experience has shown that the involvement of local specialists in these fields, even if they are less experienced than foreign specialists, will enhance the likelihood of the outcomes of the EF being accepted by local stakeholders.

Integrated basin flow assessments

A relatively new direction that has emerged in EF assessments in the last five years is towards basin-level assessments that integrate environmental, social and economic outcomes for a series of planned development and/or restoration scenarios. These integrated basin flow assessments (IBFAs) (King and Brown, 2009a) deliver scenarios that describe possible pathways into the future: multifaceted views of potential changes in, for instance, channel configuration; bank erosion; water chemistry; riparian forests; river, estuarine and near-coastal marine fisheries; rare species; pest species; human and livestock health; availability of baptism areas; household incomes; GDP; job creation; hydropower production, and more (Figure 8.1) (King and Brown, 2009a). They represent a harmonization of previously separate macroeconomic and EF assessments associated with water-resource developments, for discussion and negotiation by stakeholders.

IBFAs are similar to strategic environmental assessments (SEAs) but more contained, focusing just on flow-related issues (King and Brown, 2009a). Typically, IBFAs use either a holistic or a combination EF method able to describe the predicted consequences of flow change for many aspects of the ecosystem, that is, it is interactive or scenario-based. There are a number of reasons for this: prescriptive methods are unable to provide information on the consequences of different types of development required to facilitate negoti-ations and trade-offs, and detailed information on valued ecosystem aspects and services are required to inform the social and (often) the macroeconomic assessments. Also, aquatic ecosystems are complex, and the interplay between, *inter alia*, flow, sediment, aquatic habitat and biota (Figure 8.2) cannot be adequately described and assessed without considering the whole (or at least many aspects) of the ecosystem.

IBFAs are a major contribution to the requirements of IWRM, and offer a possible way forward for sustainable management of freshwater ecosystems in a transboundary setting (King and Brown, 2009a). They are also in line with the emerging field of sustainability science, which seeks to understand the interactions between nature and society, and to encourage those interactions along more sustainable trajectories (Kates et al, 2001).

Integrated Basin Flow Assessments in the Development and Management of Transboundary River Basins

Water-resource developments in upstream countries can impact downstream countries in numerous ways. The most obvious of these are

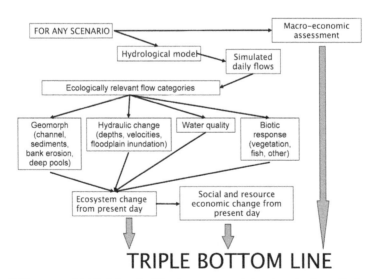

Figure 8.1   The IBFA process, highlighting the EF component and showing the flow of information that produces the three streams of information for any development (or rehabilitation) scenario

Source: King and Brown, 2009a

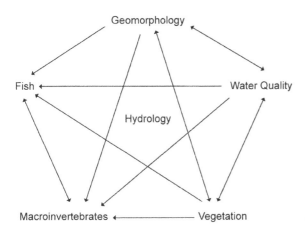

Geomorphology

Fish   Water Quality

Hydrology

Macroinvertebrates   Vegetation

Figure 8.2 Holistic EF assessments consider the interplay between various aspects of a freshwater ecosystem

changes in the volume and timing of water flowing across the border. Not only can these limit water-resource development opportunities in the downstream country, but they can also reduce the resilience and vitality of downstream freshwater ecosystems, with potential losses of resources such as wood, reeds, fish, income from tourism and hunting, potential costs associated with proliferations of pest species, and other unforeseen effects. Reduced freshwater inflow into estuaries, for instance, will affect not only the estuary, but also near-shore fisheries, which will suffer from a loss or decline in the viability of estuarine nursery areas (e.g. Sengo et al, 2005), and salt-water intrusion may affect water supply upstream of the estuary (Brockway et al, 2006). Sediment trapped in upstream dams will no longer supply riverbanks and beaches, possibly leading to bank and coastal erosion. A decline in the health of freshwater ecosystems in the upstream country will also impact on the downstream country. For instance, damage to upstream floodplains, and the resultant reduction in flood attenuation capacity, will cause increased flood peaks in the downstream reaches, with associated flood alleviation and/or reparation costs, and even loss of human life.

Fox and Sneddon (2007) argue that, in a transboundary setting, freshwater ecosystems and the flow of water through them will only be subjected to substantive international rules of law if they can be represented as simplified watercourses, demar-

cated, reduced to parts and rationally managed. IBFAs allow for such an interpretation of these ecosystems, and as such can, and should, be used to enhance technical, operational and political cooperation between neighbouring countries (after Wolf et al, 2003, and Awulachew et al, 2008; and see Jarvis and Wolf, this volume Chapter 9).

Two examples of IBFAs are for the Lower Mekong River (Thailand, Lao PDR, Cambodia and Vietnam) (MRC, 2006) in Southeast Asia, and the Okavango River (Angola, Namibia and Botswana) (King and Brown, 2009b) in southern Africa. Other examples of IBFAs, or similar, include: Pioneer River, Australia (Department of Natural Resources and Mines, 2005); Pangani Basin, Tanzania (PBWO/IUCN, 2009); and the Olifants–Doring Basin, South Africa (Dollar et al, in press). They are also in the planning stage for the Zambezi River (Angola, Botswana, Zambia, Zimbabwe, Malawi and Mozambique) and the Orange River (Lesotho, South Africa, Namibia and Botswana). The transboundary IBFAs for the Mekong and the Okavango Basin were done under the auspices of their respective river basin organizations (RBOs), the Mekong River Commission (MRC) and the Okavango River Basin Commission (OKACOM).

The Mekong IBFA focused on the Lower Mekong from the China border to the sea, and was a demonstration assessment of the possible impacts of three hypothetical levels (low, medium and high) of basin development. The multidisciplinary team comprised pairs of discipline specialists, one international and one riparian for each discipline (MRC, 2006). Over eight months, the team developed hydrological time-series for points along the main stem to describe the flow changes that would occur with the three levels of water-resource development, and used these, together with available knowledge on the system, to predict the likely changes to a suite of biophysical and social indicators for each scenario. The indicators included ecosystem variables such as sediment supply and transport, infilling of deep pools, bank erosion, viability of secondary channels, and fish and fisheries; and social issues such as hydropower production, irrigated agriculture, water supply, navigation, river gardens, salinity intrusion, and shrimp and rice farming. In addition, the present-day monetary value of the basin, in terms of a comprehensive suite of ecosystem services offered by the Lower Mekong was estimated, as well as its future value under each scenario. The results illustrated the potential threats posed by upstream water-resource developments in China and Lao to wetlands and other resources in the Lower Mekong River, in particular those to the Tonle Sap Great Lake in Cambodia, which underpins Cambodia's national fishery. They also showed that, under the highest development scenario considered, the economic gains from basin water-resource development would be far more modest than expected if the loss of ecosystem services, such as fisheries and functioning wetlands, was factored in. The assessment enhanced awareness among the Mekong Basin's collaborating governments of the two sides of development, and the negative effects that development in one country could have on a neighbouring one.

The Okavango IBFA was a preliminary assessment aimed at promoting basin-wide communication and collaboration, and building capacity in collaborative IBFAs in all disciplines in each of the three basin countries, Angola, Namibia and Botswana (King and Brown, 2009c). This was done by appointing a full biophysical and socioeconomic team from each of the three countries, with planning, coordination and training done by an international process management team. Over 12 months, the combined team provided a first-level assessment of the ecological, social and economic consequences of three possible future development paths (e.g. Figure 8.1) that were chosen through a process of government consultation (King and Brown, 2009c). In general, the hydrological team developed basin hydrological models and prepared scenarios; the biophysical team identified biophysical indicators and developed the relationships between flow and ecosystem components, using flow response curves; and the socioeconomic team identified social indicators and the links with the biophysical indicators (see Kranz and Mostert, this volume Chapter 7, for definitions of kinds of stakeholders). A decision support system (DSS) captured these relationships and used them to produce

predictions of ecological and social change for each scenario. This output was combined with the predicted macroeconomic benefits of each scenario, which emanated from a linked exercise, to provide a comprehensive cost–benefit analysis of the range of development options (King and Brown, 2009c). The results clearly illustrated the country-specific benefits associated with each development path, as well as the downstream, often transboundary, ecological and social costs associated with each. They also made explicit the costs to one country for benefits accrued in another. For example, the ecotourism benefits of the Okavango Delta accrue to Botswana, whereas (under current arrangements) the benefits of water-resource developments on the Cubango and Cuito Rivers whose waters sustain the Delta, would accrue to Angola (King and Brown, 2009c).

The IBFA provided OKACOM with a DSS, calibrated through cooperation between technical experts from each of the basin countries, which can be used to test variations of the three chosen scenarios. These could include changing the location, size or operations of particular water-resource schemes, or foregoing particularly damaging ones, in order to optimize the overall benefits to the basin, and inform on trade-offs and decision making. The information could also be used to develop international agreements that consider, *inter alia*, benefit sharing and/or payment for ecosystem services.

Negotiations and Decision Making

Assessing the implications of different development options, while technically complex, is probably more straightforward than the next stage: deciding on the appropriate trade-off between water-resource development and environmental protection (Brown and Watson, 2007). Furthermore, because it is a decision that will influence access to and the benefits accruing from water resources for all sectors of society, it has considerable equity and socioeconomic implications, which makes it an inherently social, political, economic and legal process (Dollar et al, in press).

For any point in the basin, as upstream water-resource development increases, the quantity and distribution of water available to downstream ecosystems changes and these ecosystems and the services they provide will decline (Figure 8.3a). A useful way forward for decision makers would be to determine the 'water-resource development space' (King and Brown, 2009a), which can be defined as the difference between current flow conditions and the furthest level of development found acceptable to stakeholders (including governments). Depending on the existing level of development and the nature of the ecosystems, this space may be extensive (Figure 8.3b) or very small (Figure 8.3c). In heavily developed basins, the changes to the freshwater ecosystems may have already moved beyond a point that is acceptable to stakeholders, in which case, the water-resource development space for the basin would be negative, and the level of water-resource develop-

ment would need to be reduced so that an appropriate EF can be provided to restore the ecosystem (Figure 8.3d).

This has occurred in, *inter alia*, the Murray–Darling Basin in Australia (e.g. Blackmore, 1999), in the Colorado River in the USA (e.g. Richter and Thomas, 2007) and in the Savannah River in the USA (e.g. Richter et al, 2006). To the best of our knowledge, however, although there are circumstances where water-resource developments in one country have caused a drastic decline in ecosystem condition in a downstream country (for example, the Aral Sea), there are few, if any, examples of reversal through provision of an EF. Such examples may appear in the future, but their current rarity underscores the importance of negotiating and implementing EFs in those transboundary basins where some water-resource development space still exists before they too move into deficit. Negotiations could also apportion the

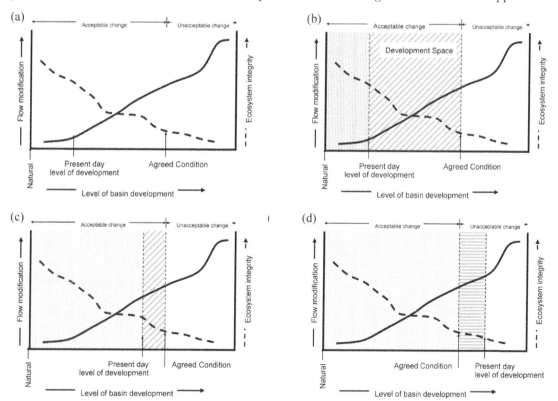

Figure 8.3   The concept of water-resource development space, which is defined by present-day conditions and the negotiated limit of ecosystem degradation as water-resource development proceeds

Source: King and Brown, 2009a

See text for explanation of a–d

recognized water-resource development space among countries in an agreed way that still allows slower-developing countries to have their share as and when desired (King and Brown, 2009a) or to develop in ways that do not require much water. This would allow them to choose to use their share of the water resources of the shared basin to maintain the associated freshwater ecosystem condition(s) at a higher level than the point of unacceptable change (King, 2009). The concept of equitable use is discussed in Chapters 5 and 9 of this volume by McIntyre and by Jarvis and Wolf, respectively. The IBFA process produces scenarios that detail the costs and benefits associated with various points along the flow modification/integrity curve to help countries locate the point of unacceptable change. The question marks in the arrows in Figure 8.4 denote that the fact that their position on the graph cannot be determined technically, but rather is a societal judgement.

The concept of water-resource development space (King and Brown, 2009a) illustrates two important principles:

1   For truly sustainable development, development planning should first identify and agree on the point beyond which ecosystem degradation should not be allowed to proceed and then to work backwards, considering how to

live and share within those limits. This is somewhat contrary to 'absolute territorial sovereignty and integrity' (McIntyre, this volume Chapter 5), where present needs take priority and future needs cannot be reserved, but unless some limit on aquatic ecosystem degradation is drawn and humanity lives within that limit, then sustainable development will remain elusive (King, 2009). By example, the UNECE Water Convention (Helsinki 1992) commits European countries to not harming freshwater ecosystems.

2   Removal of water from aquatic ecosystems and/or manipulation of their flow regimes will always be a trade-off between loss of ecosystem function and resilience, on the one hand, and benefit derived from the use of that water elsewhere, on the other. There is no minimum flow requirement, other than the full flow regime, that will maintain an aquatic ecosystem in its natural condition. This necessitates that the flow regime needed for freshwater ecosystem maintenance is determined at the end of a negotiation process, not allocated in some way before it.

Monitoring and Adaptive Management

No matter how many data are collected, or how deep the understanding of an ecosystem, it is impossible to accurately predict its every response to changing flow, climatic and other conditions. Sensitive management of these systems requires well-planned monitoring programmes supported by suitable adaptive management practices, so that management goals may be continually assessed and activities adjusted where appropriate.

EFAs produce baseline information and predictions of ecosystem and social change at points in the river basin for any scenario of interest that are an excellent basis for monitoring. Their use could begin with the countries reaching basin-wide agreement on a development scenario, which automatically defines a target condition for the river and its users. The EFA predictions for that scenario can then to be treated as hypotheses that can be tested with post-development monitoring.

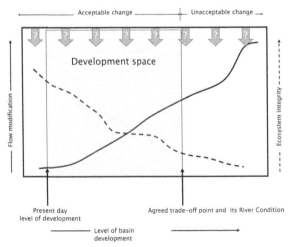

Figure 8.4   The concept of water-resource development space

Source: King and Brown, 2009a

Box 8.1

Online support for EF initiatives

Online support and guidance for those wishing to embark on EF work is provided by several organizations. These include eFlowNet (www.eflownet.org), which aims to make the EF concept accessible to managers in river basins, policy makers who shape legislation on water allocation, NGOs, governmental and international agencies and to a wider public. The network provides dialogue on EFs, assistance with technologies related to EFs and access to literature and other resources. Several eFlowNet member organizations also provide assistance and support on their websites and through their activities.

Failure to achieve the predictions can be addressed by either adjusting the desired condition or the desired amount of water needed to maintain the system; again, this is the choice of society. The costs of continual basin-wide monitoring of this nature could be funded through an initial investment factored into the project development costs, and would be a highly valuable means of facilitating technical collaboration between the countries of the basin. Such an initial investment is likely to be less than 0.5 per cent of total project costs.

Conclusion

The problem of degrading freshwater ecosystems is intractable if it is approached piecemeal. It is clear that the various needs for water, for both people and freshwater ecosystems, must be addressed collectively in order for sustainable development to become a reality (Baron et al, 2002). In transboundary settings, it also requires that countries work together towards a mutually acceptable future. Recent scientific advances have shown that the ecological and subsistence impacts of water-resource developments can be quantified. This can be done in basin-wide EFAs that provide information on the changes in ecosystem services likely to arise under different development-driven future flow regimes, as well as the more typical economic and social assessments of expected benefits.

The technical skills and knowledge to do such EFAs exist, and there is a growing number of

examples of successful EFAs, of decision-making processes used to consider EFA outputs, and, albeit fewer, of delivery of EFs for ecosystem maintenance. Organizations exist that will support and guide such EF initiatives for countries wishing to embrace more equitable and sustainable approaches to water-resource use (Box 8.1). The challenge now lies with the countries and their willingness to consider the full suite of costs and benefits before making their decisions.

References

Acreman, M. C., Farquharson, F. A. K., McCartney, M. P., Sullivan, C., Campbell, K., Hodgson, N., Morton, J., Smith, D., Birley, M., Knott, D., Lazenby, J., Wingfield, R. and Barbier, E. B. (2000) *Managed Flood Releases from Reservoirs: Issues and Guidance*, Report to DFID and the World Commission on Dams, Centre for Ecology and Hydrology, Wallingford, UK

Arthington, A. H. (1998) 'Comparative evaluation of environmental flow assessment techniques: review of holistic methodologies', LWRRDC Occasional Paper 26/98, Land and Water Resources Research and Development Corporation, Canberra, Australia

Arthington, A. H., King, J. M., O'Keeffe, J. H., Bunn, S. E., Day, J. A., Pusey, B. J., Bluhdorn, D. R. and Tharme, R. E. (1992) 'Development of an holistic approach for assessing environmental flow requirements of riverine ecosystems', in J. J. Pigram and B. P. Hooper (eds) *Proceedings of an International Seminar and Workshop on Water Allocation for the Environment*, Centre for Water Policy Research, University of New England, Armidale, Australia

Arthington, A. H., Brizga, S. O., Choy, S. C., Kennard, M. J., Mackay, S. J., McCosker, R. O., Ruffini, J. L. and Zalucki, J. M. (2000) *Environmental Flow Requirements of the Brisbane River Downstream from Wivenhoe Dam, Brisbane, Australia*, South East Queensland Water Corporation and Centre for Catchment and In- Stream Research, Griffith University, Brisbane, Australia

Arthington, A. H., Baran, E., Brown, C. A., Dugan, P., Halls, A. S., King, J. M., Minte-Vera, C. V., Tharme, R. E. and Welcomme, R. L. (2007) 'Water requirements of floodplain rivers and fisheries: existing decision-support tools and pathways for development', International Water Management Institute, Comprehensive Assessment Research Report 17, Colombo, Sri Lanka

Awulachew, S. B., McCartney, M., Steenhuis, T. and Mohamed, A. (2008) 'A review of hydrology, sediment and water resources use in the Blue Nile Basin', IWMI Working Paper 131, IWMI, Colombo, Sri Lanka

Baron, J. S., Poff, N. L., Angermeier, P. L., Dahm, C. N., Gleick, P. H., Hairston, N.G., Jackson, R. B., Johnston, C. A., Richter, B. D. and Steinman, A. D. (2002) 'Meeting ecological and societal needs for freshwater', *Ecological Applications*, vol 12, no 5, pp1247–1260

Berkamp, G., McCartney, M., Dugan, P., McNeely, J. and Acreman, M. (2000) *Dams, Ecosystem Functions and Environmental Restoration, Thematic Review II.1*, prepared as an input to the World Commission on Dams, Cape Town, http://oldwww.wii.gov.in/ eianew/eia/dams%20and%20development/kbase/ thematic/tr21main.pdf, accessed 17 May 2010

Blackmore, D. J. (1999) 'The Murray-Darling Basin Cap on diversions – policy and practice for the new millennium', *National Water*, June, pp1–12

Brisbane Declaration (2007) 10th International River Symposium and Environmental Flows Conference, Brisbane, Australia, September 2007, www.eflownet.org, accessed 12 May 2010

Brockway, R., Bowers, D., Hoguane, A., Dove, V. and Vassele, V. (2006) 'A note on salt intrusion in funnel-shaped estuaries: application to the Incomati Estuary, Mozambique', *Estuarine, Coastal and Shelf Science*, vol 66, no 1–2, pp1–5

Brown, C. A. and King, J. M. (2003) 'Water resources and environment technical note C1. Environmental flows: concepts and methods', in R. Davis and R. Hirji, (eds) *The World Bank Water Resources and Environment Technical Note Series*, The World Bank, Washington, DC, pp28

Brown, C. A. and King, J. M. (2006) 'Implications of upstream water uses on downstream ecosystems and livelihoods', *International Journal of Ecology and Environmental Sciences*, vol 32, no 1, pp99–108

Brown, C. A. and Watson, P. (2007) 'Decision support systems for environmental flows: Lessons from Southern Africa', *International Journal of River Basin Management*, vol 5, no 3, pp169–178

Bunn, S. E. and Arthington, A. H. (2002) 'Basic principles and ecological consequences of altered flow regimes for aquatic biodiversity', *Environmental Management*, vol 30, pp492–507

Carter, V., Bedinger, M. S., Novitzki, R. P. and Wilen, W. O. (1979) 'Water resources and wetlands', in P. E. Greeson, J. R. Clark and J. E. Clark (eds) *Wetland Functions and Values. The State of our Understanding*, Water Resources Association, Minneapolis, MN, pp344–376

Costanza, R., d'Arge, R., de Groot, R., Farber, S., Grasso, M., Hannon, B., Limburg, K., Naeem, S., O'Neill, R. V., Paruelo, J., Raskin, R. G., Sutton, P. and van den Belt, C. (1997) 'The value of the world's ecosystem services and natural capital', *Nature*, vol 387, pp253–259

Department of Natural Resources and Mines (2005) *Pioneer Valley Resource Operations Plan: Community Consultation Report*, Department of Natural Resources and Mines, Brisbane, Australia

DEWHA (Department of Environment, Water, Heritage and Arts) (2007) *Water Act No. 137*, Office of Legislative Drafting and Publishing, Attorney-General's Department, Canberra, Australia, pp543

Dollar, E. S. J., Nicolson, C. R., Brown, C. A., Turpie, J. K., Joubert, A. R., Turton, A. R., Grobler, D. F. and Manyaka, S. M. (in press) 'The development of the South African Water Resource Classification System (WRCS): a tool towards the sustainable, equitable and efficient use of water resources in a developing country', *Water Policy*

DWAF (Department of Water Affairs and Forestry) (1998) *National Water Act*, Government Gazette, Pretoria, South Africa

European Union (2000) *Establishing a Framework for Community Action in the Field of Water Policy* ('Water Framework Directive'), Directive 2000 /60 /EC

Fox, C. A. and Sneddon, C. (2007) 'Transboundary river basin agreements in the Mekong and Zambezi basins: enhancing environmental security or securitizing the environment?', *International Environmental Agreements*, vol 7, pp237–261

Gippel, C. J. and Stewardson, M. J. (1998) 'Use of wetted perimeter in defining minimum environmental flows', *Regulated Rivers: Research and Management*, vol 14, pp53–67

Gleick, P. H. (2002) *The World's Water: The Biennial Report on Freshwater Resources (2002–2003)*, Island Press, Washington, DC

GWP (Global Water Partnership) (2003) 'Poverty reduction and IWRM', *GWP Technical Committee Background Paper 8*, Prepared for the Water and Poverty Initiative, GWP, Stockholm, 16 January 2003

Heeg, J. and Breen, C. (1982) *Man and Pongola Floodplain*, Cooperative Scientific Programs, Council for Scientific and Industrial Research, Pretoria, South Africa

Hirji, R. and Davis, R. (2009) *Environmental Flows in Water Resources Policies, Plans, and Projects. Findings and Recommendations*, The International Bank for Reconstruction and Development/The World Bank, Washington, DC, pp189

Hughes, D. A. (2001) 'Providing hydrological information and data analysis tools for the determination of ecological instream flow requirements for South African rivers', *Journal of Hydrology*, vol 241, pp140–151

Hughes, D. A. and Hannart, P. (2003) 'A desktop model used to provide an initial estimate of the ecological instream flow requirements of rivers in South Africa', *Journal of Hydrology*, vol 270, pp167–181

Jowett, I. G. (1989) *River Hydraulic and Habitat Simulation, RHYHABSIM Computer Manual*, Fisheries Miscellaneous Report 49, New Zealand Ministry of Agriculture and Fisheries, Christchurch, pp39

Kamweneshe, B. and Beilfuss, R. (2002) *Population and Distribution of Wattled Cranes and Other Large Waterbirds on the Kafue Flats, Zambia*, Working Paper 1, Zambia Crane and Wetland Conservation Project, International Crane Foundation, Baraboo, WI, US

Kates, R. W., Clarke, W. C., Corell, R. Hall, J. M., Jaeger, C. C., Lowe, I., McCarthy, J. J., Schellnhuber, H. J., Bolin, B., Dickson, N. M., Faucheux, S., Gallopin, G. C., Grübler, A., Huntley, B., Jäger, J., Jodha, N. S., Kasperson, R. E., Mabogunje, A., Matson, P., Mooney, H., Moore, B., O'Riordan, T. and Svedin, U. (2001) 'Environment and development: sustainability science', *Science*, vol 292, pp641–642

Karenge, L. and Kolding, J. (1994) 'Inshore fish population changes in Lake Kariba, Zimbabwe', in T. J. Pitcher and P. J. B Hart (eds) *Impact of Species Changes in African Lakes*, Chapman and Hall, London

King, J. M. (2009) *The Environmental Dimension of Transboundary Freshwater Governance and Management*,

Discussion Document, UNEP High-Level Ministerial Conference on Strengthening Transboundary Freshwater Governance, Bangkok, 20–22 May 2009

King, J. M. and Brown, C. A. (2006) 'Environmental flows: striking the balance between development and resource protection', *Ecology and Society*, vol 11, no 2, p26

King, J. M. and Brown, C. A. (2009a) 'Integrated basin flow assessments: concepts and method development in Africa and South-East Asia', *Freshwater Biology*, vol 55, no 1, pp127–146

King, J. M. and Brown, C. A. (2009b) *Environment Protection and Sustainable Management of the Okavango River Basin: Preliminary Environmental Flows Assessment. Scenario Report: Ecological and Social Predictions*, Report no 07/2009, Project no UNTS/RAF/010/GEF, June 2009, pp93

King, J. M. and Brown, C. A. (2009c) *Environment Protection and Sustainable Management of the Okavango River Basin: Preliminary Environmental Flows Assessment*, Report no 02/2009, Project no UNTS/RAF/010/GEF, June 2009, pp45

King, J. M. and Louw, D. (1998) 'Instream flow assessments for regulated rivers in South Africa using the Building Block Methodology', *Aquatic Ecosystems Health and Restoration*, vol 1, pp109–124

King, J. M., Tharme, R. E. and Brown, C. A. (1999) 'Definition and implementation of instream flows', contributing paper for World Commission on Dams, Thematic Review 11.1 Ecosystems, www.dams.org

King, J. M., Brown, C. A. and Sabet, H. (2003) 'A scenario-based holistic approach to environmental flow assessments for regulated rivers', *Rivers Research and Applications*, vol 19, pp5–6, pp619–640

LHDA (Lesotho Highlands Development Authority) (2003) *Lesotho Highlands Water Project. Phase 1. Policy for Instream Flow Requirements*, LHDA, Maseru, Lesotho

Matthews, R. C. and Bao, Y. (1991) 'The Texas method of preliminary instream flow determination', *Rivers*, vol 2, no 4, pp295–310

Mendelsohn, J. and el Obied, S. (2004) *Okavango River: The Flow of a Lifeline*, Struik, Cape Town, South Africa

Metsi Consultants (2000) *Final Report: Summary of Main Findings, Lesotho Highlands Water Project, Contract LHDA 648: Consulting Services for the Establishment and Monitoring of Instream Flow Requirements for River Courses Downstream of LHWP Dams*, Report no LHDA 648-F-02, Maseru, Lesotho

MRC (Mekong River Commission) (2006) *Integrated Basin Flow Management Report Number 8: Flow-Regime*

Assessment, Mekong River Commission, Vientiane, Lao PDR, pp119

Nixon, S., Olsen, S., Buckley, E., Fulweiler, R. (2004) *'Lost To Tide' – The Importance of Fresh Water Flow to Estuaries*, Coastal Resources Center, University of Rhode Island, Narragansett, RI, pp15

O'Keeffe, J., Hughes, D. and Tharme, R. E. (2002) 'Linking ecological responses to altered flows for use in environmental flow assessments: the flow stressor-response method', *Verh Int Ver Limnol*, vol 28, pp84–92.

PBWO/IUCN (2009) *Final Scenario Report: Report 4: Pangani River Basin Flow Assessment*, Moshi, Tanzania

Petts, G. E. and Gurnell, A. M. (2005) 'Dams and geomorphology: Research progress and future directions', *Geomorphology*, vol 17, pp27–47

Poff, L. N. and Ward, J. V. (1990) 'Physical habitat template of lotic systems: recovery in the context of spatiotemporal heterogeneity', *Environmental Management*, vol 14, pp629–645

Poff, L. N., Richter, B. D., Arthington, A. H., Bunn, S. E., Naiman, R. J., Kedny, E., Acreman, M., Apse, C., Bledsoe, B. P., Freeman, M. C., Henriksen, J., Jacobson, R. B., Kennen, J. G., Merrit, D. M., O'Keeffe, J. H., Olden, J. D., Rogers, K., Tharme, R. E. and Warner, A. (2009) 'The ecological limits of hydrologic alteration (ELOHA): a new framework for developing regional environmental flow standards', *Freshwater Biology*, vol 55, no 1, pp147–170

Postel, S. and Richter, B. (2003) *Rivers for Life: Managing Water for People and Nature*, Island Press, Washington, DC

Richter, B. and Postel, S. (2004) 'Can we save earth's rivers?', *Issues in Science and Technology*, vol 20, no 3, pp31–36

Richter, B. D. and Thomas, G. A. (2007) 'Restoring environmental flows by modifying dam operations', *Ecology and Society*, vol 12, no 1, p12, www.ecologyandsociety.org/vol12/iss1/art12/

Richter, B. D., Baumgartner, J. V., Powell, J. and Braun, D. P. (1996) 'A method for assessing hydrological alteration within ecosystems', *Conservation Biology*, vol 10, no 4, pp1163–1174

Richter, B. D., Baumgartner, J. V., Wigington, R. and Braun, D. P. (1997) 'How much water does a river need?', *Freshwater Biology*, vol 37, pp231–249

Richter, B. D., Warner, A. T., Meyer, J. L. and Lutz, K. (2006) 'A collaborative and adaptive process for developing environmental flow recommendations', *Rivers Research and Applications*, vol 22, pp297–318

Sengo, D. J., Kachapila, A., van der Zaag, P., Mul, M. and Nkomo, S. (2005) 'Valuing environmental water pulses into the Incomati estuary: key to achieving equitable and sustainable utilization of transboundary waters', *Physics and Chemistry of the Earth*, vol 30, no 11–16, pp648–657

Stalnaker, C. B. (1982) 'Instream flow assessments come of age in the decade of the 1970s', in: W. T. Mason and S. Iker (eds) *Research on Fish And Wildlife Habitat*, EPA-600/8-82-022, Office of Research and Development, US Environmental Protection Agency, Washington, DC, pp119–142

Stalnaker, C. B. and Arnette, S. C. (1976) *Methodologies for the Determination of Stream Resource Flow Requirements: An Assessment*, US Fish and Wildlife Services, Office of Biological Services Western Water Association, pp199

Stalnaker, C., Lamb, B. L., Henriksen, J., Bovee, K. and Bartholow, J. (1994) *The Instream Incremental Methodology: A Primer for IFIM*, Biological Report 29, US Department of the Interior, National Biological Service, Washington, DC

Tabacchi, E., Correll, D. L., Hauer, R., Pinay, G., Planty-Tabacchi, A. and Wissmar, R. C. (1998) 'Development, maintenance and role of riparian vegetation in the river landscape', *Freshwater Biology*, vol 40, pp497–516

Tennant, D. L. (1976) 'Instream flow regimens for fish, wildlife, recreation and related environmental resources', *Fisheries*, vol 1, no 4, pp6–10

Tharme R. E. (1996) *Review of International Methodologies for the Quantification of the Instream Flow Requirements of Rivers*, Water Law Review Final Report for Policy Development, Department of Water Affairs and Forestry, Freshwater Research Unit, University of Cape Town, Pretoria, South Africa

Tharme, R. E. (2003) 'A global perspective on environmental flow assessment: emerging trends in the development and application of environmental flow methodologies for rivers', *River Research and Applications*, vol 19, pp397–442

United Nations (2007) *The Millennium Development Goals Report 2007*, United Nations, New York

URT (United Republic of Tanzania) (2002) *Water Act*, Government Gazette, Dar es Salaam, Tanzania

Watson, P. (2006) *Manage the River as Well as the Dam: Assessing Environmental Flow Requirements – Lessons Learned from the Lesotho Highlands Water Project*, World Bank Directions in Development Series, World Bank, Washington, DC

WCD (World Commission on Dams) (2000) *Dams and Development a New Framework for Decision-Making. The Report of the World Commission on Dams*, Earthscan, London

Welcomme, R. L. (1979) *Fisheries Ecology of Floodplain Rivers*, Longman, London, pp317

Welcomme, R. L. (2001) *Inland Fisheries Ecology and Management*, FAO (UN), Blackwell Science, London

Whitfield, A. K. and Wooldridge, T. H. (1994) 'Changes in freshwater supplies to southern African estuaries: some theoretical and practical considerations', in K. R. Dyer and R. J. Orth (eds) *Changes in Fluxes in Estuaries: Implications from Science to Management*, pp41–50, Olsen and Olsen, Fredensborg

Wolf, A. T., Yoffe, S. and Giordano, M. (2003) 'International waters: Identifying basins at risk', *Water Policy*, vol 5, no 1, pp29–60

Young, W. J., Scott, A. C., Cuddy, S. M. and Rennie, B. A. (2003) *Murray Flow Assessment Tool – A Technical Description*, Client Report, 2003, CSIRO Land and Water, Canberra, Australia

Zakhary, K. (1997) 'Factors affecting the prevalence of schistosomiasis in the Volta Region of Ghana', *McGill Journal of Medicine*, vol 3, pp93–101

9

Managing Water Negotiations and Conflicts in Concept and in Practice

Todd Jarvis and Aaron Wolf

- Negotiations over water depend on the characteristics of the resource. Surface water negotiations typically focus on allocations and flows; negotiations over groundwater typically focus on storage. Both resources are hydrologically linked despite governance instruments that manage the resources independently.
- Water conflicts and negotiations are not easily handled without the assistance of specialists who can interpret causal chains. Because science remains at the core of water conflicts, it might be a source of conflict as well as the means of resolving conflicts. Groundwater situations are susceptible to the 'duelling experts' syndrome and a source of tension between the political and technical.
- Good practice in water negotiations focuses on overcoming the 'duelling expert' syndrome, often through leadership and facilitation, collaborative learning, joint fact finding, emphasizing equity over rights and attempting to create a superordinate identity such as 'we are all in this together'.

Introduction

Water management is, by definition, conflict management. Water, unlike other scarce, consumable resources, is used to fuel all facets of society, from biology and economics to aesthetics and spiritual practice. Moreover, it fluctuates wildly in space and time; its management is usually fragmented and is often subject to vague, arcane, and/or contradictory legal principles. Within a nation, the chances of finding mutually acceptable solutions to the conflicts among water users drop exponentially as more stakeholders are involved. Add international boundaries to water resources, both seen at the surface and hidden or invisible beneath the ground surface, and the chances decrease exponentially still further (Wolf, 2009).

Background

Surface and groundwater crossing international boundaries present increasing challenges to regional stability, because hydrologic needs can often be overwhelmed by political considerations. There are 270 international river basins and over 270 transboundary aquifers (Wolf and Giordano,

2002; Bakker, 2009; IGRAC, 2009). The basin areas that contribute to these rivers comprise approximately 47 per cent of the land surface of the earth, include 40 per cent of the world's population and contribute almost 60 per cent of freshwater flow.

The distinction between the use of 'international' and 'transboundary' when discussing surface water versus groundwater focuses on the fact that the river basin 'watershed' or 'catchment' boundaries can be mapped with a high degree of certainty with respect to political boundaries, whereas the geography of the groundwater system and the relationship to political boundaries is known at only a reconnaissance level.

Analogically a large percentage of the world population also resides in lands underlain by transboundary aquifers. Shiklomanov and Rodda (2003) estimated that almost 90 per cent of all accessible freshwater is found in aquifers. Groundwater use is increasing because it is a 'common' resource, available to anyone with the financial resources to drill, equip and power a well (Moench, 2004); and pumping of groundwater is among the most intensive human-induced changes in the hydrologic cycle. With dramatic changes in drilling technology, pumping technology and the availability of electrical and diesel power over the past 60 years, the number of wells has increased exponentially in many parts of the world. (Moench, 2004; Shah, 2009). According to Zekster and Everett (2004) and Shah (2009), groundwater is the world's most extracted raw material, with withdrawal rates approaching 800–1000km^3 per year through millions of water wells. As a consequence of this, the global economy is becoming increasingly dependent on groundwater, and conflicts over transboundary aquifers are becoming increasingly common.

Environmental flows and ecosystem services are more dependent on groundwater than previously thought. Gautier (2008) estimates that 36 per cent of river runoff comes from groundwater, yet water management has long suffered from a case of 'hydroschizophrenia' – the creation of separate surface water and groundwater governance and policies despite the recognition of the hydraulic connection between both hydrologic regimes (Llamas, 1975). The majority of the world's cities rely on groundwater to some degree for their urban water supplies, and Giordano (2009) posits that developed groundwater contributes to the global urbanization under way today.

Garduño et al (2004), Puri et al (2005), and Puri and Struckmeier (this volume Chapter 6) suggest the differences between river systems and aquifers in terms of governance and potential conflict is that river systems are dominated by flow, whereas groundwater systems are dominated by storage. For example, in 2009 the State of Mississippi filed a lawsuit against the city of Memphis, Tennessee for capturing groundwater stored in the Memphis Aquifer underlying Mississippi, and is seeking US$1 billion in damages (Cameron, 2009). Likewise, the states of Utah and Nevada settled a dispute in 2009 over water stored in a shared fractured rock aquifer which will serve as part of the municipal water supply for the city of Las Vegas, Nevada which will be conveyed through a pipeline 350 miles in length and costing nearly US$4 billion.

Within each international basin and overlying each transboundary aquifer, the demands from environmental, domestic and economic users, and the inputs of pollution, increase annually, while the amount of freshwater in the world remains roughly the same as it has been throughout history. Given the scope of the problems and limited resources available to address them, avoiding water conflict is vital because conflict is expensive, disruptive and interferes with the efforts to relieve human suffering, reduce environmental degradation and achieve economic growth.

Key Theoretical Background

What is conflict?

Jarvis (2008) addressed this question using Thomasson (2005) who cites Wallensteen (2002) as providing the best working definition of conflict: 'conflict … (is) a social situation in which a minimum of two actors (parties) strive to acquire at the same moment in time an available set of scarce resources'. 'Resources' can mean any kind of resource, including material as well as political. Whereas the classic disputes over resources are over the territorial integrity of a state and the extent of government control as described by Anderson (1999), Thomasson (2005) posits that it is incompatibilities over resources that create grievances or conflicts and that 'scarce' does not have to mean that the resource is limited. Incompatibilities often arise over the use and equitable, or inequitable, distribution of a resource. Other good working definitions include that by Brogden (2003) followed by Sword (2006) who define conflicts as:

> complex adaptive systems, which means they are dynamic and nonlinear. In nonlinear systems cause and effect are not always directly linked, proportionate or predictable … (and)…Complexity science deals at a systems level, rather than the constituent parts of systems.

The debate on conflict and water

In water-related conflicts, sometimes there is a physical scarcity, but conflicts over distribution are more common (Thomasson, 2005). Public discourse about the role of resource scarcity or abundance is the common thread of discussions regarding conflict at the intrastate as well as international scale. For example, Thomasson (2005) describes one foundation for conflict as the neo-Malthusians pointing to the role of population growth and finite resources as creating scarcities of resources. Yet he argues that resource scarcity serving as a foundation for conflict has been challenged from at least three different perspectives. The 'Cornucopians' argue that natural resources are abundant and can be traded, substituted through technological innovation, recycled, or rationed through market mechanisms when there is a shortfall of the resource. Desalinization of seawater is often referenced as an example. The 'curse of resources' school of thought posits that resource abundance is more important than resource scarcity in creating conflict. Actors fight over material resources, with emphasis on resources that require a moderate investment in extraction and transportation, such as diamonds,

minerals, oil biomass and timber. The trade disputes between the United States and Canada over lumber serve as a good example. The 'liberal institutionalists' offer research on international shared water resources, revealing that there are a larger number of cooperative treaties than of conflicts over water. For example, the history of international water treaties regarding surface water is robust; over 400 treaties have been inventoried with the earliest dating back to 2500 BCE following the last documented war over water in Mesopotamia along the Tigris River (McCaffrey, 1997, 2001; Wolf and Giordano, 2002). The history of cooperation over groundwater resources is much less robust, with only one treaty specifically addressing transboundary groundwater and with only a small percentage of the international water treaties having provisions for groundwater (Matsumoto, cited in Delli Priscolli and Wolf, 2009). However, not all cooperation is 'pretty' and more times than not conflict and cooperation co-exist (Zeitoun and Mirumachi, 2008).

What causes conflict over water?

Jarvis (2008) combined the conceptual models of the causes of conflict developed by Moore (2003) and Rothman (1997) as depicted in Figure 9.1. While Moore's (2003) 'circle of conflict' was developed primarily for assessing disputes between individuals, it can be related to larger groups (clans, villages, towns, cities) as well as between nation states. For example, Zeitoun and Warner (2006) as well as other contributors to this volume (e.g. Cascao and Zeitoun, Chapter 3) provide examples of structural conflicts in their assessment of 'hydro-hegemony' at the river basin level in the Nile River Basin, the Jordan River Basin, and the Tigris–Euphrates River Basin.

Regardless of whether the conflict initially appears to be interest-based or resource-based, it is important to acknowledge that identity is one of the foundations for nearly all conflicts, especially those related to communal resources such as water (Rothman, 1997; van Vugt, 2009). Rothman indicates that many conflicts are poorly diagnosed since identity conflicts are usually misrepresented as disputes over tangible resources, and proposes

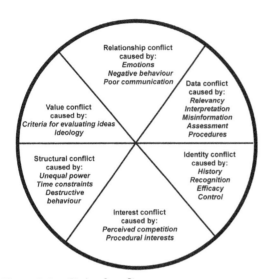

Figure 9.1   Circle of conflict

Source: adapted from Jarvis, 2008

that identity-based disputes have foundations in people's need for 'dignity, recognition, safety, control, purpose, and efficacy'. He offers the suggestion that identity-based conflicts are destructive, but once identified and processed with the right approach, these disputes can evolve into creative outcomes with significant opportunities for 'dynamism and growth'. Fitzhugh and Dozier (2001) indicate that the lack of respect afforded to technical professionals working on water-related issues also led to problems in negotiating a water-use dispute in the state of Vermont in the northeastern US. And one has to look no further than the situation of 'duelling experts' as described in later sections, to better appreciate the intensity and destructiveness of an identity-based dispute in a water conflict.

For the Basins at Risk study described by Wolf et al (2003), they investigated the data-based conflicts by testing some of the traditional indicators for future tensions over water, including climate, water stress, population, level of development, dependence on hydropower, dams or development *per se*, and 'creeping' changes such as general degradation of quality and climate change-induced hydrologic variability. Their analysis determined that these parameters are only weakly linked to disputes.

Conversely, the 'infosphere' of Lonsdale (1999) is perhaps the most important dimension of strategy in common pool resources such as groundwater because 'people want to make sense of their natural environment' (van Vugt, 2009). While groundwater monitoring networks have been in place in many intensively exploited regions of the world, few have been expanded since their inception (Moench, 2007). Groundwater data collection and information dissemination have created a large industry of 'knowledge entrepreneurs' as described by Conca (2006), but the problem with groundwater is that the 'size' of the resource is poorly known, even in the most studied basins in the world. Only reconnaissance-level estimates of how much groundwater is stored, much less how much is recoverable, exist for groundwater basins located throughout the developed world. Imagine the controversy over information on groundwater resources in the developing world; and van Vugt (2009) reports that people were more successful in managing conflicts over a resource when it was fixed in size than when its size fluctuated.

Renevier and Henderson (2002) suggest that the interests and options in water conflicts are not easily defined without the assistance of specialists who can interpret causal chains. Thus, science remains at the core of water conflicts, so the first part of a conflict may rest with 'science', and according to Ozawa (2005), 'the best science is the science whose meaning is agreed upon by the participants in a decision making process'. Yet 'water experts' come in many forms, ranging from scientists and engineers with training or experience in groundwater resources, well drillers, and water witches – each expert serving as a catalyst in the conflict, seasoned a little differently by their relationship with each other. Hydrologic professionals often develop conflicting conceptual models on how water is stored and flows in the subsurface. These models reflect the biases of the water professionals' institutional, disciplinary and personal interests – in other words, how they value water (Renevier and Henderson, 2002); and conflict is good business for conflict beneficiaries such as the knowledge entrepreneurs or the scientists who are paid for their services. Sometimes it is in their best interests to keep a conflict over water in motion.

It is clear from examining the chronologies of conflict complied by Gleick (2009) that many conflicts commence at a local level, eventually escalating into an international one involving more than one nation, as described by Trondalen and Munasinghe (2004). They describe the traditional types of international conflicts over water as arising from:

- incompatible goals related to access to, control over and unsustainable use of international water systems through water diversions, dams and reservoirs;
- problems created by utilizing the international water systems such as soil salination from irrigation, a change in water flow as a result of regulation, pollution from industry using the water, sewage from cities and communities;
- effects from other activities affecting the river systems such as eutrophication, pollution from industries which do not use the water resource in the production process, soil erosion and silting of watercourses following deforestation or overgrazing.

Conflicts over water require a holistic approach to address multidisciplinary and multimedia issues (Renevier and Henderson, 2002). Adler (2000) indicates that disputes over water are:

often large in scale, broad in impacts, and laden with values that are at odds with each other. They are emotional to both 'conscience' and 'beneficiary constituents'. At issue in many cases are matters of culture, economics, justice, health, risk, power, uncertainty, and professional, bureaucratic, and electoral politics.

At local scales, conflicts may arise between parties because of the land–water nexus and the large investments required to purchase and develop the land, while trying to weigh the value of maintaining a quality of life through open space initiatives and preserving the local water quality. In both developing and developed countries, conflicts also arise due to the plethora of beliefs surrounding the occurrence of water under the land held by the various parties.

Strategic dimensions to water disputes: competition versus collaboration

According to Daniels and Walker (2001), environmental conflicts involve strategic choices as disputants try to settle conflicts. Negotiations and decision making typically follow two dominant strategic orientations: competitive/distributive and collaborative/integrative. Competition in conflict is typically associated with distributive negotiations; collaborative approaches typify integrative negotiations as summarized in Table 9.1.

Competitive/distributive strategies are frequently employed when disputants consider the resources under negotiations as 'limited'. Each party employs competitive strategies to maximize their outcome. Parties take a 'position' and seek power and control. Conversely, collaborative/integrative negotiations emphasize alternative solutions through creative problem solving and 'mutual gains'. The alternatives to either approach focus on the 'walk-away possibilities' that each party has if an agreement is not reached – neither party should agree to something that is worse than its 'best alternative to a negotiated agreement' (BATNA). Delli Priscoli and Wolf (2009, p49) suggest that the lesson learned over nearly 40 years of experience in negotiations over shared water resources is to 'negotiate and solve problems by satisfying interests rather than capitulating to positions'.

Online negotiations are increasingly being used in disputes over environmental issues such as shared water resources (Freid and Wesseloh, 2002; Hammond, 2003). Ford (2003) indicates that it is likely that online negotiations 'will in time replace face-to-face conflict resolution and where it does not, will be used to complement the face-to-face process'.

Cultural competence skills are increasingly important for international negotiations over shared resources not only conducted face-to-face, but also online, as computers and operating systems become more available in developing countries. Negotiators need to be more sensitive to

Table 9.1   Comparative analysis of competitive versus collaborative strategic orientations

Concept/feature	Competitive/distributive	Collaborative/integrative
Party's goals	Maximize own share of benefits (individual gain)	Increase benefits for both (mutual gain)
Theory base	Game theory, economics, collective bargaining	Human relations, systems thinking, problem solving, communication
Utility orientation	Individual	Joint
Motivation	Self-interest	Mutual interest
Relationship worth	Minimal, present focus	High, future oriented
Relationship perception	Adversary, rival, competitor	Collaborator, partner
Trust	Limited, guarded	High
Communication	Controlled, selective, purposeful, tactical	Open
Dissemination of information	Cautious, intentional	Full relevant disclosure
Position sought relative to the other party	Superiority; gain or maintain advantage	Respected equal
Power	Individual-centred; coveted, sought	Shared; in relationship, process
Norm of justice	Equity	Equality
Issue focus	Positions	Interests
Deception	Accepted as inherent, justified	Inappropriate, unnecessary
Intangible issues (e.g. face, respect)	Manipulated for advantage	Addressed openly
Rules, procedures	Dictated by conflict structure or imposed	Generated by the parties

Source: adapted from Daniels and Walker, 2001

the differences between negotiating strategies for individualist cultures versus collectivist or cooperative cultures (Cohen, 2002). For example, Curry (1999) suggests that Europeans are more cooperative and are concerned not only with their own welfare, but also with that of the other party. Conversely, Brett (2001) indicates that Israelis are more individualistic and search for agreements that fulfil their interests.

The disciplinary approaches to negotiations over shared waters transcend economics, game theory, international law, international relations, geography and hydrology. The interested reader is referred to Dinar et al (2007) for a comprehensive discussion of these approaches. While international and domestic water laws are an important part of negotiations, as discussed by McIntyre (this volume Chapter 5), one of the many points of contention associated with implementing these regimes for water resources includes the fact that few have monitoring provisions, and nearly all

have no enforcement mechanisms (Chalecki et al, 2002).

Tensions between the political and technical: the hidden water conflict

According to Delli Priscoli (2004) few issues mesh the technical and political as much as water management and governance. The traditional model views the separation between the political, typically associated with legislative action, and the technical, typically associated with implementing executive agencies. He suggests that complex water management decisions often break down the distinction between the two factions, and that the implementation and administration of laws makes the distribution of impacts more clear. It is during implementation when the political benefits become clear, and when administrators of technical agencies 'begin to appear as bestowers or deniers of political benefits' (Delli Priscoli, 2004).

The development of a water management plan in the Umatilla Basin in the state of Oregon, USA, as described in the case studies in this volume (Chapter 13), provides a good example of the tension between the political and technical arenas. Even before the draft water management plan was completed, the first source of tension among some of the 'science-based' stakeholders focused on the conceptual hydrogeologic model of the deep basalt aquifers which had experienced water level declines approaching 150m over the past 50 years. Researchers at Oregon State University (OSU) synthesized the water level data maintained by the state water resources programme and integrated these data with the geological data compiled by the state geological survey. The state water resources department disagreed with the conceptual model prepared by OSU, indicating that it was too simplistic. Nevertheless they hesitated to offer an alternative conceptual model for reasons that remained unsaid, but were more than likely linked to politics, as the aquifer is shared with the state of Washington, the neighbouring state to the north, and the Confederated Tribes of the Umatilla Indian Reservation. Such a position would not be unusual given the current competitive climate regarding transboundary groundwater in the US. For example, the hydrologic analyses completed by the US Geological Survey and the University of Memphis served as the foundation for the lawsuit between the state of Mississippi and the City of Memphis, Tennessee, over groundwater shared by both states.

Conflicting conceptual models are part of the technical training of hydrogeologists focusing on the intellectual method of 'multiple working hypotheses', introduced in the late 1890s by the first hydrogeologist in the United States, Thomas Chamberlain (Chamberlain, 1897). The structure of the method of multiple working hypotheses coupled with 'affirmative action science' as described by Renevier and Henderson (2002, p124) revolves around the development of several hypotheses to explain the phenomena under study and giving voice to under-represented sectors (public, private or educational) which would 'further the political credibility of science and make for better science itself'. However, the state's criticism of the conceptualization of the groundwater system could also be considered a 'ruling theory' as described by Wade (2004) – the antithesis of multiple ways of knowing. Or the conceptual model could have been perceived as 'outside the current paradigm' as described by Shomaker (2007) who was proposing a new groundwater development scheme in New Mexico.

'The findings of water science play a role in the politics of allocation and management. But they will generally be subordinate to politics. After all, water flows uphill to money and power' (Stockholm Water Prize Winner Professor J. A. 'Tony' Allan quoted by Blenckner, 2008). According to Ozawa (2005), part of the problem is that the scientific enterprise 'in its purest form' is incompatible with the practical demands of public decision making. How science subordinates to politics is a subject of rich debate and beyond the scope of this chapter; however, Adler et al, (2000) and Ozawa (2005) provide excellent introductions to this topic for the interested reader. The reasons for the failure of rational discourse on the conceptualization may also have something to do with the power of the *status quo* as posited by Hamman (2005), who stated, 'Those individuals, communities and institutions that benefit from the current allocation or perceive they will suffer from a change have great power to defend the status quo.'

The special case of 'duelling experts' syndrome: 'hydrostitution'

Groundwater is a resource that is found everywhere. In many parts of the world, groundwater is part of the public trust and is a classic 'common pool resource' – a resource which is available to all, difficult to exclude users from gaining access to, and is 'extractable', becoming unavailable to other users after some of the resource has been used or 'mined' (Adams et al, 2003). Groundwater use is increasing because it is available to anyone with the financial resources to drill, equip, and power a well (Moench, 2004). Clearly, disputes over groundwater resources are particularly susceptible to the 'duelling experts' syndrome given that the database of information on groundwater resources is less than ideal due to the hidden nature of the resource, and that the underlying premise of the field of hydrogeology is based on the concept of multiple working hypotheses. Some knowledge entrepreneurs knowingly or unknowingly become zealous advocates for their conceptual models. Glennon (2006) has referred to some of these experts as 'hydrostitutes'.

According to Wade (2004), common causes of conflict focus on missing information, inaccurate data and procedures of data analysis. Wade (2004) suggests that data conflicts can be settled by 'the dispassionate opinions of one or more alleged experts about history, causation, or the future'. However, the 'darker' side of using experts to settle conflicts rests with 'helping' disputants 'solve' their problem, and the conflict escalates because the expert(s) become part of the problem instead of part of the solution.

Wade (2004) describes the disputant's behavior in the 'duelling experts' syndrome as having the following characteristics:

- employs an expert ('ours is the best in the field');
- hires an expert who has a reputation for favouring the disputant's preferred outcome ('reputational partiality');

- tells different stories to own expert ('garbage in, garbage out');
- makes expressed or implied hints at the advice he/she wants from the expert.

'Duelling experts' apparently exhibit the following characteristics:

- do not consult with each other ('delusionary isolation');
- tell client what they want to hear ('you get what you pay for');
- refrain from providing 'best to worst alternatives' ('delusionary certainty');
- produce long, incoherent reports ('mysterious complexity');
- refuse to share draft reports ('no early doubts of compromises').

Greek mythology describes a nine-headed water beast, the Hydra, that dwelt in a marsh near Lerna, Greece; Hercules was sent to kill the serpent as the second of his 12 labours. 'Hydra' is the term applied in this chapter to the complex situations or problems that continually pose compounding difficulties in developing or administering groundwater policy. As depicted on Figure 9.2, the duelling experts joust with a 'hydrohydra' of issues associated with groundwater development and governance.

It is clear that when a dispute evolves into the 'duelling experts' syndrome, parties in water management disputes pursue their conflicting interests and overlook common interests (Quirk, 2005, p207). While the 'duelling expert' syndrome is good business for conflict beneficiaries, trying to resolve a dispute through collaboration may save money. For example, in a study of hundreds of lawsuits, Glater (2008) reports that:

the vast majority of cases do settle – from 80–92 per cent by some estimates. On average, getting it wrong cost plaintiffs at about $43,000; the total could be more because information on legal costs was not available in every case. For defendants, who were less often wrong about going to trial, the cost was much greater: $1.1 million.

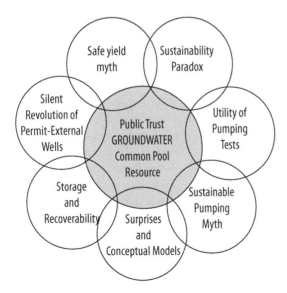

Figure 9.2   The 'Hydrohydra' of issues fuelling the duelling expert syndrome

Good Practice

Conflicts over water can best be described as a 'wicked' planning problem that has uncertain boundaries, defies absolute solutions and can be a symptom of larger problems (Rittel and Webber 1973). Yet, Delli Priscoli and Wolf (2009) state that water compels us to think regionally, and that the following are just a few reasons why water conflicts may be well suited to conflict resolution through collaboration:

- The price for control over an agreement over water is sharing ownership and cooperating in both the process and outcome of the agreement.
- The transaction costs are escalating beyond traditional management methods.
- The available money to identify needs is contracting.
- The public awareness of water resources is growing and changing.
- The traditional legal systems are unable to cope with change.

As Wolf (2008) eloquently puts it:

Water ignores all separations and boundaries save for those of the watershed itself. As such, it offers a vehicle to bring those who share it together. Since it touches all we do and experience, water creates a language through which we may discuss our common future.

Overcoming the 'duelling experts' syndrome

Wade (2004, p423) describes two approaches to responding to the 'duelling expert' syndrome: (1) reframing the expert conflict into a problem that can be addressed cooperatively; and (2) systematically identifying options to resolve the expert dilemma.

To begin the process of resolving the vexing question of 'What can be done about the current differing views of the experts', Wade (2004) proposes a 12-step approach ranging from joint meetings between the disputants and their experts, to tossing a coin. Determining who would initiate the 12-step programme remains problematic.

There are few practising mediators who are trained in water science and engineering. This is not a new problem – scientists and engineers typically do not receive training in professional communication in their technical curricula at colleges and universities. But this deficiency also leads to conflicts between technical professionals regarding the discussion and solution of environmental problems. The goals for resolving a 'duelling expert' situation are to help bridge this gap by identifying the causes of conflict between technical professionals. Reflecting on his work as a hydropolitical scientist spanning the past 40 years, the 2008 Stockholm Water Prize winner (Blenckner, 2008) suggests that 'Scientists need to understand and respect political processes, the role of advocacy and the significance of learning how to advocate responsibly.'

Technical experts need to understand that their networks as epistemic communities have the power to move governments towards cooperation (Conca, 2006). Lopez-Gunn (2009) highlighted the steering of states that took place indirectly, rather than directly, through instruments of legal regulation, such as the work on the UN Resolution on the Law of Transboundary Aquifers (see Chapter 6 in this volume), or regional or bilateral agreements. However, Conca (2006) also points out that technical experts can frame the understanding of environmental problems in ways that either facilitate or hinder cooperation as 'knowledge entrepreneurs'.

In the case of negotiations over groundwater, a common practice is to have the experts prepare a jointly signed explanation. Within the western United States, the dispute over the carbonate rocks composing the Great Basin Aquifer serve as an excellent example of difficulties over a transboundary aquifer. The Great Basin Aquifer underlies the states of Utah, Nevada, Idaho and Oregon, and is targeted for development by the Southern Nevada Water Authority to provide groundwater supplies approaching $0.21km^3$ per annum to Las Vegas due to the apparent lack of hydraulic connection between groundwater flows in the nearly 260 surface watersheds that overlie the Great Basin Aquifer System. While Bredehoeft and Durbin (2009) acted as consultants to opposing sides of the groundwater development project, both parties prepared a peer-reviewed technical paper describing the groundwater modelling predictions.

Collaborative and cooperative learning

The collaborative learning approach described by Daniels and Walker (2001) is well suited for natural resource, environmental and community decision-making situations that include (1) multiple parties; (2) multiple issues; (3) scientific and technical uncertainty; and (4) legal and jurisdictional constraints. The advantages of collaborative learning approaches to conflict management over water resources include the following:

- They are learning-based and involve public participation.
- Stakeholders learn from one another.
- Agencies interact as stakeholders.
- Technical/scientific and traditional/local knowledge are respected.
- Public participation activities are accessible and inclusive.

Collaborative learning draws upon systems thinking, conflict management, and alternative dispute resolution (Daniels and Walker, 2001). The use of collaborative learning requires a focus not only on developing a good working relationship between the disputants, but also that they begin the process of getting high-quality data and information on the table. Adler (2000) indicates that it is impossible to meet this challenge for disputes over water without good process and good working relationships.

Collaborative approaches to learning from and between water experts are documented by Kendy and Bredehoeft (2006) and Ziemer et al (2006) for conflicts over hydraulic interactions between surface water and groundwater pumping in Montana. Fenemor et al (2008) describe expert collaboration over biophysical and socioeconomic models for integrated catchment management in New Zealand; and Tidwell and van den Brink (2008) describe a 'cooperative' process for the numerical modelling of groundwater systems in a

regional water planning study in New Mexico and a groundwater protection project in The Netherlands.

Collaborative efforts take time. Tidwell and van den Brink (2008) report that the cooperative modelling process was held over 9 months for the groundwater protection project in The Netherlands, and 18 months for the regional water quality project in New Mexico. The collaborative learning process used in the Umatilla Basin in the Pacific Northwest of the US as described in the case studies summarized in this book (Chapter 13) required a few years. Jarvis (2008) reported that it took two years to reach an agreement between nearly ten groundwater professionals working as volunteers for a groundwater protection project in Wyoming. A few of the contested issues have resurfaced ten years later, some in part due to cross-jurisdictional politics between a municipal and county government; others due to the 'duelling experts' syndrome. Despite the most well-intentioned efforts at cooperation and collaboration, some conflicts cannot be resolved through this method.

Leadership and facilitation

Wondolleck and Yaffee (2000, p144) indicate successful collaborative efforts have one or two individuals who 'lead by example'; sometimes the leader is a forward-looking agency official. Leach and Sabatier (2003) also indicate that local leadership is important, and counsel that as water issues rarely fit within political boundaries, the level of agreement reached can be enhanced if the leadership can increase participation of government officials who hold permitting authority, technical expertise and the funding that allows collaborative efforts and project implementation to continue. Learning a new leadership language that better fits the complex and adaptive world of water resources and the role of public learning in these initiatives might be better served by collaborative 'leaders' who can:

- inspire political and personal commitment and action;
- function as peer problem solvers;

- build broad-based involvement in the collaborative enterprise;
- work to sustain hope and encourage participation in the consensus-building process.

While Leach and Sabatier (2003, p167) indicate that it is better for facilitators to be disinterested, they indicate that non-disinterested facilitators can be effective if they follow through on an in-kind basis or as volunteers, because paid facilitators 'may evoke feelings of resentment among those whose affairs are being facilitated for financial gain … (and that) … [w]atershed stakeholders may also be sceptical of overly polished or excessively managed processes'. Scher (2000) suggests that it is wise to seek out advisers such as scientists or engineers with process skills and technical expertise to assist with complex environmental disputes.

Rather than attempting to reach agreement on the many issues facing a dispute over water resources, Adler (2000) suggests that it is more important to get to 'maybe' rather than getting to 'yes' when discussing the upsides and downsides to all potential solutions, and once 'maybe' is reached, then what follows is mutually focused thinking and productive talk. While each party will more than likely be oriented towards its self-interests, the 'sequencing' approach to group facilitation as described by Kaner (1996) might work well. For example, when faced with a 'duelling expert' situation the topic of the sequencing might focus on one topic by the neutral board(s) such as 'what does safe yield mean to you?' or 'what does sustainability mean to you in terms of groundwater resources?'

Incentives and joint fact finding

While the top-down administrative decision making typical of many water management regimes pushes groups to accentuate their differences rather than searching for common ground, as described by Wondolleck and Yaffee (2000, pp52 and 224), the incentives for all parties to avoid litigation will not solve the problems of competition over limited natural and financial resources. Conversely, the lack of readily available funding for data collection to supplement both technical and public learning may also serve as an incentive, where participants complete some of the required data gathering on their own. For example, O'Rourke and Macey (2003) describe how citizen groups organized as 'bucket brigades' fill in the gap between monitoring and data collection for enforcement activities in neighbourhoods surrounding industrial facilities. While O'Rourke and Macey (2003) report that the bucket brigades are instigated by community members and facilitated and intermediated by NGOs, they have been only partially accepted by government agencies.

Susskind (2005, p147) indicates that joint fact finding (JFF) encourages stakeholders to specify the information that they desire to collect because 'JFF puts a premium on expanding the technical capacity of participants through facilitated face-to-face dialogue, with the support of technical experts'. For example, Redfern (2009) describes the ongoing joint investigation of the 1 million km^2 Northwest Sahara aquifer system by Algeria, Tunisia and Libya for the development of a mathematical model managed by the Sahara and Sahel Observatory. Ozawa (2005) recommends JFF, because it acknowledges 'surprises' and forces resource managers or elected decision makers to issue public statements clarifying the basis for the discrepancies when analyses or reports point to opposing policy prescriptions. Public confusion can discredit decision makers and lead to the rejection of all scientific work and a retreat to other forms of decision making. Likewise, periodic review and reassessment of monitoring and data collection are essential and eminently practical as industries, firms, private individuals and municipalities are sensitive to capital investment costs and rely on a certain degree of stability and predictability in decisions (Ozawa, 2005). Other incentives to the JFF approach also include the political risk of ignoring consensual proposals and data collection efforts – any elected body that ignores the consensual proposals of a truly representative stakeholder group that has done its homework does so at substantial political risk (Susskind, 2005, p143).

The Four Worlds and the 4i framework

Rothman (1997) proposes the 'ARIA' framework for the negotiator's toolkit. The ARIA Framework focuses on a process of antagonism, resonance, invention and action. Rothman explains that antagonism brings out the problem, and the anger that has been festering, so that they can be discussed. Resonance is the process of emerging harmony brought forth by the deep learning and speaking of what is going on within each disputant, and focuses on the 'why' and 'who' of the conflict, or what he calls 'reflexive reframing'. Rothman promotes the inventing phases as the process of brainstorming mutually acceptable options for addressing the conflict, fostering a 'how to' approach to cooperatively resolve the conflict. He suggests that the action phase builds upon the previous stages by focusing on the implementation of 'what' should be done and 'why', by 'whom', and 'how' it will be completed.

Building upon the ARIA framework developed by Rothman (1997), Wolf (2008) and Delli Priscoli and Wolf (2009) describe a four-stage process in water conflict transformation based on the observations of these workers over decades of professional experience. Wolf (2008) describes the idea of the Four Worlds and the use of transformative processes to move through negotiations. In many faith traditions, our relationship to the world can be experienced through four types of perception: physical, emotional, knowing and spiritual. He maps how each of the Four Worlds can be adapted to water negotiations as summarized in Table 9.2. The core motive influencing decision making, or the 4i framework as proposed by social psychologist van Vugt (2009), is paired with the Four Worlds model to show what motivates decision making within each of the four stages of conflict transformation.

Wolf (2008) describes Stage I as an adversarial setting, with regional geopolitics often overwhelming the capacity for efficient water resources management. Van Vugt's (2009) 4i framework indicates the core motive for decision making is institutions, and while institutions are designed to manage limited resources, they are also designed to build trust and invoke fairness. For example, in

the early phases of their respective negotiations, India claimed absolute sovereignty over the Indus Waters Treaty, and Palestine over the West Bank aquifer. Downstream riparian countries often claim absolute integrity, claiming rights to an undisturbed system or, if on an exotic stream, historic rights based on their history of use. Spain insisted on absolute integrity regarding the Lac Lanoux project, while Egypt claimed historic rights to the Nile, first against Sudan, and later against Ethiopia (see Cascão and Zeitoun, this volume Chapter 3).

In Stage II (described as Changing Perceptions: Basins Without Boundaries) the adversarial stage plays out, and the strict rights-based, country or state/province-based positions of each side may play out, but in reality the actual water negotiations during this stage can last decades. In the 4i framework, information gathering promotes listening by each party and joint fact finding. Van Vugt (2009) indicates that with better information, parties face less uncertainty and can move towards making more sustainable choices. Allocations of the Rio Grande/Rio Bravo and the Colorado between Mexico and the USA are based on Mexican irrigation requirements. Similarly, a 1975 Mekong River agreement among the four lower riparian states of Laos, Vietnam, Cambodia and Vietnam defined 'equality of right' not as equal shares of water, but as equal rights to use water on the basis of each riparian country's economic and social needs.

Stage III is defined as Enhancing Relations and Benefits: Beyond the River, where participants have moved from thinking about rights to thinking about needs, the problem-solving capabilities that are inherent to most groups which can begin to foster creative, cooperative solutions (see also Daoudy, this volume Chapter 4). Van Vugt (2009) indicates that incentives are the motivators for appealing to people's desires to enhance themselves through seeking pleasure and avoiding pain. For example, as part of the 1961 Columbia River Treaty, the United States paid Canada for the benefits of flood control and Canada was granted rights to divert water between the Columbia and Kootenai for hydropower purposes. The result is an arrangement by which power may be exported out of the basin for gain, but the water itself may not be. But the 'benefit-sheds' can be global in extent extending from the local scale, to the nation, region and world economy (Kolossov and O'Loughlin, 1998). For example, more than 90 per cent of the groundwater pumped from the Ogallala Aquifer located in central North America is used to irrigate crops for food and fibre worth more than US$20 billion per annum to global markets (Little, 2009); and bottled water from springs is shipped thousands of kilometres fuelling a global industry worth nearly US$50–100 billion per year (Glennon, 2006).

Finally, although tremendous progress has been made within the first three stages, both in terms of group dynamics and in developing cooperative benefits, Stage IV (Bringing it all Together: Institutional and Organizational Capacity and Sharing Benefits or the 'action stage') helps with tools to guide the sustainable implementation of the plans and to ensure that the benefits are distributed equitably among the parties. The scale at this stage is now regional where, conceptually, we need to put the political boundaries back on the map, reintroducing the political interest in seeing that the baskets that have been developed are to the benefit of all. To circumvent the potential for conflict once the political boundaries are reintroduced, the collaborative learning emphasis focuses on capacity-building, primarily of institutions. At first glance, it would seem that the core motive influencing decision making under the 4i framework are institutions. But van Vugt (2009) indicates that identity works towards action by connecting groups of competitors to move towards action. Van Vugt (2009) indicates it is important to create superordinate

identities such as regions by thinking of ways to 'blur group boundaries' by referencing 'we are all Europeans', implying we are all in this together. In the same light, Shah (2009) references the concept of 'aquifer communities' where aquifer users in a locality are aware of their mutual vulnerability and mutual dependence in the use of a common aquifer. For example, the 2009 draft agreement between the US states of Utah and Nevada over the allocation of the groundwater stored in the transboundary Carbonate Aquifer blurs the boundaries between both states:

> *The States agree to work cooperatively to (a) resolve present or future controversies over the Snake Valley Groundwater Basin; (b) assure the quantity and quality of the Available Groundwater Supply; (c) minimize the injury to Existing Permitted Uses; (d) minimize environmental impacts and prevent the need for listing additional species under the Endangered Species Act; (e) maximize the water available for Beneficial Use in each State; and (f) manage the hydrologic basin as a whole.* (Utah Department of Natural Resources, 2009).

Conclusions

There is room for optimism, notably in the global community's record of resolving water-related disputes along international waterways and over transboundary aquifers. For example, the record of acute conflict over international water resources is overwhelmed by the record of cooperation. Despite the tensions inherent in the international setting, riparian countries have shown tremendous creativity in approaching regional development,

Table 9.2   The four stages of water conflict transformation and the 4i framework

Negotiation stage	Common water claims	Collaborative skills	Geographic scope	Core motive influencing decision making
Adversarial	Rights	Trust-building	Nations	Institutions
Reflexive	Needs	Skills-building	Watersheds	Information
Integrative	Benefits	Consensus-building	'Benefit-sheds'	Incentives
Action	Equity	Capacity-building	Region	Identity

Source: adapted from Wolf, 2008 and van Vugt, 2009

often through preventive diplomacy, and the creation of 'baskets of benefits' which allow for positive-sum, integrative allocations of joint gains. Vehement enemies around the world have negotiated water-sharing agreements, and once cooperative water regimes are established through treaty, they turn out to be impressively resilient over time, even as conflict rages over other issues. Shared interests along a waterway seem to consistently outweigh the conflict-inducing characteristics of water. Moreover, international organizations such as the UN International Law Commission, through its work on shared natural resources (see Chapter 6 in this volume), the International Law Association and even the Government of Germany have hosted efforts to offer guidelines for the legal resolution of international water issues.

References

Adams, W. M., Brockington, D., Dyson, J. and Vira, B. (2003) 'Managing tragedies: understanding conflict over common pool resources', *Science*, vol 302, pp1915–1916

Adler, P. S. (2000) 'Water, science, and the search for common ground', www.mediate.com/articles/ adler.cfm, accessed 12 September 2009

Adler, P. S., Barrett, R. C., Bean, M. C., Ozawa, C. P. and Rudin, E. B. (2000) 'Managing scientific and technical information in environmental cases: principles and practices for mediators and facilitators', www.mediate.com/articles/wjc.cfm, accessed 12 September 2009

Anderson, E. W. (1999) 'Geopolitics: international boundaries as fighting places', in C. S. Gray and G. Sloan (eds) *Geopolitics, Geography, and Strategy*, Frank Cass, Portland, OR, pp125–136

Bakker, M. H. N. (2009) 'Transboundary river floods: examining countries, international river basins and continents', *Water Policy*, vol 11, pp269–288

Blenckner, S. (2008) 'Do the right thing a little badly ...', *Stockholm Water Front*, July, no 2, pp8–9

Bredehoeft, J. and Durbin, T. (2009) 'Ground water development – the time to full capture problem', *Ground Water*, vol 47, pp506–514

Brett, J. M. (2001) *Negotiating Globally: How to Negotiate Deals, Resolve Disputes, and Make Decisions across Cultural Boundaries*, Jossey-Bass, San Francisco, CA

Brogden, M. (2003) 'The assessment of environmental outcomes', in R. O'Leary and L. B. Bingham (eds) *The Promise and Performance of Environmental Conflict Resolution*, Resources for the Future, Washington, DC, pp277–300

Cameron, A. B. (2009) 'A study in transboundary ground water dispute resolution', *Sea Grant Law and Policy Journal, 2009 Symposium, Water Quantity: Ongoing Problems and Emerging Solutions*, http://nsglc.olemiss.edu?SGLPJ?presentations_09/cameron.pdf, accessed 10 September 2009

Chalecki, E. L., Gleick, P. H., Larson, K. L., Pregenzer, A. L., and Wolf, A. T. (2002) 'Fire & Water: An examination of the Technologies, Institutions, and Social Issues in: Arms Control and Transboundary Water-Resources Agreements', Pacific Institute, Oakland, CA, www.pacinst.org/reports/ fire_and_water/fire_and_water.doc, accessed 18 October 2009

Chamberlin, T. C. (1897) 'The method of multiple working hypotheses', *Journal of Geology*, vol 5, pp837–848

Cohen, R. (2002) *Negotiating across Cultures: International Communication in an Interdependent World*, United States Institute of Peace, Washington, DC

Conca, K. (2006) *Governing Water: Contentious Transnational Politics and Global Institution Building*, MIT Press, Cambridge, MA

Curry, J. E. (1999) *International Negotiating: Planning and Conducting International Commercial Negotiations*, World Trade Press, San Rafael, CA

Daniels, S. E. and Walker, G. B. (2001) *Working through Environmental Conflict: The Collaborative Learning Approach*, Prager, Westport, CT

Delli Priscoli, J. (2004) 'What is public participation in water resources management and why is it important?', *Water International*, vol 29, no 2, pp221–227

Delli Priscoli, J. and Wolf, A. (2009) *Managing and Transforming Water Conflicts*, Cambridge Press, New York, NY

Dinar, A., Dinar, S., McCaffrey, S. and McKinney, D. (2007) *Bridges Over Water: Understanding Transboundary Water Conflict, Negotiation and Cooperation*, World Scientific Publishing, Singapore

Fenemor, A., Deans, N., Davie, T., Allen, W., Dymond, J., Kilvington, M., Phillips, C., Basher, L., Gillespie, P., Young, R., Sinner, J., Harmsworth, G., Atkinson, M. and Smith, R. (2008) 'Collaboration and modelling – tools for integration in the Motueka catchment, New Zealand', *Water SA*, vol 34, no 4, pp448–455

Fitzhugh, J. H. and Dozier, D. P. (2001) 'Finding the common good: Sugarbush water withdrawal', www.mediate.com/articles/dozier.cfm, accessed 12 September 2009

Ford, J. (2003) 'Integrating the internet into conflict management systems', www.mediate.com/articles/ford10.cfm, accessed 18 October 2009

Freid, T. L. and Wesseloh, I. (2002) 'Integrating information technology into environmental treaty making' in L. Susskind, W. Moomaw and K. Gallagher (eds) *Transboundary Environmental Negotiation – New Approaches to Global Cooperation*, Jossey-Bass, San Francisco, CA

Garduño, H., Foster, S., Nanni, M., Kemper, K., Tuinhof, A., and Koudouori, P. (2004) 'Groundwater dimensions of national water resource and river basin planning', Sustainable Groundwater Management Concepts & Tools, Briefing Note Series no 10, The World Bank, Washington, DC, http://siteresources.worldbank.org/EXTWAT/Resources/4602122-1210186362590/GWM_Briefing_10.pdf, accessed 14 September 2009

Gautier, C. (2008) *Oil, Water, and Climate: An Introduction*, Cambridge University Press, New York, NY

Giordano, M. (2009) 'Global groundwater? Issues and solutions', *Anns Rev Environ Resource*, vol 34, pp7.1–7.26

Glater, J. D. (2008) 'Study finds settling is better than going to trial', *The New York Times*, 7 August 2008, www.nytimes.com/2008/08/08/business/08law.html, accessed 12 September 2009

Gleick, P. (2009) 'Water brief no 4: water conflict chronology', in P. Gleick (ed.) *The World's Water 2008–2009*, Island Press, Washington DC, pp151–193

Glennon, R. (2006) 'Tales of French fries and bottled water: the environmental consequences of groundwater pumping', *Environmental Law*, vol 37, no 3, pp3–13

Hamman, R. (2005) 'The power of the status quo', in J. T. Scholz and B. Stiftel (eds) *Adaptive Governance and Water Conflict: New Institutions for Collaborative Planning*, Resources for the Future, Washington, DC, pp125–129

Hammond, A. G. (2003) 'How do you write "yes": a study on the effectiveness of on-line dispute resolution', *Conflict Resolution Quarterly*, vol 20, pp261–286

IGRAC (International Groundwater Resources Assessment Centre) (2009), 'Transboundary aquifers of the world', www.igrac.net/publications/320, accessed 12 September 2009

Jarvis, W. T. (2008) 'Strategies for groundwater resources conflict resolution and management', in C. J. G. Darnault (ed.) *Overexploitation and Contamination of Shared Groundwater Resources: Management, Biotechnological, and Political Approaches to Avoid Conflicts*, Springer and NATO Public Diplomacy Division, United Nations Educational, Scientific and Cultural Organization, Paris, pp393–414

Kaner, S. (1996) *Facilitator's Guide to Participatory Decision-Making*, New Society Publishers, Gabriola Island, British Columbia, Canada

Kendy, E. and Bredehoeft, J. D. (2006) 'Transient effects of ground-water pumping and surface-water irrigation returns on streamflow', *Water Resources Research*, vol 42, W08415, doi:10.1029/2005WR004792

Kolossov, V. and O'Loughlin, J. (1998) 'New borders for new world orders: territorialities at the *fin-de-siècle*', *GeoJournal*, vol 44, pp259–273

Leach, W. and Sabatier, P. (2003) 'Facilitators, coordinators, and outcomes', in R. F. Durant, D. Fiorino and R. O'Leary (eds) *Environmental Governance Reconsidered: Challenges, Choices, and Opportunities*, MIT, Cambridge, MA, pp148–171

Little, J. B. (2009) 'The Ogallala Aquifer: saving a vital U.S. water source', *Sci Amer Earth 3.0*, vol 19, pp32–39

Llamas, M. (1975) 'Non-economic motivations in ground water use: hydroschizophrenia', *Ground Water*, vol 13, pp296–300

Lonsdale, D. J. (1999) 'Information power: strategy, geopolitics, and the fifth dimension', in C. S. Gray and G. Sloan (eds) *Geopolitics, Geography, and Strategy*, Frank Cass, Portland, OR, pp137–157

Lopez-Gunn, E. (2009) 'Governing shared groundwater: the controversy over private regulation', *The Geographical Journal*, vol 175, no 1, pp39–51

McCaffrey, S. C. (1997) 'Water scarcity: institutional and legal responses', in E. H. H. Brans, E. J. deHaan, A. Nollkaemper and J. Rinzema (eds), *The Scarcity of Water, Emerging Legal and Policy Responses*, Kluwer Law International, Boston, MA, pp43–58

McCaffrey, S. C. (2001) *The Law of International Watercourses, Non-navigational Uses*, Oxford University Press, New York, NY

Moench, M. (2004) 'Groundwater: the challenge of monitoring and management', in P. Gleick (ed.) *The World's Water 2004–2005*, Island Press, Washington DC, pp79–100

Moench, M. (2007) 'When the well runs dry but livelihood continues: adaptive responses to groundwater depletion and strategies for mitigating the associated impacts', in M. Giordano and K. G. Villhoth (eds) *The Agricultural Groundwater Revolution: Opportunities and Threats to Development*, CABI International, Cambridge, pp173–192

Moore, C. W. (2003) *The Mediation Process*, Jossey-Bass, San Francisco, CA

O'Rourke, D. and Macey, G. P. (2003) 'Community environmental policing: assessing new strategies of public participation in environmental regulation', *Journal of Policy Analysis and Management*, vol 22, no 3, pp383–414

Ozawa, C. (2005) 'Putting science in its place', in J. T. Scholz and B. Stiftel (eds) *Adaptive Governance and Water Conflict: New Institutions for Collaborative Planning*, Resources for the Future, Washington, DC, pp185–195

Puri, S., Gaines, L., Wolf, A. and Jarvis, T. (2005) 'Lessons from intensively used transboundary river basin agreements for transboundary aquifers', in A. Sahuquillo, J. Capilla, L. Martinez-Cortina and X. Sanchez-Vila (eds) *Groundwater Intensive Use*, International Association of Hydrogeologists Selected Papers No. 7, A. A. Balkema Publishers, Leiden, The Netherlands, pp137–145

Quirk, P. J. (2005) 'Restructuring state institutions', in J. T. Scholz and B. Stiftel (eds) *Adaptive Governance and Water Conflict: New Institutions for Collaborative Planning*, Resources for the Future, Washington, DC, pp204–212

Redfern, B. (2009) 'Securing supplies for the future', *MEED*, 4–10 September, pp30–31

Renevier, L. and Henderson, M. (2002) 'Science and scientists in international environmental negotiations', in L. Susskind, W. Moomaw and K. Gallagher (eds) *Transboundary Environmental Negotiation – New Approaches to Global Cooperation*, Jossey-Bass, San Francisco, CA

Rittel, H. and Webber, M. M. (1973) 'Dilemmas in a general theory of planning', *Policy Sciences*, vol 4, pp155–169

Rothman, J. (1997) *Resolving Identity-Based Conflict in Nations, Organizations and Communities*, Jossey-Bass, San Francisco, CA

Scher, E. (2000) 'Using Technical Experts in Complex Environmental Disputes', www.mediate.com/articles/expertsC.cfm, accessed 12 September 2009

Shah, T. (2009) *Taming the Anarchy: Groundwater Governance in South Asia*, Resources for the Future Press, Washington, DC

Shiklomanov, I. A. and Rodda, J. C. (eds) (2003) '*World Water Resources at the Beginning of the 21st Century*',

UNESCO International Hydrology Series, Cambridge University Press, Cambridge

Shomaker, J. (2007) 'What shall we do with all of this ground water?', *Natural Resources Journal*, vol 47, no 4, pp781–791

Susskind, L. (2005) 'Resource planning, dispute resolution, and adaptive governance', in J. T. Scholz and B. Stiftel (eds) *Adaptive Governance and Water Conflict: New Institutions for Collaborative Planning*, Resources for the Future, Washington, DC, pp141–149

Sword, D. (2006) 'Complexity science analysis of conflict', www.mediate.com//articles/swordL1.cfm, accessed 12 September 2009

Tidwell, V. C. and van den Brink, C. (2008) 'Cooperative modeling: linking science, communication, and ground water planning,' *Ground Water*, vol 46, no 2, pp174–182

Thomasson. F. (2005) 'Local conflict and water: addressing conflicts in water projects', Swedish Water House, www.swedishwaterhouse.se/opencms/en last, accessed 10 September 2009

Trondalen, J. M. and Munasinghe, M. (2004) 'Water and ethics: ethics and water resources conflicts', Series on Water and Ethics, Essay 12, United Nations Educational, Scientific and Cultural Organization, Paris, France

Utah Department of Natural Resources (2009) 'Agreement for management of the Snake Valley Groundwater System', http://naturalresources.utah.gov/pdf/snake_valley_agree.pdf, accessed 10 September 2009

van Vugt, M. (2009) 'Triumph of the commons: helping the world to share', *New Scientist*, vol 2722, pp40–43

Wade, J. H. (2004) 'Duelling experts in mediation and negotiation: How to respond when eager expensive entrenched expert egos escalate enmity', *Conflict Resolution Quarterly*, vol 21, no 4, pp419–436

Wallensteen, P. (2002) *Understanding Conflict Resolution: War, Peace and the Global System*, Sage, London

Wolf, A. T. (2008) 'Healing the enlightenment rift: rationality, spirituality and shared waters', *Journal of International Affairs*, vol 6, no 2, pp51–73

Wolf, A. T. (2009) 'International water convention and treaties', in G. E. Likens (ed.) *Encyclopedia of Inland Waters*, vol 1, Elsevier, Oxford, pp286–294

Wolf, A. T. and Giordano, M. A. (2002) '*Atlas of International Freshwater Agreements*', United Nations Environment Programme, Early Warning and Assessment Report Series, RS. 02-4

Wolf, A. T., Yoffe, S. B., and Giordano, M. (2003) 'International waters: identifying basins at risk',

Water Policy, vol 5, pp29–60

Wondolleck, J. M. and Yaffee, S. L. (2000) *Making Collaboration Work: Lessons from Innovation in Natural Resource Management*, Island Press, Washington, DC

Zeitoun, M. and Mirumachi, N. (2008) 'Transboundary water interaction I: reconsidering conflict and cooperation', *Int Environ Agreements*, vol 8, pp297–316

Zeitoun, M. and Warner, J. (2006) 'Hydro-hegemony – a framework for analysis of transboundary water conflicts', *Water Policy*, vol 8, pp435–460

Zekster, I. S. and Everett, L. G. (eds) (2004) '*Groundwater Resources of the World and Their Use*', IHP-VI, Series on Groundwater No 6, Springer, Dordrecht, The Netherlands

Ziemer, L. S., Kendy, E. and Wilson, J. (2006) 'Groundwater management in Montana: on the road from beleaguered law to science-based policy', *Public Land and Resources Law Review* (University of Montana), vol 27, pp75–97

10

Identifying Business Models for Transboundary River Basin Institutions

Jakob Granit

- In building a transboundary water management and development regime an analysis of the business model to implement the regime is critical to success.
- The analysis of business models is an analysis of what to achieve collectively, and of what can be undertaken better at the national level, allowing for effective stakeholder/client participation, assessing costs, risks and consequences of cooperation.
- The more complex a basin is, and the more complex the chosen business model is, the larger the consequences will be in terms of expectations, costs, accountability and risks in meeting agreed objectives.
- A business model that is less complex with fewer management and development functions can be effective and demonstrate good results to both the clients and the customers. It can be the starting point for more cooperation and the sharing of benefits.
- Complex business models will raise high expectations, be costly and if expectations are not met could result in lack of confidence among clients including public and private agencies and other stakeholders.
- Clear objectives and a well-designed business model for meeting agreed objectives support the raising of external financing and maintain ownership in the cooperative process.

Introduction

The objective of this chapter is to present an analytical framework to support riparian governments in determining a business model for an institutional framework that would effectively deliver agreed management and development outcomes, create long-term water security and strengthen the regional political and economic integration agenda. The analysis will help in assessing the costs of different institutional business model alternatives from a functional perspective. For institutions to be efficient it is necessary to have clarity on which functions to carry out cooperatively, what are the benefits and costs, and what can be better undertaken at the unilateral level. Business models are defined as frameworks for creating economic, social and environmental value in a transboundary watercourse context (Granit and Löfgren, 2009). The institutional business model should reflect the overall cooperative agenda in which water is seen as a key, cross-sectoral input for development and growth.

The management of transboundary water resources is considered a public good as well as a source of public goods (Jägerskog et al, 2007). Typical use of water resources as a public good include recreation, aesthetics, management of specific ecosystems such as wetlands for biodiversity conservation, flood and drought risk reduction, water quality improvement through watershed management and return flow of water of a high quality from different uses. These water management goods and services are 'public' because they benefit all and the consumption of water to produce them does not limit the availability of water to other users (International Task Force on Global Public Goods, 2006) from reduced water flows or water quality deterioration.

At the same time, water transforms from being a public good to a private good when the use of water resources moves to the private domain and contributes to generate private goods and services (Granit, 2009) such as electricity from hydropower generation, and water supply for industry, domestic and agricultural use. This transformation makes water management and development complex and there is a role for public as well as for private sector actors.

International regimes for the management and development of transboundary watercourses are ideally built on negotiations among riparian countries rather than imposed by a basin hegemon or developed through a reactive process (Kibaroglu, 2002), and are not built by third parties without the full ownership of the riparian governments. In the quest to identify effective institutional frameworks as one important element of building a management and development regime on transboundary watercourses, existing cooperative governance models are being reviewed, institutional business models tested and new forms emerging.

Key Theoretical Background: Drivers, Principles and Objectives for Management and Development of Transboundary Water Resources

Three key drivers in the water resources management and development sector can be identified behind the establishment and continued development of cooperative governance models to reach an overall objective of regional integration, prosperity and stability. The regional integration objective (see 'Wider regional integration objectives', below) is in this analysis defined as an overall objective driving multi-country cooperation (for more on the contested nature of cooperation refer to Cascão and Zeitoun, this volume Chapter 3).

Drivers for collective management of transboundary water resources

The first driver for an improved regional governance framework is the constrained freshwater resource in many regions due to demand increase from growing economies, population and climate change stresses (Granit, 2009). Recent data indicate that by 2030 there will be a global 40 per cent *water supply gap* of current accessible, reliable water supply for economic development, including return and environmental flows. This figure is an aggregation of a very large number of local water gaps made up of the quantity represented as accessible, reliable, environmentally sustainable supply, which is a much smaller quantity than the absolute raw water available in nature (McKinsey & Company, 2009).

The second driver is the growing sign of environmental stress in freshwater and marine ecosystems globally and in industrialized regions, particularly where water resources have been extensively utilized for economic development for agriculture, hydropower and industrial production (Granit, 2009). Evidence of environmental stress are dead zones in coastal oceans and seas that have spread exponentially since the 1960s, due primarily to worldwide coastal eutrophication fuelled by riverine runoff of fertilizers and the burning of fossil fuels, which have led to an increase in primary production (Diaz and Rosenberg, 2008). Water pollution both from point source and non-point sources is on the rise globally (World Water Assessment Programme, 2009). This can be identified as the *water quality gap*.

The third driver is related to the current *under-investment in the water value chain* that needs to be addressed in order to close the water supply and quality gaps. Recent estimates indicate that Africa's annual financing requirements for water could be US$50 billion, for funding drinking water and sanitation, wastewater, desalination, irrigation and water management, hydropower and multipurpose storage (Winpenny et al, 2009). Underinvestment can be seen in the whole water value chain in respect of:

1   water supply services, both for basic needs in meeting domestic water supply and productive services in agriculture and industry necessary for economic growth;
2   water resource management for maintaining or restoring watersheds and development, for example through artificial water storage;
3   water governance functions such as law, regulation, pollution control, data collection, monitoring and stakeholder engagement (Winpenny et al, 2009).

These three drivers come together in sub-Saharan Africa where access to improved sanitation is only 31 per cent and access to improved drinking water sources is 51 per cent (UNICEF and WHO, 2008). Only about 7 per cent of the technically feasible hydropower potential has been developed, and demand is outstripping supply (World Atlas and Industry Guide, 2007). Several countries are utilizing expensive fossil fuel-based thermal electricity generation that could partly be offset by the development of hydropower; this would also reduce

greenhouse gas emissions (King et al, 2008). Freshwater quality indicators are falling due to poor waste management, agricultural and industrial discharges and over-extraction (ECA, 2001). At the same time, in Africa, most water resources are transboundary and 63 transboundary river basins account for 85 per cent of the freshwater resource (Ashton, 2002). In this context transboundary institutions that are smart, efficient and output-oriented need to be maintained and developed

Wider regional integration objectives

The *World Development Report 2009* (World Bank, 2009) provides strong arguments for the benefits of regional collaboration and integration bringing economic success, prosperity and stability. When economies grow from low to high income, production becomes more concentrated spatially as some places are favoured as producers. The most successful countries implement policies that make living standards of people more uniform across space. *Economic integration* is said to be the most efficient way to get both the immediate benefits of the concentration of production and the long-term benefits of a convergence in living standards. The report further claims that challenges to integration posed by long distances, small economies and geography can be overcome by connective infrastructure and market integration reducing barriers to cooperation. In places where integration is difficult, the policy response should be comprehensive and include institutions that unite, infrastructure that connects, and interventions that target (World Bank, 2009). In a transboundary river basin context, joint institutions that promote data collection, knowledge sharing and joint monitoring are a first step towards cooperation. Where cooperation is more advanced, hydraulic infrastructure such as multipurpose water storage, water transfer schemes and electricity transmission networks are examples of connecting frameworks.

The transboundary water governance regime that is being developed and the linked institutional framework need to be context-specific, and consider the physical, economic and social context, in order to successfully promote regional integration. The basic principles for collaboration and building a regime in a transboundary river basin context are outlined in the draft UN Convention on the Law of the Non-navigational Uses of International Watercourses 1997 (GA, 2005). Even though this UN Convention is not ratified, the principles are regarded as customary international law and guide riparian states in their quest for common governance frameworks. The principles (see, for example, McCaffrey, 2001; Kibaroglu, 2002; Dombrowsky, 2007; and McIntyre, this volume Chapter 5) address:

- equity in utilization;
- the obligation not to cause harm;
- the obligation to cooperate;
- the exchange of information (in the absence of agreement no use enjoys priority over another use);
- planned measures, to be notified to all riparians;
- the holding of consultations on planned measures;
- the obligation to undertake pollution control and ecosystem protection.

Transboundary water resources management and development can play an important role towards promoting regional political and economic integration. This is illustrated by the Baltic Sea Strategy that has been developed as a new *macro region strategy* by the EU Commission during 2008 and 2009, in consultation with the member states and Russia. This is the first regional strategy the EU Commission is developing to address the environment with a focus on making the Baltic Sea region a prosperous, accessible, attractive, safe and secure place. With this new integrated economic governance regime, provided by the EU, there is a good opportunity to tackle the challenge of legal compliance and integrate the environment into mainstream regional economic development planning, and moving beyond voluntary agreements (Granit, 2009).

The EU Water Framework Directive (European Union, 2000) stipulates that member states shall take an integrated water resources

management (IWRM) perspective and achieve good water status in all their freshwater bodies by 2015. By 2009 river basin management plans and governance frameworks were to be in place for all national and transboundary basins in the European Union. Generally agreed IWRM principles and concepts are embedded in a binding regulatory instrument which provides for:

- a sustainable approach to manage an essential resource: links between water, ecosystems, economy and human health;
- holistic ecosystem protection: rivers, lakes, estuaries, coastal waters and groundwater are covered;
- ambitious objectives with flexible means: satisfying human needs, ecosystem functioning, and biodiversity protection and priority setting;
- integration of planning: water quality, and quantity and sectoral integration;
- the right geographical scale: the catchment, including transboundary basins;
- the principle of 'polluter pays' (water pricing);
- participatory processes;
- better regulation and streamlining.

Significant progress has been made in restructuring administrations, compiling information and assessments, and public awareness-raising campaigns. This provides a good starting point for preparing river basin management plans. However, part of the difficulty in implementing river basin plans and setting up catchment management authorities relates to the long and strong tradition of local government in Europe. A key issue that is emerging is how decision making regarding water is going to be handled. In several cases new authorities are being created that will supersede local and regional authorities. The new river basin management plans need to be adopted into existing administrative planning and legislative processes, and business models for transboundary river basin institutions (TRBIs) need to be developed.

Beyond the core group of 27 member states, the EU works through, *inter alia*, its European Neighbourhood Policy, and has a vision of a ring of countries sharing the EU's fundamental values and objectives. This is important since pollution or transboundary rivers do not respect borders and can therefore be best addressed through a mix of international, regional and national action. Energy production features strongly in the EU's Neighbourhood Policy (Commission of the European Communities, 2004), and this is significant considering the role water plays in generating electricity.

The cooperative management and development of transboundary waters is about finding new opportunities to create added value beyond unilateral development. Basin states can use different mechanisms to identify new opportunities from cooperation, to demonstrate what those opportunities may be, and what the steps are to realize them. This can be through the framework of a regional economic community (REC) or in the absence of such a framework through other forums. Philips et al (2008) have developed a conceptual framework called the Transboundary Waters Opportunity (TWO) analysis, a flexible tool to aid analysis and decision making between riparian states, within RECs and by potential investors engaged in the identification and implementation of cooperative water management and development opportunities. The conceptual framework consists of a matrix with key areas of development opportunities and main categories of sources of water to realize those opportunities.

The three key categories of sources of water that are first assessed cooperatively to set the boundaries for the analysis are:

1   New water: The potential for 'new water' to be developed within the basin (i.e. for an increased volume of freshwater to be made available through, for example, desalination or water transfers). This is most relevant in regions with water scarcity.
2   The efficient use and management of water: Options for improving the existing efficiency in the use of freshwater by the basin states in transboundary basins (through, for example, institutional strengthening, joint management regimes, physical infrastructure and demand side management).

3   Other sources: In river basins that are not closed, blue water resources can be allocated or reallocated to specific uses. Green water provides a major source of water to be considered, especially in the agriculture sector.

The five key areas of development opportunity factors in the analytical framework are:

1   Hydropower production and power trading: The link between water management and hydropower for electricity production and power trade, to support economic development. Regional power trading is a central element of benefit sharing and an example of a growth strategy where water resources as a source of fuel can provide low-cost electricity able to balance other generation centres that depend on other sources of fuels.
2   Primary production: Options related to improvements in primary production using green and blue water resources, such as in agriculture for food and bio-energy production, and in forestry; this may be of particular importance in developing countries.
3   Urban and industrial development: The potential for an inter-sectoral reallocation of freshwater from uses with low economic returns to applications with higher returns, involving urban growth and industrial development.
4   Environment and ecosystem services: Ensuring key environment and ecosystem services for future generations. Two specific forms of economic ecosystem services addressed here relate to fisheries and tourism.
5   Other opportunities: Every transboundary river basin is unique in terms of its physical, economic and social context. Drought proofing at the regional scale or conjunctive use of existing water infrastructure for flood management and increased hydropower production are examples of other opportunities.

After having identified sources of water and different development opportunities, the next step is to assess trade-offs from development and identify a prioritized set of opportunities for further studies towards implementation. The TWO analysis helps to clarify the role water can play to promote social, economic and environmental benefits in a transboundary river basin. It points out the functions a transboundary institution must have to achieve the identified opportunities. The TWO framework highlights the benefits of inter-sectoral planning beyond the river basin and within RECs where that is possible (for an in-depth discussion of benefit sharing see Daoudy, this volume Chapter 4).

Water as a transboundary, scarce and vulnerable resource is suffering from under-investment along the whole water value chain. Unclear transboundary governance regimes with unclear business models for the institutional framework facilitating cooperation are partly to blame for this situation. Figure 10.1 provides an overview and link between identified drivers for collaboration, principles of transboundary water resources management based on customary international water law and IWRM principles. It includes the central role of having effective transboundary institutions that advance agreed management and development objectives identified through a REC framework or through a TWO analysis. The conceptual model points towards the long-term objective of regional integration as evidenced theoretically by the World Bank (2009) and practically by the European Union (2000).

Institutional Functions for Management and Development of Transboundary River Basins

A review of the functions of national and transboundary river basin organizations (RBOs) and their evolution (Granit and Löfgren, 2009) indicates ambiguity of the functions of RBOs in a rapidly changing regional and global political and economic context. Current institutional models for the governance of transboundary water resources as a public good seem to adhere to the UN Convention and IWRM principles only to a certain degree, and do not tackle under-investment in water efficiently in order to fill the water supply and quality gaps. The role of RBOs in promoting regional and economic integration is not clearly spelled out. However, the review also notes that the roles and the water management and development functions the RBOs carry out differ significantly depending on the physical, political and socioeconomic context (Granit and Löfgren, 2009). No RBO is a copy of another. There are examples of less complex bilateral RBOs that focus on hydropower generation, for instance, and of other RBOs with complex mandates and specific functions operating multipurpose structures with co-ownership (Granit and Löfgren, 2009).

Water management and development functions

CAPNET (2008b) has analysed key performance indicators of RBOs (primarily at the national level)

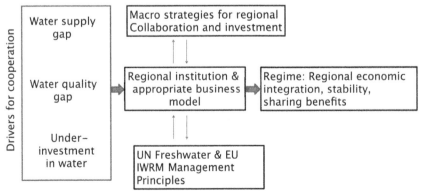

Figure 10.1 A conceptual model illustrating drivers for cooperation, water management and development principles and the overall objective of regional integration

that link to a set of operational functions. The functions and associated indicators are regarded as the minimum actions required for an RBO to meet basic water management objectives. CAPNET (2008a) claims that good water resources management and development at the transboundary level depends on strong systems and structures in-country. RBO functions and their objectives, according to CAPNET (2008b), are: water allocation; pollution control; monitoring; basin planning; economic and financial planning; information management; and stakeholder participation.

Hooper (2006) emphasizes the evolution of an RBO at the national level and stresses 'learning by doing'. He sees three stages in the evolution of an RBO – the 'initial RBO', the 'emerging auto-adaptive RBO' and the 'mature and auto-adaptive RBO'. Five functions are then linked to these stages as the RBO matures, develops and takes on more responsibilities. Water allocation is a central function of an RBO, yet this function is not usually performed until the second stage of its evolution when it has evolved into an 'emerging auto-adaptive RBO' (Hooper, 2006). Policy and monitoring functions require a 'mature RBO'. The stages and functions are summarized in Table 10.1 (Granit and Löfgren 2009).

A spirit of cooperation will be a prerequisite for the evolvement of the RBO. This can be within the framework of a broader REC such as the EU, where the role of water in providing key goods and services is valued, or it can be in an emerging cooperative framework driven by the identification of potential benefits as defined by riparian countries through, for example, the TWO analysis that will highlight development opportunities to both public and private stakeholders. The Nile Basin Initiative (NBI) is an example of a transboundary river basin cooperative framework that includes all the ten riparians and which is being developed in the absence of an REC (Granit, 2009). However, a subset of countries around the Lake Victoria Region, part of the NBI group, is developing such broader cooperative framework through the East African Community (EAC) (Granit, 2009).

Granit and Löfgren (2009) made a review of six transboundary watercourse institutions to further determine the functions these institutions carried out. The review assessed the International Commission for the Protection of the Danube River (ICPDR), the NBI, the Mekong River Commission (MRC), the Organisation Pour la Mise en Valeur du Fleuve Sénégal (OMVS), the Estonian–Russian Joint Commission on the protection and rational use of transboundary waters (ERJC), and the Zambezi River Authority (ZRA). The assessment pointed towards the following water management and development functions undertaken by these institutions and listed in Table 10.2. None of the institutions assessed in the study performed all the functions shown in the table.

Based on this analysis Granit and Löfgren (2009) compiled a comprehensive list of functions that could be considered in the context of TRBIs.

Table 10.1 Functional stages in the evolution of an adaptive RBO at the in-country level

Functions	Initial RBO	Emerging auto-adaptive RBO	Mature and auto-adaptive RBO
1 Water (and natural) resource data collection and processing, modelling, planning, stakeholder consultation and issue clarification	✗	✗	✗
2 Project feasibility, design, implementation, operation and maintenance, raising funds, community consultation and awareness raising	✗	✗	✗
3 Allocating and monitoring water shares; establishing cost-sharing principles		✗	✗
4 Policy and strategy development for economic, social and environmental issues			✗
5 Monitoring water use, pollution and environmental conditions; oversight and review role for projects promoted by RBO partners			✗

Source: Hooper, 2006, cited in Granit and Löfgren, 2009

Table 10.2   Functions carried out by six assessed transboundary institutions at various level of effort

Function	Objective
1 Visioning and trust building	To create a common framework for action and build trust between parties moving forward
2 Strategic planning and policy development	To ensure multi-sector linkages and prioritize key cooperative management and development activities
3 Pre-investment work	To undertake cooperative strategic assessments, and pre-investment and feasibility studies, building the business case for investment
4 Joint infrastructure management/development	Cooperative operation and management of joint infrastructure assets such as multipurpose hydropower facilities, flood and drought protection
5 Conflict resolution	To ensure a structured approach to problem solving in case of diverging views opposing management and development objectives, and clearing house for major in-country and multi-country investments. Prevention is better than conflict resolution
6 Corporate management	To ensure professional executive management, governance, resource mobilization, capacity and harmonization of administrative systems across the transboundary river basin organizations (TRBOs); financial management monitoring and evaluation, human resources management and decision support systems (World Bank, 2008, Nile Basin Initiative Project Appraisal Document)
7 Capacity-building	To level the playing field among riparian countries working together in a river basin context (World Bank, 2008, Nile Basin Initiative Project Appraisal Document)

Source: Granit and Löfgren, 2009

The list identifies two core business functions in a transboundary context: first, *management* of the transboundary water resources to ensure ecosystem services and second, *development* to promote the achievement of livelihood objectives and regional integration (Table 10.3). These two groups of business functions represent the core elements of IWRM that stress the river basin as the single management unit and the integration of fresh-water using sectors and stakeholders across society (Granit, 2009). The two groups of functions also illustrate how transboundary water resources management and development can support regional integration.

Role of stakeholders, clients and customers

Governments are the clients of a TRBI. As clients, they determine the specific objectives they have in

Table 10.3   A comprehensive list of cooperative water management and development functions for transboundary river basin institutions

Management function	Development function
1 Corporate management	9 Strategic basin planning (input to regional planning)
2 Financial management – cost recovery	10 Policy and strategy development (economic, social and environmental issues)
3 Monitoring and modelling (water and natural resources data and socioeconomic and legal developments)	11 Water allocation (to sectors and/or users)
4 Pollution control/monitoring	12 Pre-investment work at multi-country level
5 Information and communication	13 Support to in-country development planning
6 Stakeholder engagement	14 Transaction advisory services
7 Conflict resolution	15 Operation and management of joint infrastructure
8 Visioning and trust building	

Source: Granit and Löfgren, 2009

Stakeholders in a transboundary basin

Figure 10.2 Stakeholders and their roles in transboundary river basin management and development

mind and the role transboundary waters will play in promoting cooperation. The customers who benefit from services delivered by the TRBI include citizens, private sector entities, public institutions and international financial institutions (IFIs). The customers will implement change and therefore ensure that the TRBI adds value to water management and use in the river basin (Figure 10.2). The TRBI must facilitate the day-to-day business of the customers (Granit and Löfgren, 2009). The business model should be determined by both the customers and the clients, the latter being the end users of the services provided by the TRBI.

The mandate of the TRBI should ideally come from a negotiated process by the political clients. In parallel the business model of the institution should be assessed with input from the customers. The clients operate in a national and international context and should take both customary international water law and national governance frameworks into account. In regional economic communities such as the European Union, treaty texts and directives will guide the behaviour of governments. At the national level there is an increasing recognition of the need to integrate water resource planning and management into national economic development frameworks (Rogers, 1997).

When clients and customers assert issues to be catered for by the TRBI, they negotiate future outcomes, including how the basin will be managed and developed in a regional context. What they can expect to get in terms of livelihood improvements, the political implications of permitting some in-country decision shifting to collective management, and the development benefits they gain are all part of the regime-building process. In the absence of a REC the process of collectively identifying joint management and development opportunities will be critical to promote behavioural change among decision makers. The NBI is an example of that strategy.

Good Practice – Determining Functions in the Business Model for a Transboundary River Basin Institution

The design of a business model for TRBIs should be based on an assessment of its potential functions, the needs of clients and customers, the linked costs and benefits of organization, and the functions carried out by other cooperative frameworks such as RECs if in existence. A TRBI can have a strategic character backed up by a small organization or it can have a more operational character backed up by a larger management organization. It is possible to choose from business models scaled from simple to complex. The business model will have to adapt to the level of functions chosen and the level of basin complexity. The TRBI can move from being an advisory body to one that operates and manages joint infrastructure on behalf of its clients. The more management and development functions added to a TRBI the more complex the business model will be (Granit and Löfgren, 2009). Table 10.3 lists 15 operational functions split into management and development clusters. Some of the management functions are corporate and others relate to monitoring, pollution control, forecasting and vision building. The latter functions are more complex. In the development cluster the functions range from strategic basin planning, policy and transaction advice to operation of joint infrastructure. For each function added the costs of organization will grow in parallel with the expectations of the customers.

The basin-level complexity is defined in its simplest term by the number of countries in the TRBI. Complexity grows with the number of countries. Basin complexity may also be a function of management and development issues, economic differences between countries, the presence of civil strife or in-country organizational capacity. It can include managing complex issues such as the impacts of droughts and floods and adapting to climate change. Complexity will also be linked to major asymmetries in economies and the different sizes of countries in the transboundary basin.

Complexity will also grow with any increase in the number of functions to be carried out and as the number of countries increases. The relationship is not linear: a complex basin that has many countries can have a simple business model, and a basin with few countries can have a complex business model in terms of the number of functions it is mandated to carry out. The degree of sovereignty that countries are prepared to hand over to a regional function increases complexity (Granit and Löfgren, 2009). The countries involved in defining or redefining the business model for a TRBI would be capable of assessing the level of complexity when assessing the political economy of the countries involved and other cooperative frameworks that influence cooperation.

Figure 10.3 illustrates the relationship between simple and more complex business models in different transboundary basins with different levels of complexity. As the complexity of the business model grows, so do the expectations of the clients and customers. Organizational costs will increase along with responsibility and risks (Granit and Löfgren, 2009).

As the review in Granit and Löfgren (2009) asserts, there is no correlation between a successful TRBI and the complexity of its business model. A TRBI using a simpler business model may find it easier to secure resources to carry out its objectives, and expectations of stakeholders would be lower and easier to meet (Granit and Löfgren, 2009). This may be a more desirable situation than

creating high expectations by selecting a very complex business model that may not be able to deliver. The riparian governments/clients will assess how much of their national responsibilities they are prepared to hand over to a cooperative function, and they need to trust that the business model they choose will support their interest beyond management of the public good.

Conclusion – Clarifying Cooperative Objectives and the Institutional Business Model for Success

An analytical framework to assist riparian governments in assessing appropriate business models for TRBIs, and which supports an overriding goal of regional cooperation and integration, is presented. The starting point is an analysis of three context-specific drivers for cooperation: the water supply gap, the water quality gap and under-investment in water along the water value chain. The strength of these drivers will differ depending on the geography and size of the basin, the number of riparians and the political economy. An effective business model should be designed to allow for effective stakeholder participation by both customers and client, delivering economic, social, and environmental value. The business model should unlock a mix of technical and policy lever approaches creating incentives for investment.

There are strong arguments for regime building that supports regional collaboration and economic integration. As the regime develops it is critical to build the institutional framework and select a business model that supports the regime. This entails building solid transboundary watercourse institutions that are capable of handling functions clustered into management and development, and range from corporate management, monitoring and allocation of water to the operation of joint infrastructure.

It is concluded that in a transboundary watercourse context it is important to clarify the objectives of the regime the governments intend to build based on a detailed analysis of the basin's complexity and the long-term objectives of

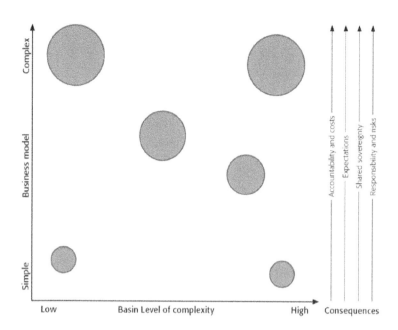

Figure 10.3 A model for illustrating the complexity of different business models for a transboundary river basin institution by assessing functions to carry out, their costs and benefits. Large circles illustrate river basin institutions with many functions. An institution does not have to carry out many functions to be regarded as successful; the number of functions is linked to its mandate. Additional functions carry costs and raise expectations among the customers

Source: Granit and Löfgren, 2009, modelled after a CSIR/SIWI workshop, Durban, South Africa, September 2008

cooperation. In parallel with this, business models for institutional cooperation should be explored. Cooperating partners should strive towards simplicity in terms of the business model chosen. Complex business models mean giving up some of a nation's sovereignty for the benefit of cooperative management of the public good.

The analytical framework presented here is designed to clarify what functions could be carried out jointly and which functions can be undertaken at the national level, coupled with a risk analysis. This will help in determining effective TRBIs which strive for clarity in the institution-building process and the ability to serve the clients for change. In the case that external financial or technical support to the regime-building process is received, this analysis will help in determining areas that need support and in maintaining ownership of the cooperative process. Clarity on

common objectives, commitment based on mutual understanding and a roadmap to achieve them are the secrets to success.

Acknowledgements

The analysis of business models is based on research by Jakob Granit and Rebecca Löfgren (from SIWI, the Stockholm International Water Institute) as input to a research consultancy carried out by the Council for Scientific and Industrial Research (CSIR), South Africa, and the Stockholm International Water Institute (SIWI), Sweden, to develop tools and approaches to strengthen RBOs in the Southern Africa Development Community (SADC–GTZ, 2009). Thanks go to Dr Marius Claassen and his team for their invaluable input to the business model concept.

References

Ashton, P. (2002) 'Avoiding Conflicts over Africa's Water Resources', *AMBIO*, vol 31, no 3, pp236–242

CAPNET (2008a) *Implementing Integrated Water Resources Management at River Basin Level*, UNDP

CAPNET (2008b) *Performance and Capacity of River Basin Organisations – Cross-case Comparison of Four RBOs*, UNDP

Commission of the European Communities (2004) *European Neighbourhood Policy Strategy Paper*, COM (2004) 373 final, European Commission, Brussels, pp564–570

ECA (2001) *State of the Environment in Africa*, ECA FSSDD/01/06, Economic Commission for Africa, Addis Ababa, Ethiopia

Diaz, R. and Rosenberg, R. (2008) 'Spreading dead zones and consequences for marine ecosystems', *Science*, vol 321, no 5891, pp926–929

Dombrowsky, I. (2007) *Conflict, Cooperation and Institutions in International Water Management: An Economic Analysis*, Edward Edgar, Cheltenham, UK

European Union (2000) *Establishing a Framework for Community Action in the Field of Water Policy*, Directive 2000/60/EC of the European Parliament and of the Council, Official Journal of the European Union, Luxembourg

GA (2005) *Convention on the Law of the Non-navigational Uses of International Watercourses*, Adopted by the General Assembly of the United Nations 21 May 1997, (not yet in force), General Assembly resolution 51/229, annex, Official Records of the General Assembly, Fifty-first Session, Supplement no 49 (A/51/49), General Assembly of the UN, United Nations Department of Economic and Social Affairs (DESA)

Granit, J. (2009) *Reconsidering Integrated Water Resources Management: Promoting Economic Growth and Tackling Environmental Stress*, H. G. Brauch (ed.), Hexagon Series on Human and Environmental Security and Peace (HESP), Springer-Verlag, Heidelberg, Germany

Granit. J. and Löfgren R. (2009) *A Framework for Exploring Business Models for Shared Watercourse Institutions*, Stockholm International Water Institute, Stockholm, Sweden

Hooper, B. (2006) *Key Performance Indicators of River Basin Organizations*, The Institute for Water Resources, US Army Corps of Engineers, Alexandria, VA

International Task Force on Global Public Goods (2006) *Meeting Global Challenges: International Cooperation in the National Interest*, Final Report, International Task Force on Public Goods, Stockholm, Sweden

Jägerskog, A., Granit, J. Risberg, A. and Yu, W. (2007) *Transboundary Water Management as a Regional Public Good. Financing Development – An Example from the Nile Basin*, Report no 20, SIWI, Stockholm, Sweden

Kibaroglu, A. (2002) *Building a Regime for the Waters of the Euphrates–Tigris River Basin*, Kluwer Law International, London/The Hague/New York

King, M., Noël, R. and Granit, J. (2008) 'Strategic/sectoral social and environmental assessment of power development options in the Nile Equatorial Lakes region – lessons in cooperation', paper for the *African Development Bank First Water Week*, Tunis, Tunisia, 26 March 2008

McCaffrey, S. (2001) *The Law of International River Basins – Non-navigational Uses*, Oxford University Press, Oxford

McKinsey & Company (2009) *Charting our Water Future – Economic Frameworks to Inform Decision Making*, 2030 Water Resources Group, US

Rogers, P. (1997) *Strategic Role of Water in Agriculture, Industry, and the Health of Humans and Ecosystems*, United Nations Commission on Sustainable Development, United Nations Department of Economic and Social Affairs (DESA)

Phillips, D., Allan, A., Claassen, M., Granit, J., Jägerskog, A., Kistin, E., Patrick, M. and Turton, A. (2008) 'The TWO analysis: introducing a methodology for the transboundary waters opportunity analysis', SIWI Report no 23, Stockholm, Sweden

UNICEF and WHO (World Health Organization) (2008) *Progress on Drinking Water and Sanitation – Special Focus on Sanitation*, WHO/UNICEF Joint Monitoring Programme for Water Supply and Sanitation, WHO Press, World Health Organization, Geneva, Switzerland

Winpenny, J., Bullock, A., Granit, J. and Löfgren, R. (2009) 'The global financial and economic crisis and the water sector', internal policy report to Sida, www.siwi.org, accessed 23 May 2010

World Atlas and Industry Guide (2007) *The International Journal on Hydropower and Dams*, Aqua Media International, Sutton, Surrey, UK

World Bank (2008) *Nile Basin Initiative Institutional Strengthening Project*, World Bank Project report no 46432 – AFR

World Bank (2009) '*World Development Report 2009 – Reshaping Economic Geography*, International Bank for Reconstruction and Development/World Bank, Washington, DC

World Water Assessment Programme (2009) *The United Nations World Water Development Report 3: Water in a Changing World*, UNESCO, Paris; and Earthscan, London

PART 3

CHALLENGES AND OPPORTUNITIES

11

Sustainability of Transnational Water Agreements in the Face of Socioeconomic and Environmental Change

Malin Falkenmark and Anders Jägerskog

- Transboundary water agreements are often not flexible or adaptive. They do not have an inbuilt flexibility to address natural as well as anthropogenically induced variations in water flow.
- Transboundary water agreements are often informed by political considerations, and trade-offs and sub-optimal solutions (from a scientific perspective) are the result. This is often not well understood by water managers but should be, since this is the reality in which they have to operate.
- In order for states sharing water to be able to address coming environmental challenges, such as increased climate variability and socioeconomic challenges, there is a need for more adaptive and flexible agreements.
- While agreements are now generally reactionary, they run the risk of becoming obsolete and triggering political conflicts in basins where population and therefore water demands grow, and where climate changes and consumptive water use increasingly depletes the river flow.
- Proactive agreements may become flexible by introducing a focus on development opportunities, the better use of infiltrated rain/green water and trade/import of water-consuming goods such as food.

Introduction

The main aim of the chapter is to secure a realistic sense of the implications of quite rapid changes in time in terms of pressure and possibilities, manoeuvring space, and options for benefit sharing. We put the focus on water agreement basics and typical threats which drive cooperation out of balance. We discuss criteria for sustainable agreements that are flexible enough to remain practical, as water stress increases with growing populations and water requirements, and the relations between upstream and downstream countries alter. The purpose is to highlight the time dimension behind riparian country relations in transnational river basins. Particular stress is put on the importance of seeking durability in basin agreements, so that they can absorb shifting relations between basin countries in response to the work of out-of-the-box driving forces. The chapter starts by taking a water perspective of the basin, seeing it as a physical system for water delivery both to humans and ecosystems, full of interactions between the inhabitants internally, and between them and the ecosystems, terrestrial as well as aquatic. We highlight country relations from a legal and political feasibility perspective, including constraints originating from history and culture, from the dominant discourse in riparians as well as existing power relations (asymmetries). We discuss the resulting room for manoeuvre as the stress on the shared water resource intensifies, and the possibilities of TWO (Transboundary Water Opportunity) analysis to alter the resource by incorporating green water as well, in order to moderate requirements, buy time, and so on. The possible openness to compromise-building and towards more general benefit sharing is also taken into account.

The Prospect of Transnational Water Dilemmas

Around half the world's population lives in areas fed by a transnational river, meaning that the water available for supporting a country's water-dependent socioeconomic development may originate from rain falling outside its borders. As the global population expands, the challenge of reaching agreement between such countries increases in importance, and has implications for international peace and security, as well as for development and poverty reduction. Such agreements have to regulate the principles by which the countries may together benefit from use of the water resource that they share with co-basin countries.

Agreements between states sharing a water resource are, however, usually static and seldom take into account the climatic variability of the shared water systems. With both increased

pressure on limited water resources due to socio-economic factors such as population growth, as well as potential climatic changes, the pressure on those static agreements will increase. The inflexibility of transboundary water agreements needs to be addressed if the problems the world is facing with regards to management of shared waters are to be efficiently addressed.

Water agreement basics

In general water agreements between countries sharing a river, lake or aquifer have either of two focuses. *First*, countries have an option to agree on volumetric allocations to be divided between countries. If this option is chosen the agreement should take into consideration variations in flow both based on seasonal changes as well as variations between years. However, often agreements do not adequately reflect the natural variations that occur, and which are likely to increase because of climate change (Jägerskog and Phillips, 2006). Thus, the flexibility needed to be well prepared for climate change is often not in place in many agreements. *Second*, countries may opt to agree on broader arrangements that involve not only the sharing of water but also the sharing of benefits that may be derived from using the water in an optimal manner. These benefits may come in different types and forms (see Daoudy, this volume Chapter 4, and Granit, this volume Chapter 10, for a description of various types of benefits).

There exist no rules for how states should engage in negotiations over their shared waters. In the anarchic system of international relations there are, however, some key guiding principles at hand. These are laid down in the Helsinki Rules as well as in the UN Convention on the Law of the Non-navigational Uses of International Water Courses, adopted in the UN General Assembly in 1997.[1] In most cases states base their agreement to a larger or lesser extent on these international principles, while others have entered into negotiations and eventually settled on a division of their resources. In addition, water agreements are often not solely focused on water, but are part of larger agreements between states, and trade-offs between different issues take place as a result.

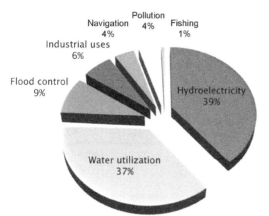

Figure 11.1 Areas of principal focus of 145 international agreements on transboundary water resources

Source: Wolf and Hamner, 2000

Research at the Oregon State University has shown that unexpectedly few agreements on transboundary waters include specific references to volumetric divisions of available water. Only 37 per cent of an analysis of 145 international treaties included volumetric allocations to the riparian states (Wolf and Hamner, 2000). The agreements deal with a range of other issues as well, as shown in Figure 11.1.

As the use-to-availability ratio will be growing in transnational river basins in semi-arid hot-spot regions with rapid socioeconomic development and population growth, alternative principles for sharing the benefits of water passing down the river are under development.

Since evidently neither absolute territorial sovereignty (allowing every country to operate independently), nor absolute territorial integrity (allowing every country the right to remain undisturbed from water-related activities in the other countries) are realistic, the agreement will have to specify compromise principles, by which they can peacefully share the resource with limited consequences for the others (Dinar et al, 2007).

In arriving at such compromises, it will be essential to be clear about how different types of water uses interact with the river system, and how the water cycle connectivity links different human activities so that one activity will be impacting other activities, and will be influencing the natural

ecosystems, which are also dependent on the water flow through the basin and its seasonal fluctuations and quality. Not least in view of predicted climate change, with its direct and indirect effects on the water cycle, many agreements are likely to need to be revisited.

Basic Water Management Challenges

Biophysical systems with internal connectivity

Looked at from a water perspective, the basin is to be seen as a physical system for water delivery, characterized by water mobility and connectivity between all water-dependent activities, phenomena and ecological systems. The human population is fundamentally water-dependent for many different uses. It makes use of the liquid (blue) water while it passes by in the streams and rivers or is accessible from wells. Similarly, use is made of the wettening of the soil in terms of naturally infiltrated rain (green) water in the soil, for food production, timber, fuelwood, fibre, and so on. By the connectivity linked to the water cycle, one human action impacts on other human actions as well as on the ecosystems that are part of the life support system – the terrestrial ecosystems dependent on green water and the aquatic ecosystems dependent on blue water.

The ideal way of seeking compromises in a shared river system would be adaptation to its geographical characteristics, such as storage in existing lakes or depressions, hydropower production in steep areas and food production in fertile parts. However, in the real world the ideal solutions are seldom forthcoming. In reality, compromises may be found around the sharing of water-related benefits, founded on differential biophysical opportunities – for instance sharing water; storage, food production and hydropower potential; flow control opportunities, and so on.

The hydroclimatic conditions differ a good deal between basins in different climatic zones. Figure 11.2 visualizes the typical differences between water balance in the temperate climate zone, the semi-arid tropics and the humid tropics. Runoff generation is basically the outcome of

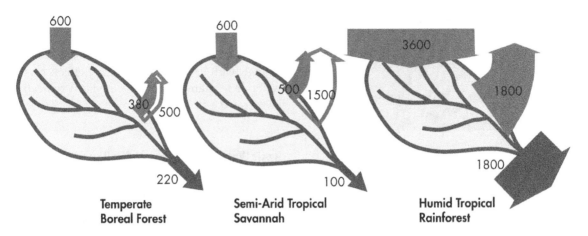

Figure 11.2 Livelihood contrasts in terms of annual precipitation, potential and actual evaporation and runoff generation (mm/yr)

Source: Falkenmark and Rockström, 2004

three phenomena: precipitation, evaporative demand of the atmosphere (potential evapotranspiration), and land use/vegetation pattern determining actual consumptive water use (evapotranspiration). The most intriguing situation is found in the semi-arid tropics and subtropics, where climatic vulnerability coincides with poverty, undernutrition and rapid population growth. Transnational basins in this region are characterized by basin-level hydroclimatic contrasts between runoff generation zones with water surplus, and water deficit zones, where the dominating water use is often irrigated agriculture with large consumptive water use. Streamflow is easily impacted by land-use changes, and deforestation may influence moisture feedback into the atmosphere and therefore precipitation.

Foreseeable changes in climate and water stress

The combination of population growth and increasing wealth is increasing the demand for food security, safe household water and energy, all of which involve water resources development. The outcome will be an increase of both use-to-availability ratio and water crowding (Falkenmark et al, 2007).

Climate change

As climate changes through global warming, so freshwater availability also tends to change, resulting in a reduction of the stability and predictability of the climate systems, and complicating water resources planning. In the past, stability in terms of hydroclimate has been the pillar on which our water resources planning capacity has been built, benefiting from hydrological statistics (Milly et al, 2008). We have relied on hydroclimatic experiences from the past repeating themselves in the future.

Changing precipitation and temperature conditions will, however, alter the distribution and availability of freshwater resources. Climate change adaptation will therefore mainly be an issue of adaptation to altered freshwater conditions. As a result of global warming, snowpack and mountain glaciers are now declining, and incidents of extreme climate events are projected to increase in frequency. Multi-year droughts and more disastrous floods are already being experienced in the most vulnerable basins, such as those in Australia and the western United States.

Socioeconomic development

As development proceeds in a transnational basin, more water will have to be put to use, the water resource system will need to be adapted by physical infrastructures and the water quality becomes increasingly polluted by more wastewater returned to the river after use. As populations continue to grow, more and more people will have to share every flow unit of the river, and water crowding increases – see Figure 11.3, horizontal axis (Falkenmark et al, 2007). When more and more people have to share each unit of available water, the possible per capita use decreases (Figure 11.3, diagonal lines). As more and more water is being mobilized and put to use, use-to-availability ratio rises (Figure 11.3, vertical axis), increasing the water stress and complicating compromise building between riparian countries. In the Niger River Basin, for instance, water use is low compared to the amount available, and the number of people per unit of water is also comparatively low (Falkenmark et al, 2009b). At the other end of the

spectrum the water in the Nile Basin is over-allocated and the number of people depending on its water is very high, indicating severe water shortage and environmental problems related to water scarcity. The Amu/Syr Darya Basin is also over-allocated, driven by cotton irrigation, but the population pressure is less. It is clear that from a water availability point of view the Niger Basin still has options available, while the Nile Basin as a whole has little room left to manoeuvre.

The challenge of river basin closure

As socioeconomic change in a river basin proceeds, the basin will pass through successive *river basin development* phases, moving from a low to medium to high use-to-availability ratio (Molle et al, 2007). In *early development phases* plenty of water is accessible to

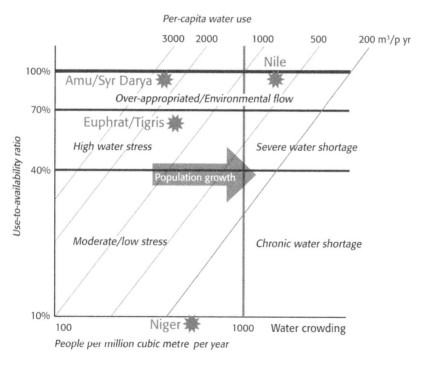

Figure 11.3 Two dimensions of present and imminent water scarcity risks: demand-driven water stress and population-driven water shortage. Where the ratio of use-to-availability is high (Y-axis), blue water resources are stressed to the point where no more water is accessible for use and the basin is said to be closed. Beyond 70 per cent, water can be seen as over-appropriated since the last 30 per cent of the resource should preferably be reserved for aquatic ecosystems as 'environmental flow'. Where many people depend on a limited amount of water (shown as high 'water crowding' on the X-axis), severe water shortage occurs

Source: Falkenmark et al, 2007

share between upstream and downstream countries. As the development along the vertical axis in Figure 11.3 proceeds with time, the water resource development gets more and more advanced and passes from low water stress during early development phases, to medium stress during the accelerating development phase, to high water stress when the basin approaches the level of development limited by environmental flow constraints and the state of a closed river basin (Figure 11.4). When water is used for irrigation, the resource gets successively more and more depleted due to evapotranspiration of the plants, and availability decreases (Falkenmark and Lannerstad, 2005; Falkenmark and Molden, 2008), further complicating compromise-building, since *less and less water remains* for additional use. Finally *the basin is closed* and no more water is accessible for additional use – only committed water remains (Molle et al, 2007). These successive development phases *complicate interaction* between countries sharing the same transnational river system. Agreements made in early development phases will be losing relevance as use-to-availability proceeds and the streamflow depletion increases.

When no more water remains to be allocated and the river basin is closed for additional withdrawals, one way to meet further expansion of water needs is by water reallocation. The sign of a closing river basin is that outflow decreases, first during the dry season, and later all the year round.

Water-sharing agreements between upstream and downstream states will need flexibility in order to meet increasing water requirements, not only for increased population and rising socioeconomic development in general but for the even more water-consuming production of food. The alterations of water stress discussed above will alter the balance in terms of water requirements and degree of interaction with the shared resource between the riparian countries. Earlier, relatively undeveloped countries will be increasing their water demands on the shared water resource as they strive towards poverty and hunger alleviation and try to achieve the rest of the Millennium Development Goals (Figure 11.5).

As illustrated, the emerging challenges will be rather different in the case of a river where the

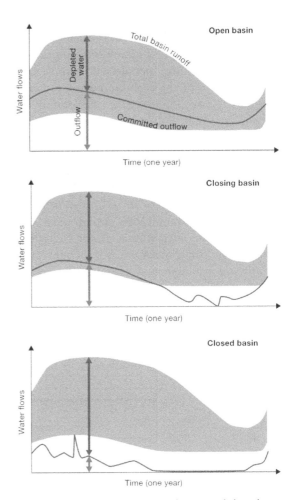

Figure 11.4 Rivers under stress: closing and closed river basins

Source: Molle et al, 2007

upstream country is already well developed and the water requirements increase downstream (Case A, symbolized by the Tigris–Euphrates); a river where all the countries are more or less developing in parallel (Case B, symbolized by the Niger); and the most challenging situation (Case C, symbolized by the Nile) where upstream riparians need more water in a river where the downstream riparian is already well developed, but may meet – depending on the outcome of negotiations – a shrinking water resource as the river flow gets increasingly depleted and/or polluted.

Upstream

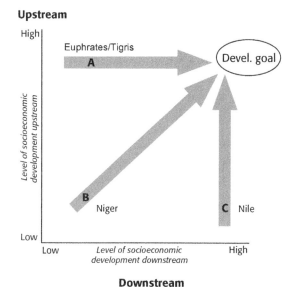

Figure 11.5 Three contrasting types of relations between upstream and downstream riparians. Horizontal axis shows water resources development in downstream part of the transnational basin; vertical axis the same for the upstream part. The arrows indicate foreseeable changes in relations between riparian countries as socioeconomic development proceeds

Source: Falkenmark et al, 2007

The Jordan Basin – precursor of the challenges[2]

In the Jordan Basin one can note a situation of *upstream consumptive use with subsequent downstream streamflow depletion.* As such it can be seen as an example of what could happen in transboundary basins that are not effectively addressing climate variability. Israel diverts large quantities of water into its National Water Carrier upstream from Lake Tiberias, and transports water out of the basin for irrigation, as well as for other uses. Syria, upstream of the Yarmouk River (a tributary to the Jordan River) diverts water for irrigation upstream, and when the remaining water has moved into Jordan the bulk of it is diverted into the East Ghor Canal before it can enter the Jordan River, and is used mainly for irrigation in the Jordan Valley (Jägerskog, 2003).

The case of Israel and Jordan can be said to be a precursor to the challenges that future climate change as well as socioeconomic changes may

bring. This is because the Jordan Basin is already characterized by significant climate variability, with rainfall in the region changing drastically from year to year. Earlier analysis (Jägerskog, 2003) has shown that risks (primarily climatic variability) were not addressed in the treaty that the parties reached in 1994. On the contrary, the climatic variations were conveniently ignored and the agreement based on multi-year averages. As such, the case illustrates the problems this chapter intends to bring forth – that is, the challenges and problems that riparians will face when their agreements are placed under increasing pressure; they are not constructed so as to meet such challenges.

The challenges that countries face as a result of their agreements not addressing fully the issue of climatic variations have strained political relations between them, although it may have been possible for the difficulties to have been managed through peaceful means. When there was a drought (which happens at semi-regular intervals in the Jordan Basin) in 1998–2000, Israel did not want to supply Jordan with the full amount stipulated in their agreement, and a political crisis erupted. Israel argued that the parties should 'share the deficit'. Jordan, on the other hand, argued that Israel was obliged to supply them in accordance with their peace agreement. The crisis was eventually solved when Israel supplied the water (Khatib, 1999a; Haaretz, 1999).

The problem however remains for the future. The former Jordanian Minister of Water and Irrigation, Kamal Mahadin, who was minister at the time of the crisis, stated that in the deliberations with the Israelis it was clear that they understood the problems (K. Mahadin, pers comm, 9 March 2002). Still, as Shmuel Cantour, an Israeli expert involved in the Israeli–Jordanian water discussions acknowledged, the issue is not settled once and for all, as discussions on the issue are still ongoing within the Israeli–Jordanian Joint Water Committee (JWC) which was installed through the agreement (S. Cantour, pers comm, 30 April, 2001).

Serious political conflict resulting from disagreement on allocation in times of drought could actually have been avoided had the parties not mutually de-emphasized the issue of reliable and non-reliable water in the negotiations over the agreement. Reliable water is available even when there is a drought and non-reliable water is the water that exceeds the reliable amount. It might be argued that the negotiators took a calculated risk when they did not include the issue of drought in the agreement and instead argued that it should be solved within the JWC. Indeed, the Israeli water negotiator Uri Shamir argues that drought policies should be dealt with in the JWC (U. Shamir, pers comm, 30 April 2001). This seems to be confirmed by a Jordanian negotiator and legal expert from Jordan, Al-Khasawneh, who also states that from a Jordanian perspective it is not bad that there are no provisions for droughts in the treaty (A. Al-Khasawneh, pers comm, 27 November 2001). However, other water experts argue that the non-inclusion of drought provisions was a serious mistake (N. Kliot, pers comm, 2 May 2001). Perhaps the uncertainty regarding the possibility of droughts occurring was de-emphasized in the communication between experts and the decision makers.

This illustration shows some of the challenges that riparians sharing a river or aquifer may face due to external challenges in the forms of increased climatic variability as well as changes in the socioeconomic context. While there may be – politically – perfectly rational explanations for de-emphasizing the risk of non-reliable water (resulting from the non-inclusion of provisions for what to do with seasonal and climatic variations), the end result may well come back to haunt the riparians at a later stage.

Broadening the Approach

As outlined above, the new challenges that transboundary water agreements are facing are considerable. With predicted future socioeconomic and environmental changes, agreements on transboundary water will face increasing pressure. The starting point for this is the predicted climatic changes. On top of this is population growth (meaning increased pressure on the water resources), increased economic development (craving more water resources) and

in certain basins political challenges may add a further dilemma to the equation. Thus, there is a need to chart a way forward that will be able to accom-modate the future challenges. This will need approaches that move outside of the (blue) water sector while at the same time are seen as politically feasible. In some less politically charged situations, this may be relatively easy, while in others it will be highly challenging. To chart a way forward towards sustainable agreements in such a situation is complicated, but some suggestions will be offered. It is argued that it is necessary in devising a way forward to find incentives that are attractive to all parties in a riparian setting. A number of components are seen to be important to this end.

Shift in thinking

Benefit sharing

In recent years, debate has repeatedly been stressing the need to seek benefits *outside of the (blue) water box* (e.g. WWAP, 2009). In line with this, the stress on benefit sharing rather than on water allocation has increased. Four different perspectives are currently used in the international debate, distinguishing between benefits that can be shared (Sadoff and Grey, 2002; Daoudy, this volume Chapter 4):

- benefits from the river: additional water allocation, storage potential, hydropower production, food production, navigation, recreation;
- benefits to the river: clean up, altered river regime, healthier ecosystems, fish production;
- costs because of the river: floods, droughts, water deficit, tensions;
- benefits beyond the river: peace, economic integration.

As already indicated, today less than half the agreements studied involve volumetric blue water allocations, while the rest involve hydropower, fishing, pollution, navigation, flood control and so on. Thus, states do occasionally agree on areas outside of the water box, not focusing only on division of water rights and allocation (Jägerskog and Phillips, 2006).

Compensation

In the case of river pollution, an upstream country may compensate the downstream state, in accordance with the 'polluter pays' principle. For example, the Permanent Court of Arbitration decided that France had to financially compensate their downstream riparian neighbour, The Netherlands, for excessive discharges of chloride into the Rhine River (PCA, 2004). The Convention on the Protection of the Rhine (1998) is actually directly based on a principle of compensation and the 'polluter pays' principle, among others (Article 4).

However, sometimes the downstream riparian utilizes side-payments to motivate abatement by the polluting upstream state, a so-called 'victim pays regime'. Side-payments could also be utilized simply to encourage actions of the upstream riparian that would be beneficial for the downstream partner. In the 1961 treaty on the Columbia River, the US (the downstream state) agreed to financially compensate Canada for building dams for flood control which through improved streamflow would benefit the US as well. Canada was also granted rights to downstream power benefits and to redirect water for hydropower (Dinar et al, 2007; Treaty between the United States of America and Canada, 1961, Article 5 and 6).

Just as the majority of the transboundary water agreements do not describe allocations in detail (Wolf and Hamner, 2000), the method of compensation is not always specified. As for the Dniester River for instance, the agreement between Moldova and Ukraine stipulates that the method of compensation shall be decided on a case-by-case basis and damage shall be evaluated by a joint expert group (Agreement between the Government of the Republic of Moldova and the Government of Ukraine, 1994, Article 5).

Broader water resource conceptualization

Green/blue water approach

The increasing stress on food production, the water costs of reforestation and ecosystems more generally, in the last few decades, have caused hydrology-oriented expertise to expand the water resource concept to include green water in the soil (naturally infiltrated rainwater). This makes it possible to incorporate among other water uses the huge amounts of water consumed in crop and timber production (Falkenmark and Rockström, 2005). It is seldom realized that on the global scale green water is responsible for 83 per cent of the water used for current food production, while blue water added through irrigation is responsible for only some 17 per cent (Rost et al, 2009).

However, based on experiences from South Africa and Australia, hydrologists have also drawn attention to the *runoff generation process* and the fact that, in tropical and subtropical climates, *runoff tends to change as a consequence of land-use change* that alters the partitioning of the incoming precipitation (between evaporation/consumptive water use and runoff generation). South Africans refer to this by describing forest plantation as a 'streamflow impacting activity' (GWP, 2003).

Opportunity analysis

Another effort in broadening 'out-of-the-box' conceptualization is the transnational water expertise development of the so-called Transboundary Water Opportunity (TWO) analysis (Phillips et al, 2008). The TWO analysis takes as its base the benefit-sharing approach primarily championed by Sadoff and Grey (2002) and further develops that thinking. In preparing for long-term sustainability of transnational agreements, different types of components may be incorporated: both so-called 'in-the-water-box' water components and various types of 'out-of-the-water-box' components. The TWO analysis structure includes different types of what they call water-related 'use', thereby distinguishing between ways to *utilize more water* (evaporation-loss reduction, green water in the soil, recharge of groundwater, inter-sectoral transfers/reallocation and ecotourism), and ways to *use the accessible water better* (inter-seasonal storage, increased irrigation efficiency, bioenergy production, water pollution abatement by organic waste or nutrient 'treatment' in wetlands, waste water reuse, increased water-use efficiency in urban areas, electricity generation, flood protection and aquaculture).

Attention to virtual water opportunities

A *problematique* expected to play a dominant role in the coming decades is global food security for a growing world population. When combining projections of hydroclimatic differences due to climate change with population growth (UN medium projection), hunger alleviation efforts and altered diet expectations, a global pattern of increasing virtual water flows emerges (Falkenmark et al, 2009a).

It may be anticipated that – unless current water productivity can be radically increased – countries with only some 30 per cent of the world population would be expected to radically increase their food production for export in order to complement water shortage-related food deficits in the rest of the world. For national food security in low-income countries, it will be of major importance to give priority to socioeconomic development for generating the purchasing power needed for import of the necessary food. Unless the economic development in these countries can be accelerated, some 40 per cent of the world's population – mainly in the semi-arid tropics – might have to reduce their dietal expectations and/or depend on complementary food aid from the water surplus countries (Rockström et al, in press).

Being Aware of Water's Political Role

Power, discourse and the ways things do (or don't) work

As if the natural and social aspects of development in a basin (climate variability and change, population growth, etc.) were not challenges enough in themselves, political and security considerations come into play when discussing challenges and opportunities in transboundary basins (Phillips et al, 2006). These challenges may often be even larger than the challenge to address the naturally induced occurrence of, for example, climatic variability. While the development opportunities outlined above may seem rational, they need also to find acceptance within the states which they should ultimately benefit. As outlined in the Introduction to this volume (Earle, Jägerskog and Öjendal, Chapter 1), the linkages between the research community (bringing new concepts), the water resource communities (bringing practical solutions on the ground) and the politicians (mastering compromises while allocating values in society) are not always easy to understand. While there may be scientific and seemingly rational ways to master water resources in the face of socioeconomic as well as climate changes, the interaction between the different sectors are somewhat messy and often hard to grasp.

Understanding water policies and discourse-related limitations

How can one understand water policies pursued by states at the international level? What are the linkages between the national and international level? Historically most studies that deal with international relations tends to treat every state/nation as an unproblematic singular unit. This approach is a gross simplification. This is because the policies pursued internationally are likely to be a reflection of the domestic discourse (Jägerskog, 2003). Thus, the need for governments to be in line with their respective domestic discourses in their pursuit of international policies is imperative to understanding foreign policy, including how states pursue their interests with regards to water internationally (Gross Stein, 1988; Allan, 2001).

The discourse sets limits within which policies have to be pursued; that is, it indicates what avenues may be politically feasible (Jägerskog, 2003). It represents what may be said, who may say it and also how it is to be interpreted. The rationale for explaining events in one way or another is often a result of the surrounding social context and the particular discourse that is dominating. Ben-Meir (pers comm, 29 April 2001) argues that the water agreements do not solve the problem of water scarcity. The reason for the non-inclusion of measures that would be helpful in this regard is that they are politically stressful. For instance, it was not seen as politically viable for Israel and Jordan to agree on properly including the issue of droughts (climatic variability) in their agreement at the time it was made (Jägerskog, 2003). This results in them presently having to face a situation where today's climatic variability is putting a strain on their relations that is likely to increase, with larger climatic variations and increased socioeconomic development putting greater pressure on their shared waters.

Development opportunities

Another area of importance when analysing trans-boundary waters, in particular when approaching the problem constellation outlined in this chapter, is what happens outside of the (blue) 'water box'. The national discourse in a state is made up of perspectives and decisions that have a large influence on the water sector. However, many of those decisions are taken by actors outside of the water sector. Similarly, the perspectives prevalent in a society, for example the notion that national security is on top of the agenda, affect the discourse strongly. This makes it imperative to analyse non-water issues and their linkage to the water agenda as well, both at a national and international level (Jägerskog, 2003; WWAP, 2009). In the case of the peace negotiations between Israel and Jordan, for instance, which to a significant degree were focused on water, trade-offs between water and other issue areas were made (Jägerskog, 2003). In a more development-oriented and benefit-sharing approach, linkages outside of the water sector may thus offer incentives for increased riparian cooperation in that they present a range of *development opportunities* that states may realize if they cooperate over their shared waters (Sadoff and Grey, 2002; Phillips et al, 2008).

Power asymmetries

While the national discourse is helpful in discerning the respective national policies of states in international negotiations, another political layer to the analysis is the understanding of how power relations, and the relative difference in power, are decisive for the understanding of why riparian relations look as they do (see Cascão and Zeitoun, this volume Chapter 3). Often the stronger power (hegemon) in a basin is able to secure a better outcome for itself than what could be deemed to be its equitable share (Zeitoun, 2008; Cascão and Zeitoun, this volume Chapter 3).

Zeitoun and Jägerskog (2009) have outlined an approach to deal with power asymmetries. The first option is to *influence* the situation. This could be done through identification of positive-sum outcomes or win–win situations that can satisfy all parties in a basin. This is closely related to the thinking on benefit-sharing and the identification of *opportunities* elaborated above. The second option is to *challenge* power asymmetry – either by trying to level the 'players' or by levelling the playing field. To help level the 'players', capacity-building programmes for the weaker parties can improve negotiating and administrative capacities, for instance. When making the playing field more level, the international community may try to strengthen international water law for example, which would benefit the weaker parties.

Conclusions – Moving Towards More Sustainable Agreements

The aim of this chapter has been – with an understanding of the global water crisis in mind and in the face of approaching socioeconomic and environmental changes – to address the issue of transboundary water agreements. It is concluded that much needs to be done in order for riparians to be better prepared to address the challenges that both environmental changes (primarily increased climatic variability) and changes in the socioeconomic context will bring.

The challenge faced by politicians and practitioners in many parts of the world is that more and more river basins are moving towards being 'closed', that is to a situation when there is no more water to allocate from it. While this is a challenge in itself, that challenge is compounded by the frequent lack of mechanisms to address future changes in a basin. Drawing on the case of Israel and Jordan – used as an example for what may come – it is concluded that while there may be politically convenient reasons to keep climate variability out of an agreement on water, that may come back and haunt the countries later, when for example a drought is hitting the region. As such it may be seen as an example of what may come in other parts of the world, with agreements that are static, and therefore not able to address complex and changing situations.

The chapter has stressed the need for a shift in thinking to secure agreements that are sustainable and involve a proactive rather than a reactive approach. They should properly face the increasing water stress caused by the ongoing global change in terms of climate, demographic and economic changes. Approaches that may be considered when devising new agreements and contemplating ways for development in shared river basins include increased focus on green water and compensation, as well as a focus on development opportunities (for example the TWO analysis).

In the face of coming climatic changes and related environmental uncertainties, it is likely that many water agreements will become obsolete. Many do not have an inbuilt flexibility to deal with variations but are rather based on annual averages, which may already be a problem, and with larger variations will be likely to increase the difficulties. Indeed, as been pointed out by Falkenmark et al (2009b), 'legal arrangements in transnational basins are, however, principally reactionary in character, often developed to resolve existing conflicts through compromise'. While this situation is unlikely to change overnight, it is still argued that such a development is needed, and some key components and perspectives are outlined below. As shown in the discussions of the Israeli–Jordanian agreement, climatic variability has 'conveniently' been left out of the agreement for political reasons, only to resurface when there is a drought in the Jordan Basin. Crises have been solved in the political arena, but this has resulted in strained political relations. While aiming for more sustainable and inclusive agreements on transboundary waters, which arguably are needed, it is

important to keep in mind that agreements between riparians are not always informed by the best knowledge available but are devised through political negotiations. Thus, our list (below) of necessary features which will allow agreements to be sustainable is not always going to be reached easily, precisely because agreements are political in nature.

Nevertheless we believe it is useful to outline what would improve agreements and make them more sustainable:

1   Agreements have an inbuilt flexibility, and allocations are not written in stone. Nevertheless, as noted by Drieschova et al (2009), the potential trade-off following increased flexibility can reduce certainty about the actual flows which may increase negotiation costs.

2   Agreements are based on an appreciation for the political context in which they are to function. This includes the cultural, historical and discursive composition of the states in the basin. These largely determine the 'room for manoeuvre'.

3   Agreements focus on development opportunities that benefit all parties in a basin. These opportunities are often found outside the (blue) water sector (cf. TWO analysis) such as land use/green water, trade, energy and tourism. Increased interaction between the research community (producing innovative knowledge), the water resources managers (with the task to implement policies and strategies) and the politicians is expected to be key in this respect.

4   Power asymmetries in the basin need to be addressed either through efforts to level the playing field or by levelling the players. The international system can play an important role here by capacitating the weaker parties to interact on more equal terms.

These recommendations would improve the situation and preparedness to address the coming socioeconomic as well as potential climate changes. Some of them may be less feasible than others but it is nevertheless important to highlight the measures that, ideally, should be taken.

In sum, political agreements on shared waters need to be based on an understanding of the political context in which they are to operate. They should be more flexible than they are today, in order to be sustainable. Furthermore, broader agreements, including a focus on development opportunities, also represent a way forward that would open up more benefits that could be shared by the riparians. If agreements are not more flexible than most are today, they run the risk of triggering political conflicts.

Notes

1   The convention has not yet entered into force since only 17 countries (30 are needed) have signed and ratified it. However, a growing consensus in the water world is emerging that argues that the convention should be seen as part of customary international law.
2   This section is based on Jägerskog (2003).

References

Agreement between the Government of the Republic of Moldova and the Government of Ukraine on the Joint Management and Protection of the Cross-Border Waters, (1994) http://dniester.org/en/legislation/ moldavsko-ukrainskoe-moldovan-ukrainian, accessed 12 August 2009

Allan, J. A. (2001) The Middle East Water Question: Hydropolitics and the Global Economy, I. B. Tauris, London

Convention on the Protection of the Rhine (1998) http://ocid.nacse.org/tfdd/tfdddocs/193ENG.htm, Accessed 12 August 2009

Dinar, A., Dinar, S., McCaffrey, S. and McKinney, D. (2007) Bridges Over Water: Understanding Transboundary Water Conflict, Negotiations and Cooperation. World Scientific, Singapore

Drieschova, A., Giordano, M. and Fishlander, I. (2009) 'Climate change, international cooperation and adaptation in transboundary water management', in W. N. Adger, I. Lorenzoni and K. O'Brien (eds) Adapting to Climate Change: Thresholds, Values, Governance, Cambridge University Press, Cambridge

Falkenmark, M. and Lannerstad, M. (2005) 'Consumptive water use to feed humanity – curing a blindspot', Hydrology and Earth System Sciences, vol 9, pp15–28

Falkenmark, M. and Molden, D. (2008) 'Wake up to realities of river basin closure', Water Resources Development, vol 24, pp201–215

Falkenmark, M. and Rockström, J. (2004) Balancing Water for Humans and Nature: The New Approach in Ecohydrology, Earthscan, London

Falkenmark, M. and Rockström, J. (2005) Rain: The Neglected Resource, Swedish Water House Policy Brief, no 2, Stockholm International Water Institute, Stockholm, Sweden

Falkenmark, M., Berntell, A., Jägerskog, A., Lundqvist, J., Matz, M. and Tropp, H. (2007) On the verge of a new water scarcity, SIWI Policy Brief, SIWI, Stockholm, Sweden

Falkenmark, M., Rockström, J. and Karlberg, K. (2009a) 'Present and future water requirements for feeding humanity', Food Security, vol 1, pp59–69

Falkenmark, M., Vick, M. and de Fraiture, C. (2009b) 'Global change in four semi-arid transnational river basins: analysis of institutional water sharing preparedness', Natural Resources Forum, vol 33, pp310–319

Gross Stein, J. (1988) 'International negotiation: a multidisciplinary perspective', Negotiation Journal, vol 4, pp 221–231

GWP (2003) Water Management and Ecosystems: Living With Change, TEC Background paper no 9, Global Water Partnership, Stockholm, Sweden

Haaretz (1999) 'Israel eases back on water cut for Jordan, but alters schedule', Haaretz, 22 April

Jägerskog, A. (2003) 'Why states cooperate over shared water: the water negotiations in the Jordan River Basin', PhD dissertation, Linköping Studies in Arts and Science, Linköping University, Linköping, Sweden

Jägerskog, A. and Phillips, D. (2006) Managing Transboundary Water for Human Development, Human Development Report Occasional Paper, UNDP, New York, http://hdr.undp.org/en/reports/global/ hdr2006/papers/jagerskog%20anders.pdf, accessed 11 May 2010

Khatib, A. (1999a) 'Tensions ease in the Jordan–Israel water dispute, officials say', Jordan Times, 7 April

Khatib, A. (1999b) 'Israel agrees to provide Jordan with full share of water share', Jordan Times, 21 April

Milly, P. C. D., Betancourt, J., Falkenmark, M., Hirsch, R. M., Kundzewicz, Z., Lettenmaier, D. P. and Stouffer, R. J. (2008) 'Stationarity is dead: whither water management?' Science, vol 319, pp573–574

Molle, F., Wester, P., Hirsch, P., Jensen, J. R., Murray-Rust, H., Paranjpye, V., Pollard, S. and van der Zaag, P. (2007) 'River basin development and management', in D. Molden (ed.) *Water for Food, Water for Life: A Comprehensive Assessment of Water Management in Agriculture*, Earthscan, London; IWMI, Colombo, Sri Lanka, pp585–625

PCA (Permanent Court of Arbitration) (2004) 'Case concerning the auditing of accounts between the Kingdom of the Netherlands and the French Republic pursuant to the additional protocol of 25 September 1991 to the Convention on the Protection of the Rhine Against Pollution by Chlorides of 2 December 1976, Arbitral award of 12 March 2004', www.pca-cpa.org/upload/files/Neth_Fr_award_English.pdf, accessed 12 August, 2009

Phillips, D., Daoudy, M., McCaffrey, S., Örjendal, J. and Turton, A. (2006) 'Trans-boundary water cooperation as a tool for conflict prevention and broader benefit-sharing', *Global Development Studies* no 4, Ministry for Foreign Affairs, Stockholm, Sweden

Phillips, D. J. H., Allan, J. A., Claassen, M., Granit, J., Jägerskog, A., Kistin, E., Patrick, M. and Turton, A. (2008) *The TWO Analysis: Introducing a Methodology for the Transboundary Waters Opportunity Analysis*, Report no 23, SIWI, Stockholm, Sweden

Rockström, J., Karlberg, L. and Falkenmark, M. (in press) 'Global food production in a water-constrained world: exploring green and blue water challenges and solutions', in *Water Resources, Planning and Management: Challenges and Solutions*, Cambridge University Press, Cambridge

Rost, S., Gerten, D., Hoff, H., Lucht, W., Falkenmark, M. and Rockström, J. (2009) 'Global potential to increase crop production through water management in rainfed agriculture', *Environmental Research Letters*, vol 4, 044002

Sadoff, C. and Grey, D. (2002) 'Beyond the river: the benefits of cooperation on international rivers', *Water Policy*, vol 4, pp389–403

Treaty between the United States of America and Canada Relating to Cooperative Development of the Water Resources of the Columbia River Basin (1961) http://ocid.nacse.org/tfdd/tfdddocs/116ENG.pdf, accessed 12 August 2009

Wolf, A. and Hamner, J. (2000) 'Trends in transboundary water disputes and dispute resolution', in Green Cross International, *Water for Peace in the Middle East and Southern Africa*, Green Cross International, Geneva

WWAP (World Water Assessment Programme) (2009) *The United Nations World Water Development Report 3: Water in a Changing World*, UNESCO, Paris; Earthscan, London

Zeitoun, M. (2008) *Power and Water in the Middle East: The Hidden Politics of the Palestinian–Israeli Water Conflict*, I. B. Tauris, London

Zeitoun, M. and Jägerskog, A. (2009) 'Confronting power: strategies to support less powerful states', in A. Jägerskog, M. Zeitoun and A. Berntell (eds) *Getting Transboundary Water Right: Theory and Practice for Effective Cooperation*, SIWI report no 25, Stockholm International Water Institute, Stockholm, Sweden

12

Enhanced Knowledge and Education Systems for Strengthening the Capacity of Transboundary Water Management

Léna Salamé and Pieter van der Zaag

- Independently from the nature of the water body or its geographical setting, and whatever the legal regime to which it is subject, the development and the sharing of knowledge constitute:
  - a basis for the understanding of the water body in question and for establishing cause–effect relationships in its management;
  - an essential element to build confidence and enhance trust among the stakeholders concerned at the technical level;
  - a sound and scientific basis for joint decision making at the political level.
- When the various technical and scientific players are on the same page, they may offer a solid and scientific knowledge base for the political players who will ultimately engage in political and legal agreements for the management of a shared water body.
- A 'new' type of water manager, planner and decision maker is required. They will be asked to act as a 'problem manager' rather than a 'problem solver', and to act as a 'first-line conflict preventer' by resolving problems before they arise.
- Collaborative capacity-building exercises are the most efficient and sustainable.
- Since people have to cooperate in real life, they might as well start within the framework of a training exercise.

Introduction

As competition over the use of transboundary water resources grows, more efforts to anticipate, prevent and manage water conflicts are required. Political will and financial resources are essential ingredients in both effective transnational water management and conflict prevention and mitigation. However, while necessary, these two factors alone are not sufficient. Financial resources and political will cannot create effective water management when institutions lack the capacity to assess water resources as well as to make and implement appropriate decisions. Transboundary water management (TWM) is knowledge and data intensive. It demands that managers be more than technical problem-solving engineers, more than negotiators, and more than specialists able to make sense of mounds of data. Rather, water management institutions must be able to do all this and more. The demands placed upon today's and tomorrow's institutions mean that the water managers within them need effective education, that equips them with the knowledge and skills to allow specialists and generalists to come together and support one another in tackling complex problems. Institutions can only be as good as their component parts. Therefore, effective water management requires developing capacities of both individual managers and the institutions they comprise. Focusing more broadly on institutional capacity is also useful because institutions can often last through strained situations, providing structures and channels for communication even when relationships among individuals or states are difficult (Salamé et al, 2009), and allowing collaboration to pick up where it left off if working relationships become temporarily unfeasible.

Part I of this chapter presents an overview of the knowledge, skills and tools that water managers need in order to contribute effectively to the management of transboundary water resources. Four cases where knowledge served directly or indirectly to enhance cooperation and water management are presented. Part II describes the role that an appropriate set of knowledge, skills, and tools can play in the management of transboundary water resources. Finally, part III describes possible approaches to education that can equip water managers and others with the necessary skills, tools and knowledge that will allow tomorrow's society to address the challenges of TWM.

Part I: Knowledge, Skills and Tools in Transboundary Water Management

Appropriate knowledge, skills and tools enable water managers to better understand the basin in question, to anticipate potential conflicts that are

inherent to transboundary waters, and to work with stakeholders to develop sustainable management strategies that will prevent or mitigate conflict while creating the best outcome for all parties involved. In addition, knowledge can serve less obvious but equally important functions. Where trust among states is lacking or when the political situation precludes cooperation on certain issues, information sharing can open the door for cooperation, providing a basis upon which to build trusting relationships across borders. Such relationships will not only make cooperative water management sharing possible, but can also facilitate cooperation extending into other areas such as trade and integrated development.

This section presents four cases where knowledge and information played a role in facilitating cooperation and sustainable water management. The cases demonstrate that when the needed institutional capacity is available, knowledge is developed, shared and used, which facilitates the tasks of water managers who, in turn, become better equipped to contribute to enhancing cooperation around transboundary water resources. Each river basin is unique, so there are myriad ways that knowledge could come into play.

The choice of these cases illustrates that knowledge can play the role of a catalyst for cooperation in different transboundary settings. The examples were chosen to give a balanced overview from various perspectives:

- Kind of water bodies: the cases cover a variety of water bodies (rivers, an aquifer and a lake).
- The type of legal regime: the examples are each subject to different legal regimes (international water bodies shared by two or more sovereign states, national water bodies shared within one country by various federated states).
- The geographical setting: different continents and climatic zones.

Each case shows how concerned parties (whether sovereign or federated states) have been confronted with challenges related to the joint and cooperative management of the resources they share, and each also shows how knowledge and education played a direct or indirect role in the resolution of the problem.

It is well recognized that knowledge and education usually contribute to improve individual and institutional capacities in managing water resources. Although it is not possible to establish a direct link between a certain political cooperative action and a given training or knowledge-building activity, it remains obvious that sound water management can only be achieved if the basic capacity and the critical knowledge to implement it have progressively been put in place.

The cases presented in this chapter will show that knowledge and education contribute to the facilitation of dialogue among various parties concerned with a transboundary water body. They will show that knowledge and education also constitute at least two ingredients – among others – that contribute to the establishment of a sound and scientific basis for the development or the enhancement of cooperation. This selection of cases could therefore be seen as a starting point from which to begin thinking about creative ways to employ knowledge in order to facilitate more directly cooperative management.

The Nile, an international water body in Africa shared by sovereign riparian states[1]

The River Nile is 6500km long. The basin drains an area of over 2.9 million km^2, about 10 per cent of the African continent, where about 150 million people are living. The Basin's climate range varies between extreme aridity in the north (Egypt and Sudan in particular) to tropical rainforest in Central and East Africa and parts of Ethiopia. The richness of cultural, ethnic and religious diversities adds further layers of complexity that need to be taken into account by water managers and decision makers when developing and managing the water resources of that basin. The multifaceted heterogeneity of the Nile indeed exacerbates the classical competitiveness between riparian states of a transboundary basin (Nicol, 2003).

Various initiatives have attempted to facilitate the dialogue among the Nile states users, establish some kind of cooperative frameworks and support the development of the peoples whose lives are (inter-)dependent on the river. Some of these initiatives focused on the political level exclusively. Others focused on education, knowledge and capacity-building and have had a positive impact on the dialogue and exchanges at the political level. One such positive impact is the FRIEND project, to which we now turn.

In March 1996, the Flow Regimes from International Experimental and Network Data (FRIEND) Nile project was initiated. UNESCO invited the National Committees of the International Hydrological Programme (IHP) and the Nile Basin countries' representatives of the Technical Cooperation Committee for the Promotion of the Development and Environmental Protection of the Nile Basin (TECCO-NILE) to meet at the University of Dar es Salaam, Tanzania. Since 1996, the project has expanded to include all ten Nile riparian countries (Burundi, Democratic Republic of Congo, Egypt, Eritrea, Ethiopia, Kenya, Rwanda, Sudan, Tanzania and Uganda). The overall objective of the FRIEND Nile project is 'to improve international river basin management of the Nile through improving cooperation amongst the Nile countries in the fields of water resources management and regional-scale analysis of hydrological regimes'. However, the establishment of shared databases met with serious difficulties as some countries refused to share data, although this was certainly not in conformity with the established principles of FRIEND.

In 1999, therefore, the Steering Committee agreed to postpone the establishment of a regional database component until difficulties in data sharing could be resolved. Instead, the Steering Committee added a new component for training and capacity-building in order to provide support to researchers and operational staff of the hydrological services, thereby contributing to their capacity to better assess and manage the Nile Basin. The main content of the new component was the following three objectives:

1   to provide in-job training courses tailored to different disciplines within the research components;

2   to define the possible use of ongoing training courses in the same fields as the research components;
3   to initiate workshops that address the main issues related to the joint management of the Nile.

Since then, training and capacity-building have been given a high priority within the FRIEND Nile project. They achieved their above-mentioned objectives and this also led to the establishment of a work plan for further activities related to data sharing with well-defined objectives, methodologies, research needs, expected results and relevant budget requirements. Following these activities, and based on the trust that was then built among its various parties, the team working on solving the data sharing problems indeed progressed in its task and further adopted criteria in order to assess the objectives, feasibility and applicability of each research theme, and nominated a person from each riparian state for a special research component. Finally, since the FRIEND Nile has developed strong links with other international organizations, its technical experience and the results it reached ended up contributing to the negotiations processes between the different players in the Basin.

The FRIEND Nile case constituted an entry point that allowed the parties to learn to know and subsequently trust each other. It ended with the constitution of clear work plans that paved the way for the sharing of data, which in turn was to be used in wider cooperation processes. The difficulties in sharing valuable data between the Nile Basin actors in 1999 was indeed eased through the implementation of the different training and capacity-building activities by the Steering Committee. The cooperation processes that followed were facilitated by the existence of the data that was jointly collected by, and shared among, the various parties.

Another Nile Basin-wide initiative has also addressed the importance of knowledge generation and the sensitive issue of data and knowledge sharing, and may itself be an indirect spin-off of the earlier FRIEND initiative. In 2000, the Nile Basin Capacity Building Network for River Engineering (NBCBN-RE) was established. This network comprises universities and other knowledge institutes in the ten countries riparian to the Nile. The network focuses in particular on stimulating applied research in the specific river engineering domain. At the moment a total of 25 research groups are active under the four NBCBN research modalities:

1   Regional Cluster Research (13 groups), aiming to bring together experts sharing a common thematic passion. Each cluster is hosted by a different Network Node, whose role is to develop and sustain regional research capacity in their research domain. These domains are: including GIS and Modelling Applications in River Engineering (hosted by Egypt); River Structures (hosted by Ethiopia); Environmental Aspects of River Engineering (hosted by Uganda); Flood Management (hosted by Kenya); Hydropower (hosted by Tanzania); and River Morphology (hosted by Sudan).
2   Integrated Research (two groups), aiming to implement more integrated applied research by linking three or more clusters or research groups.
3   Local Action Research (nine groups), aiming to contribute to capacity-building at local level and to enhance the collaboration among researchers in the same country.
4   Multidisciplinary Research (one group), aiming to address research on the impact of climate change on the water resources of the River Nile.

The network has grown and by 2008 it included 400 water professionals from the Nile Basin. Some concrete results of this collaborative knowledge network are:

•   A Nile Basin Knowledge Map: a web-based information management system that offers easy access to all information related to professionals, organizations, training courses, other networks and projects related to the water sector in the Nile Basin.
•   A knowledge-sharing and collaborative platform that maximizes the efficiency of

geographically dispersed NBCBN research groups in knowledge sharing and collaborative working (www.nbcbn.com).

- The launching of the *Nile Water Science and Engineering Scientific Journal* that stimulates experts in the region to share their research results.
- NBCBN Newsletters, published three to four times a year to inform the network about research, education and training opportunities.
- Several shared databases, such as the web-based 'Reservoir Information System' for reservoirs all over the Nile Basin (developed by the River Morphology cluster) and, also online, a database of existing traditional and modern diversion systems schemes in the Nile Basin (developed by the River Structures Research cluster).

Given the sensitive nature of transboundary issues on the Nile, the establishment of technical cooperative networks and shared databases must be considered small yet significant steps, which are prerequisites for further collaboration in future. An independent evaluation in 2008 revealed that the NBCBN network is viewed as very successful by most of the interviewees and that it caters for a large majority of Nile Basin professionals as their main source of knowledge and information.

As the potential economic benefits of the ten riparian countries cooperating are huge (e.g. Whittington et al, 2005), the initiatives in knowledge networks must therefore be considered very valuable. Without these, future cooperation, and thus future potential benefits, are at risk (Luijendijk and Lincklaen Arriëns, 2009)

The Mekong River Basin in Asia, an international water body shared by sovereign riparian states[2]

The Mekong River originates from the Tibetan Plateau of the Himalayas, flowing over 4800km southward to the South China Sea. It flows through China, Myanmar, Laos, Thailand, Cambodia and Vietnam, draining an area of 795,000km^2 with 48 per cent of the basin shared between Thailand and Laos. This basin is home to over 70 million people, with over 80 per cent of the population residing in the Lower Mekong Basin (UNEP, 2009).

The Mekong region faces several critical challenges. Among the contributing factors are a rapidly eroding basin, presently affecting 21 per cent of its area; declining original forests (only 30 per cent of the forests are intact); continued regional population growth (expected to remain at 2 per cent per annum over the next decades); and a threatened and deteriorating ecosystem. Compounding these challenges are the effects of climate change. All these factors could trigger social and economic instability for Mekong residents who depend on the river system for their livelihoods.

To enhance effective regional and intergovernmental cooperation among the Lower Mekong countries (Cambodia, Laos, Thailand and Vietnam), the Mekong River Commission (MRC) has set itself the goal 'To identify potential transboundary issues for negotiation, mediation and conflict prevention, and develop mediation and conflict management capacity' (MRC, 2006). The MRC is addressing this goal with two relevant initiatives. One is a training programme at the broadest level on water conflict prevention; the other is implemented at the level of MRC's Flood Management and Mitigation Programme and concerns capacity-building to enhance cooperation in addressing transboundary flood issues. Both initiatives are briefly described here. The case study of the Mekong River Commission (this volume Chapter 13) provides further information on this institution.

Through MRC's cooperative partnering with the United States Agency for International Development (USAID) and Environmental Cooperation-Asia (ECO-Asia), water conflict prevention and management training is under way in the region. More specifically, the training covers hydropolitics and conflict management; ways of changing perception through thinking of basins without boundaries; approaches to enhancing and sharing benefits; and techniques towards building institutional capacity. The objective of this training is to build capacity at the individual and institutional levels, and inform and shape policy. This training is being addressed at three levels simultaneously:

1   Workshops aimed at building awareness and understanding of transboundary water issues, their causes and triggers. This provides principles and practices in cooperation, and the prevention and management of conflict, tailored towards specific basin stakeholders.

2   Skills-building training in facilitation and mediation aimed at the national and agency level, such as the National Mekong Committees and organizations that are expected to assist in potential transboundary issues. The skills acquired through this training are practical in nature, and provide staff with the effective command to enhance communication, negotiation, facilitation and mediation.

3   To give policy makers and senior water managers the ability to analyse the significance of institutional frameworks for water resources at multiple scales – local, national, regional and international – in terms of the potential for cooperation and conflict, and the opportunity to explore real-world case studies and event analysis.

This threefold comprehensive training approach to water conflict prevention and management is combined and made available through a cohesive package. Furthermore, there is at each level a module that moves the participants through the evolutionary process and stages of water conflict transformation, thereby bolstering the region's capacity to strategically and economically anticipate, address and mediate between competing water users.

Other activities have also focused on enhancing the capacities of water managers and decision makers in the Mekong basin. Following the request of the MRC and within its Flood Management and Mitigation Programme, the UNESCO-IHE Institute for Water Education embarked on a special capacity-building project. This had the aim of enhancing cooperation by member countries in addressing transboundary flood issues. The project developed a suite of training events for high-level decision makers and professionals in the four MRC member countries and the National Mekong Commissions, national experts, managers and academic lecturers, as well as the MRC Secretariat. The project developed and piloted three training workshops, including field visits to flood management areas in the Lower Mekong Basin. These workshops were complemented with especially designed exchange study visits to pertinent international and regional river basin organizations. A visit was made to the Yangtze River Basin and the Changjiang (Yangtze) Water Resources Commission (CWRC) in Hubei and Hunan provinces, China, so that participants from the MRC could learn from China's practical experiences of controlling floods and managing water across inter-provincial borders.

The participants of this study visit included water experts from the Lower Mekong countries who, in their daily professional lives, are involved directly or indirectly in decision-making processes related to the management of the water resources of the Mekong, and coordination with neighbouring countries. The participants overwhelmingly considered that the visit was useful to them as they had learnt from the experience of their Chinese counterparts in flood control and management. The visit did not only allow the exchange of experiences and ideas: it has also opened a door of communication between experts and decision makers from China and their counterparts in the Lower Mekong Basin. This is not trivial, as China is an upper riparian country of the Mekong, and the CWRC is the Chinese institution in charge of water management in the Upper Mekong (Lancang) basin.

These types of activities usually go beyond their official learning objectives. They offer a place where experts and decision makers learn from each others' experiences. They also provide a forum where links and relations between decision makers and professionals working on the same issues in different countries can be nurtured and improved. Such achievements of knowledge building/education events are indeed as equally beneficial and important, if not more so, than the official, announced objectives. They influence, directly or indirectly, the work of the individuals and the institutions in which they work.

It is not possible to assert which training or which knowledge-building activity has shaped the given advice of a professional to his or her senior officer, or the given position of a decision maker. It remains certain, though, that the benefits of training activities and learning opportunities such as those described in this section can enhance the MRC's capacity, and that of similar river basin organizations, to negotiate and implement transboundary agreements; deepen transboundary cooperation on issues related to floods, droughts, irrigation, hydropower, environmental impacts, climate change, fisheries and navigation; and complex combinations of these. Moreover, training events imply undergoing shared experiences, which more often than not help to build mutual respect and understanding, essential ingredients for trust and confidence-building. In short, this type of training creates an atmosphere of collaborative learning that is broad and encompassing.

The Guaraní Aquifer System, a groundwater body shared by various sovereign states in Latin America[3]

A natural resource cannot be managed in a sustainable manner if its essential characteristics remain unknown to the water managers. This is particularly clear in the case of groundwater as it is an invisible resource. Furthermore, it is not enough for a country to have its own knowledge of its groundwater resources when it shares them with one or more neighbours. Knowledge should indeed be built, shared and accepted by the riparian countries and their related communities. These were lessons learned from the execution of the Project for the Protection and Sustainable Development of the Guaraní Aquifer System (2003–2009), implemented by the governments of Argentina, Brazil, Paraguay and Uruguay, with the support of the Global Environment Facility (GEF), World Bank and the Organization of American States (OAS). Case studies in Chapter 13 of this volume also refer to the Guaraní Aquifer, and to the La Plata River Basin Treaty.

The Guaraní Aquifer System (GAS) is a deep, practically confined aquifer (and, hence, a non-renewable water resource) covering more than 1 million km^2 in four countries of South America; more than 90 per cent of its extension is covered

by basalts. The primary knowledge about the GAS came from oil companies working in Brazil and Uruguay, who during the second quarter of the 20th century dug holes looking for petroleum, and discovered water. The use of the water resources of the GAS started to play an important role in meeting domestic water demands in the urban areas of Brazil, Paraguay and Uruguay, where the aquifer was shallow. Additionally, in Uruguay, springs of thermal water were used for the development of thermal touristic centres based on the deep areas of the aquifer, where the original oil drilling was carried out. This activity expanded to the Argentinean border area. As a result, the use of the aquifer was intensified without any national or regional knowledge of it. Coordinated management of the GAS was practically non-existent in each of the countries sharing it, with the exception of some initial discussions over issues of critical concern. Information was piecemeal, haphazard and scattered among different institutions, and, needless to say, was not shared among the countries concerned.

During the 1980s and 1990s, universities of the region started to show scientific and technical interest in the GAS without really knowing its transboundary characteristics. With the support of the Canadian International Development Agency (CIDA) they started sharing their available information among themselves. They produced a document with all the information they could gather together and later prepared the 'Declaration of Paysandú' in which they commit to try and disseminate the available knowledge in order to show its importance as well as its transboundary characteristics.

Around the same period, water agencies from Brazil and Uruguay were in the late 1990s planning to start a project on their common border on the Cuareim River Basin. The project proposal of these institutions and their vision included a reference to the GAS (the transboundary characteristics of which were now known to, and recognized by, all the key players) and the need to establish a solid basis for its sustainable management. Both countries agreed to ask for the support of OAS and the GEF for the implementation of this project.

In parallel, CIDA's support of the universities ended and the Canadians called for a meeting in order to launch a new initiative to develop knowledge about GAS at a regional level. The World Bank and the OAS were present, and the idea of merging both initiatives with the financial support of the OAS, the GEF and the World Bank was agreed upon and implemented. The new initiative aimed, in a very concrete way, at building a better knowledge of the aquifer system in order to protect and use it in a sustainable manner.

These various efforts show that the continental and strategic dimension of the aquifer was recognized, as well as the lack of information related to the real dimension, borders, physical characteristics, quality and dynamics of the water. As a consequence though, countries were not always confident in working together, as they did not know what each country had to manage or negotiate. Given these circumstances, the initial common agreement for the Guaraní Project was to protect the transboundary aquifer and not to discuss its possible uses.

Furthermore, the interests of the four countries were not exactly the same. Argentina was interested in knowledge *per se*. It was indeed the last country to become aware of the existence of the GAS under its territory. It did not know much about the boundaries of the GAS to the east and south, or the quality of its resources, since the aquifer is very deep in its territory. On the other hand, tourist activities related to thermal water had already started in Uruguay based on this country's good experiences with the use of the Guaraní water to attract tourism. Uruguay was consequently interested in the protection of the resources from competing uses by other countries, and sound management was the main objective for this country. Brazil, under which territory lies the biggest portion of the aquifer and which has the most developed cities and economy in the GAS area, wanted to demonstrate that the use of the aquifer by some stakeholders can have serious impacts on others, especially when such uses are located close to international borders. Finally, Paraguay was focusing more on the use of the resource in a sustainable manner, as the GAS in this country is shallower than in other places.

Knowledge was thus the common need of all four countries and the advances in understanding the GAS gained through the project helped and was critical for the final agreement between the four.

As mentioned earlier, one of the most important components of the project was to strengthen the technical and scientific knowledge of the aquifer. As a result of the project, an improved understanding of the system enabled improved management of the aquifer's resources, and, more importantly, confidence and trust for mutual cooperation were built on a solid basis between the four countries. Such positive developments were based on the information held in common, and they resulted in the establishment of permanent working relations between the concerned institutions and communities in the riparian countries.

A better understanding of the dynamics and quality of groundwater flows, and on the characteristics of water uses obtained from boreholes, was generated. It led to the acknowledgement by water managers and decision makers concerned with this aquifer that judicious groundwater management was needed at both regional and local levels, encompassing provinces, states, countries and the region as a whole. Regional cooperation fostered under the GAS project enabled specific advances that certainly benefited the aquifer itself but also served the broader context of regional integration.

Lake Chapala, a water body shared by different federated states within one country in Central America[4]

The Lerma-Chapala river basin system is situated in the central part of Mexico, Lake Chapala being located at the downstream end. The $54,000km^2$ catchment that feeds Lake Chapala lies in five states: Querétaro, Guanajuato, Michoacán, Mexico and Jalisco. Some 17 million people depend on the basin's water resources, including 2 million inhabitants of Mexico City and a similar number in Guadalajara, the second largest city of Mexico and capital city of Jalisco state. The main water use in the basin, though, is for irrigated agriculture, which has expanded exponentially during the 20th century to a current area of nearly

800,000ha. Now, the total water use is greater than the long-term average surface and groundwater replenishment rates. This overuse is reflected not only in steadily declining groundwater levels, but also in the declining water levels of Lake Chapala. This lake (with a surface area of 1100km^2 and maximum storage capacity of 8100 million cubic metres) used to have high ecological value and is a well-known tourist attraction. It is very shallow and sustains large evaporation losses. The levels tend to fluctuate, depending on natural wet and dry cycles of rainfall and higher or lower inflows from the Lerma river, and this has been exacerbated by the increase in water abstractions mainly for irrigation in upstream tributary rivers and aquifers. Lake Chapala used to irregularly discharge into the Santiago river, but this hardly happens any more.

At the same time water quality is affected by direct sewage discharges and polluted return flows from irrigation districts. The combination has resulted in a serious deterioration of water quality in the rivers and lake in recent years. The water quantity and quality crisis of Lake Chapala brings out political issues that had otherwise been hidden for many years.

The fate of Lake Chapala has been the concern of many people and institutions. The first time the lake level fell below 40 per cent storage capacity was in 1950, and the Lerma-Chapala-Santiago Study Commission was created. This was chaired by the water ministry and consisted of representatives of all five states sharing the transboundary resource, and the capital city. It was tasked to develop an equitable water distribution through an adequate and combined operation of existing infrastructure. The commission elaborated an integrated basin development plan, and concluded that more storage capacity was needed. However, once inflows into the lake increased again and lake levels returned to normal in 1959, the urgency of the commission's coordinating activities decreased.

The second time that the lake level fell below 40 per cent capacity was in 1983. By that time irrigation in upstream states had nearly quadrupled during water surplus periods while storage infrastructure more than doubled. Water allocation to the various sectors was an *ad hoc* process

with predictable outcomes: 15 per cent of all water use was clandestine, cost recovery in irrigation districts was only 18 per cent and hardly any municipal and industrial wastewater was treated. This created tensions between various water user groups, such as large and small irrigators, municipal water supplies and environmentalists in the different states, to the extent that the federal government became seriously concerned. It was the Mexican president himself who in 1989 convened a meeting with the governors of the five states. This culminated in a signed agreement, whereby the parties committed themselves to improve water management in the long term, and more specifically to:

- allocate surface and groundwater fairly among users and regulate its use;
- improve water quality by treating municipal and industrial effluents;
- increase water-use efficiency;
- conserve the river basin ecosystem and protect watersheds.

In 1993 the Lerma-Chapala Basin Council was established. During the period leading to the creation of the council, a wide-ranging diagnostic study was conducted, on the basis of which a surface water allocation agreement was negotiated and signed in 1991 by the governors of the five riparian states and the federal government. The outcome was that irrigation rights in any given year would depend on (a) the volume of water in the lake, and (b) the surface runoff generated in the various parts of the basin during the preceding year. Since 1991 the treaty has been strictly enforced and actual water consumption has not exceeded the allocated amounts. In spite of this policy, lake levels continued to decrease. Thus, after some years of unusually low rainfall and in answer to pressures by ecologists and Jalisco state (located downstream), in 1999 the Department of Water decided for the first time to release 200 million cubic metres of water from reservoirs (stored for irrigation) for the sole purpose of replenishing the lake, an exercise repeated in 2001 and 2002. Even these extraordinary efforts failed to stop the decline of the lake, while creating much resentment among the irrigators who were forced to reduce their irrigation operations. Representatives of various large irrigation districts in three different states formed a special working group to strengthen their representation in the Basin Council, as well as to develop strategies to save water in dry years, such as planting crops that required less water.

It was now clear that the existing allocation agreement of 1991 was not solving the problem and the Council decided to revise it. As a result, new studies were conducted, and these showed that the 1991 agreement had been based on overly optimistic assumptions of water inflows, while neglecting inevitable evaporation losses from Lake Chapala. More sophisticated numerical models were developed, longer time series were considered and new environmental economic techniques were used in order to come up with a more refined water allocation agreement. User representatives of the various sectors and the state representatives consulted hydrological, environmental, economic and social experts. After many negotiation rounds the so-called 'joint political optimum' scenario for water allocation was defined. This allocation algorithm formed the basis of a non-binding voluntary agreement that was adopted by the Basin Council and converted into a legally binding treaty. Meanwhile by 2005 the lake level had largely recuperated, mainly due to abundant rainfall and exceptionally large inflows in 2003 (Mollard and Vargas Velazquez, 2004).

University experts, specialists from the Instituto Mexicano de Tecnología del Agua (IMTA), the Mexican Water Technology Institute, and experts from within the state governments all contributed critical knowledge resources. Importantly, each of the parties in the negotiation process (the five states) hired independent experts, who then debated the conditions for a resolution of the problem. Some of the protagonists argued that the development of numerical models greatly facilitated the difficult negotiation process towards a new agreement (Guitron et al, 2004; Huerta, 2004). Such models helped in reaching consensus based on facts, input data and a sound understanding of basic biophysical processes.

Part II: Necessary knowledge, skills and tools

The four cases presented above show that independently from the nature of the water body or its geographical setting, and whatever the legal regime to which it is subject (international, shared between various countries such as the Nile, the Mekong or the GAS; or national, shared by different states such as Lake Chapala), the development and sharing of knowledge constitute:

- a basis for the understanding of the water body in question and for establishing cause–effect relationships in its management;
- an essential element to build confidence and enhance trust among the stakeholders concerned at the technical level;
- a sound and scientific basis for joint decision making at the political level.

Even though all these cases do not show a direct impact of the knowledge development on the diplomatic track in negotiation and cooperation processes, they certainly reflect how knowledge facilitates communication among various protagonists in the management of a shared water body. As the FRIEND Nile initiative shows most obviously, such communication often occurs at the technical level upon which politicians and decision makers have to rely when entering into an official cooperation process. When the various technical and scientific players are on the same page, they may offer a solid and scientific knowledge base for the political players, who will ultimately engage in political and legal agreements for the management of a shared water body.

Water managers need to have sufficient knowledge to understand the issues and challenges with which they are presented. Thus, they need to have a thorough understanding of the basin at issue. First, this requires them to obtain and understand hydrological data with sufficient spatial and temporal specificity. Hydrological data include precipitation, interception, the amount of water filtering into the soil and used by vegetation, and the amount of water flowing in rivers, lakes and recharging aquifers. Second, water managers should understand the relevant ecosystems –

aquatic as well as other related ecosystems nearby – and the environmental flows required to keep them in good health. Third, water managers need to understand the social, economic and political context in which the basin is situated, and be able to quantify and explain the consumptive and non-consumptive uses of water in the basin. The types of data and knowledge needed for this purpose include, but are not limited to, those concerning population, cultures, livelihoods, water withdrawals and return flows, levels of development, legal and political institutions, and relationships among different groups living in the area. As each water basin is unique, it is important that water managers are able to identify and obtain data regarding factors that may be particularly pertinent in a given area. The cases of the Nile, Lake Chapala and the GAS show the importance of such kinds of knowledge and their impact on communication across borders.

In addition to gaining information in various areas, water managers learn to see water in different ways – for example, moving beyond 'blue water' to consider 'green water' and 'virtual water.' Such shifts in understanding can help managers to see new connections and to come up with innovative management approaches. Training courses such as those provided within the Mekong basin can help the key players gain such types of knowledge and develop new ways of thinking. The exchange of experiences that these activities facilitate constitute one of the most efficient ways to open the minds and the spirits of water managers to look outside the box, and find creative solutions for the problems they face: the transboundary nature of the water resources upon which the lives of their respective nations depend.

It is impossible for one person to be an expert in all of the issues involved. Therefore, water management requires specialists with expertise in particular areas, as well as generalists who possess some knowledge in a variety of areas and can help bring specialists together to put their expertise to its best combined use. This is why initiatives that aim to build the capacities of water managers and decision makers should be addressed to groups of people who have to work together in real life, even though they have different disciplinary backgrounds, and in

so doing actively create what is known as epistemic communities (Haas, 1992). The activities that took place in cooperation with the MRC have proved to be enriching in that sense. Diplomats, engineers, lawyers and other scientists have come together to follow the same course and were given the opportunity to combine their knowledge within the framework of the courses and outside. As Jarvis and Wolf describe in Chapter 9 of this volume, scientific specialists who can 'interpret causal chains' are a central need in water management.

Simply possessing knowledge is of little use if managers lack the skills needed to utilize it effectively. First, water managers will not always have the required data on hand – as the Nile example shows, often they will have to identify the information needed and the best techniques for obtaining it, and then be able to gather this data. This requires that they be conversant in the most current observational and monitoring techniques, including those based on remote sensing technology. Water managers do not need to be experts in such advanced technologies; it is enough if they understand them well enough to identify situations in which a particular technique would be useful, and then be able to contact the relevant experts to put the technology to use. Second, water managers may gain by having the cultural sensitivity and understanding required to collect data in a variety of contexts and settings, building relationships with local stakeholders. The GAS example shows the value added by such collaboration with pertinent institutions and communities. Third, water managers need to be able to understand and analyse relevant datasets, and should be able to use modelling software, as the example of Chapala Lake exemplifies. Modelling technologies allow a better understanding of complex systems and possible outcomes of various scenarios, and provide an additional understanding of the basin at issue, as well as an awareness of the limitations of models.

In addition, since TWM requires water managers to work across borders and often across cultures, accommodating diverse concerns and interests, they have to address three different language challenges:

- Language in its conventional sense: people do not speak the same language across borders and yet they need to communicate when they share the same resource and when they decide to jointly manage it.
- Technical language: people with different disciplinary backgrounds do not use the same words to designate the same things. Furthermore the same word does not always mean the same thing across the disciplines.
- Strategic language: people involved in a TWM process need to understand the strategic position of their counterparts. They need to be able to interpret such positions and move towards the discussion of underlying interests and the creation of options that could possibly address the conflict they face.

Water managers therefore need to possess the communication skills that would allow them to express needs and share information with other managers and stakeholders.

In addition, water managers need to have conflict management skills, which should allow them to understand the benefits and drawbacks of litigation, negotiation, facilitation, mediation, arbitration and other alternative dispute resolution techniques, and identify the approaches or combination of approaches that may work best in a given situation. In particular, it would be useful for water managers to possess strong negotiation skills. They also need to be flexible and open-minded enough to recognize when different methods of conflict resolution, such as indigenous practices, may be more appropriate.

Finally, water managers need to have access to tools that will allow their skills and knowledge to be put to its best use. Therefore, the appropriate political institutions, NGOs, or international organizations should seek to provide the technology needed for data gathering, monitoring and computer modelling, and should establish channels facilitating communication across disciplines, levels of government, sectors of society and even international borders.

In conclusion, water managers, planners and decision makers are expected to contribute to successful outcomes in the management of trans-

boundary waters. In response to increased water challenges worldwide, 'new' types of water managers, planners and decision makers are required. They will be asked to act as 'problem managers' rather than 'problem solvers', and as 'first-line conflict preventers' by resolving problems before they arise (van der Zaag et al, 2003). Therefore, water managers, planners and decision makers do not only need to have strong process-orientated technological knowledge, but a certain set of negotiating skills as well. However, when learning a skill, it is not just enough to 'know' about the skill. They have to be practised, preferably in conditions that replicate the circumstances under which the skill will be used (Cosgrove, 2003). There are formal techniques as well as informal means to manage water-related conflicts that stem from the nature of transboundary water bodies. Some even say that conflict resolution is more art than science (Salamé et al, 2009). Of course, besides the necessary set of skills, it is important that the main participants of the transboundary water conflict show a willingness to participate.

An appropriate level of knowledge, skills and tools will help to ensure that institutions have the necessary intellectual capital described by Jakob Granit in this volume, Chapter 10. For example, equipped with the knowledge of what information is useful and the skills and tools needed to gather and use that data, water managers will be able to build 'structural capital' – those 'information, monitoring, forecasting and financial management systems' that stay in place and are easily managed.

Part III: The Way Forward: Education for Water Management

Sustainable and peaceful cooperation around transboundary water resources requires expertise at various levels and across many disciplines. Most water managers today are trained as engineers, who have learned to 'focus on things' instead of interests and people, and to seek to control rather than adapt to existing processes (van der Zaag, 2009). The water managers who will ensure the future of TWM must be able to take a broader and more flexible view of problems, to recognize the complex interplay among natural phenomena, people and people's diverse interests, to address competitiveness among various users, and to manage processes cooperatively and adaptively rather than seeking to gain control. Where, then should education related to TWM begin? If we cannot simultaneously implement state-of-the-art water education for *all*, where would training programmes have the greatest impact? While arguments can be made for virtually any starting point, one approach is to train those who will become the managers, educators and decision makers of water resources in the future: today's graduate students. This rationale leads to prioritizing training at a regional master's level.

Intervention at master's level

Master's level students are expected to hold a degree in a specific field (such as agriculture, engineering, economics, political science, etc.) already. The master's programme will then build upon this knowledge, giving students a deeper level of specialization in their chosen field, but with a focus on transboundary water. In addition, the programme should give students a broader view of the multidisciplinary water field and an understanding of how various disciplines are interconnected. Thus, the master's programme should be broad in scope but still rigorous in its scientific approach – while students need to gain qualitative knowledge and process-oriented skills, they also need a strong quantitative knowledge of water sciences and the specificities of transboundary settings. The master's programme should also include an opportunity to conduct research, so that an attitude of learning and discovery is cultivated and so that the programme itself contributes to available knowledge and stays up to date, with findings from one year's students incorporated into the curriculum for the incoming class.

The required curriculum is so broad that any single university is unlikely to possess the requisite combination of expertise and interdisciplinary linkages, so it could be best provided by a consortium of universities with a coordinated programme. Connecting universities and bringing students together in this manner will also serve the purpose of spreading and sharing knowledge among different regions, institutions and communities. In addition, it will expose students to information and perspectives from different areas, climates and cultures and help to build working relationships among future water managers from different nations. Such a collaborative programme will nurture a spirit of cooperation, understanding and communication across regions or even continents, which can prepare the students to deal with neighbours across their borders. It will also serve to develop a common language within the water management community, which will facilitate international dialogue and cooperation.

Finally, the master's programme should cultivate formal and informal connections to policy makers and water management professionals. Short courses based on the master's curriculum and offered to water management professionals are one way to build such links, and will also further training objectives and generate income for the university. Annual regional meetings would also provide an opportunity for policy makers, professionals, academics and students to come together to share information and ideas, and to build relationships. This would contribute to cross-border trust and facilitate future cooperation.

This approach, of a number of different universities pooling resources and offering a joint or shared curriculum, has been piloted and tested at different spatial scales: within countries (Indonesia, Ethiopia), and in entire regions and subregions, for example in Latin America, Southern and Eastern Africa (WaterNet) and South Asia (Crossing Boundaries/SaciWaters). In each of these exercises, trainers and students from various disciplines and countries have gathered and shared their knowledge and experiences in issues related to transboundary waters.

The process constituted in itself a genuine learning experience for the trainers as well as for their students. Starting with the preparation of the training material and leading to the delivery of the first training and its replication by the trainers themselves or their students (in some cases), it offered them opportunities to create links among their institutions across international political boundaries. It allowed them to establish actual communication channels among different disciplinary perspectives. Engineers, lawyers, specialists in conflict resolution techniques, social scientists, biologists and others have understood each others' technical languages. They have learned how other disciplines can help them in their work and how their own disciplines can fit into a negotiation process. Above all, they saw the value added by this exercise, and many of them integrated it in the regular curricula of their respective institutions.

The experience described above shows how strong a collaborative capacity-building exercise can be. It shows that a broad training programme can be more easily achieved if various training institutions bring in their respective contributions: each one adds its own main theme to the basket of contributions and enriches it. Furthermore, the chances of such a broad training programme being implemented in a sustainable way are bigger if many institutions are involved in its preparation, and feel ownership for it. When dealing with the topic of transboundary waters, this becomes even more true and pertinent: since people have to cooperate in real life, they might as well start within the framework of a training exercise.

Intervention beyond the master's level

While a master's programme is one possibility for a priority programme, education related to the challenges of transboundary waters must be more broadly focused. For example, sustainable and peaceful cooperation around transboundary water resources also requires a well-informed and water-conscious public. Civil society is often the first to question proposed policies and to bring new interests, such as social and ecological concerns, to the table. Public opinion is often important to the policy makers in charge of water management, and can therefore be a powerful force. In addition, water management solutions will not be sustainable unless they have buy-in and ownership from relevant stakeholders. To accomplish this, stakeholders should be involved in water management, and the more knowledge they have of water issues, the more beneficial their participation will be. A public well versed in the importance and complexity of the issues involved in TWM will be most likely to keep abreast of water policy, to recognize potential problems and to propose creative solutions. It is not easy for average householders in Cairo to understand the consequences for their daily lives of a decision related to the management of the Nile at the international level. It is also difficult for farmers in Laos to understand the impact that their actions can have on the other side of the border in Thailand, Cambodia or even Vietnam. To this end, journalists and other media professionals should be trained in the importance of transboundary water issues, and in creative ways to communicate these issues to the public in such a way that is accessible to all.

Furthermore, an awareness of water, its transboundary aspects and connected issues, can be incorporated into primary and secondary education systems, so that a new generation understands the importance of water and of cooperating to manage natural resources. Since a variety of professions, including health and social workers, environmentalists and politicians play a part in the water sector, it is important to ensure that everyone receives basic education in transboundary water issues and the importance of cooperation. Incorporating transboundary water education into primary and secondary education will also ensure that the curriculum reaches a large number of people, and pupils may be able to have a positive impact on their parents' knowledge and behaviour.

Conclusion

TWM is not an engineering puzzle, but rather a mesh of interconnected interests and issues that continues to grow more complex. Therefore, sustainable TWM requires a new kind of water manager and new institutions to facilitate cooperation. Given the knowledge- and data-intensive nature of TWM, and the diversity of issues involved, specialists and generalists are expected to come together, able and prepared to pool their knowledge and skills, employing state-of-the-art tools and innovative thinking to understand complex situations and develop possible solutions.

Since TWM requires working across borders, cultures and socioeconomic gaps, water managers must possess not only technical skills, but also 'people skills', allowing them to build relationships, understand diverse perspectives and cooperate in order to create win–win solutions. An appropriate set of knowledge, skills and tools will facilitate such cooperation by serving as a starting point for building trusted relationships, by ensuring that everyone looking at the problem is working with the same facts, and by giving water managers negotiation and conflict-resolution tools to use when disputes arise.

Preparing water managers to meet the challenges that lie ahead requires an innovative approach to education, with a variety of foci and purposes, directed at various levels of society. Regional master's programmes that draw together expertise from various institutions would be one place to start, giving students the expertise they need to move forward and build international relationships in the process. However, other opportunities for education should also be explored, including in primary and secondary schools, so that a well-informed public can participate in the sustainable management of its water resources. Water education has to occur at all levels to equip people with the skills, knowledge and values required for positive societal responses. Only if our critical water resources are managed cooperatively and sustainably can we hope to build a peaceful and prosperous future.

Notes

1   Based on Servat and Demuth, 2006, and Gustard and Cole, 2002.
2   The first part of this case study is based on MacQuarrie et al, 2008, with additional inputs from Lynette de Silva.
3   This case study was generously contributed by Jorge Rucks.
4   Wester et al, 2001; Guitron et al, 2004; Huerta, 2004; Mollard and Vargas Velazquez, 2004; Dyrnes and Vatn, 2005; Van der Zaag et al, 2006; Wester et al, 2008)

References

Cosgrove, W. (2003) 'Water security and peace: legal approaches – a sound framework', study prepared for UNESCO programme *From Potential Conflict to Cooperation Potential*, UNESCO, Paris, p102

Dyrnes, G. V. and Vatn, A. (2005) 'Who owns the water? A study of a water conflict in the Valley of Ixtlahuaca, Mexico', *Water Policy*, vol 7, pp295–312

Guitron, A., Mollard, E. and Vargas Velazquez, S. (2004) 'Models and negotiations in water management', Workshop on Modelling and Control for Participatory Planning and Managing Water Systems, International Federation of Automatic Control, Venice, 29 September–1 October 2004

Gustard, A. and Cole, A. G. (eds) (2002), 'FRIEND – a global perspective 1998–2002', Centre of Ecology and Hydrology (CEH), Wallingford, UK

Haas, P. M. (1992) 'Introduction: epistemic communities and international policy coordination', *International Organization*, vol 46, pp1–35

Huerta, J. M. (2004) 'A systems dynamics approach to conflict resolution in water resources: the model of the Lerma-Chapala watershed', presented at the *22nd International Conference of the Systems Dynamics Society*, Oxford, 25–29 July 2004

Luijendijk, J. and Lincklaen Arriëns, W. (2009) 'Bridging the knowledge gap: the value of knowledge networks', in M. W. Blokland, G. J. Alaerts, J. M. Kaspersma and M. Hare (eds) *Capacity Development for Improved Water Management*, Taylor and Francis, London

MacQuarrie, P., Viriyasakultorn, V. and Wolf, A. T. (2008) 'Promoting cooperation in the Mekong region through water conflict management, regional collaboration, and capacity building', *GMSARN International Journal*, vol 2, no 4, pp175–186

Mollard, E. and Vargas Velazquez, S. (2004) "Liable but not guilty": the political use of circumstances in a river basin council (Mexico)', *Proceedings of the workshop on Water and Politics – Understanding the Role of Politics in Water Management*, World Water Council, Marseille, France, 26–27 February 2004, pp39–50

MRC (Mekong River Commission) (2006) *Mekong River Commission Strategic Plan 2006–2010*, MRC, Vientiane, Lao PDR

Nicol, A. (2003) *The Nile: Moving Beyond Cooperation*, UNESCO-IHP, Paris, p33

Salamé, L., Swatuk, L. and van der Zaag, P. (2009) 'Developing capacity for conflict resolution applied to water issues', in M. W. Blokland, G. J. Alaerts,

J. M. Kaspersma, and M. Hare (eds) *Capacity Development for Improved Water Management*, Taylor and Francis, London, pp116–133

Servat, E. and Demuth, S. (eds) (2006) *FRIEND – A Global Perspective 2002–2006*, UNESCO-IHP/HWRP, Koblenz, Germany

UNEP (2009) 'Hydropolitical vulnerability and resilience along international waters: Asia', UNEP, New York, p184

Van der Zaag, P. (2009) 'Sharing knowledge for water sharing', *Irrigation and Drainage*, vol 58, ppS177–S187

Van der Zaag, P., Bos, A., Odendaal, A. and Savenije, H. H. G. (2003) 'Educating water for peace: the new water managers as first-line conflict preventors', presented at the UNESCO-Green Cross 'From potential conflict to cooperation potential: water for peace' session; *3rd World Water Forum*, Shiga, Japan, 20–21 March 2003

Van der Zaag, P., Van Dam, A. and Meganck, R. (2006). 'Looking in the mirror: how societies learn

from their dependence on large lakes', Invited paper for the special session 'Large lakes as drivers for regional development', *Stockholm World Water Week*, Stockholm, Sweden, 20–26 August 2006

Wester, P., Melville, R. and Ramos Osorio, S. (2001) 'Institutional arrangements for water management in the Lerma Chapala Basin', in A. M. Hansen and M. Van Afferden (eds) *The Lerma-Chapala Watershed*, Kluwer Academic/Plenum Publishers, Dordrecht and New York, NY, pp343–369

Wester, P., Vargas-Velazquez, S., Mollard, E. and Silva-Ochoa, P. (2008) 'Negotiating surface water allocations to achieve a soft landing in the closed Lerma-Chapala Basin, Mexico', *Water Resources Development*, vol 24, no 2, pp275–288

Whittington, D., Wu, X. and Sadoff, C. (2005) 'Water resources management in the Nile basin: the economic value of cooperation', *Water Policy*, vol 7, pp227–252

13

Case Studies of Transboundary Water Management: Fifteen Initiatives from Various Parts of the World

Compiled by Virginia Hooper and Michael McWilliams

Introduction

Much of what has been discussed in the preceding chapters is based on an idealized view of how Transboundary Water Management (TWM) should be done, as well as providing a critical analysis of TWM approaches. The arguments developed should be applicable to a range of different TWM processes in various parts of the world. However, what *should* happen is not necessarily what *does* happen – the implementation of policies and strategies can pose unexpected hurdles which have to be overcome. In an effort to illustrate some of the difficulties encountered when implementing TWM initiatives the editors have collated a selection of case studies.

These cases cover a range of issues, geographic locations and types of water bodies and have been written to be short and concise. They were chosen to illustrate some of the issues raised by the chapters in the book, thus seeking to provide real-world examples of the theoretical and policy dimensions introduced. Some of the cases could be considered examples of 'good' TWM – although recognizing that in each case there is room for improvement. Several of the cases are perhaps 'less good' – certainly not bad, as every situation has its own challenges to contend with. As a whole they illustrate the complexity involved in managing water across boundaries. The approach has been not to judge the processes or cases, but rather to describe the case and focus on the key learning points each one represents. It is left to the reader to decide on the degree of success achieved!

The cases have been divided into three thematic sections, based on the key points they illustrate. These themes represent something of a progression in TWM, starting with early negotiation processes, leading into joint basin management and ultimately into collaboration over the development of projects and infrastructure. It should be emphasized that each case as it appears in the book is representative of a larger process, in which a wider range of key points could be illustrated, and which could be placed into one or more of the other thematic sections. The cases are deliberately short, thus each one is written to illustrate only a limited number of key points – not a whole range of issues. The thematic sections and cases contained in them are:

- Negotiations Process
  - 13.1 Changing nature of bargaining power in the hydropolitical relations in the Nile River Basin;
  - 13.2 The Indus Treaty: conflicting riparians and the role of a third party;
  - 13.3 The International Joint Commission: a successful treaty between Canada and the USA;
  - 13.4 La Plata River Basin Treaty;
  - 13.5 Lessons from the Johnston negotiations 1953–1955.
- Basin Management Strategy
  - 13.6 The Danube River Basin: joint responsibility for river basin management;
  - 13.7 Guaraní Aquifer: the development of a coordinated management approach;
  - 13.8 The Mekong River Commission;
  - 13.9 Stakeholder participation in the Orange-Senqu River Basin, Southern Africa;
  - 13.10 Orange-Senqu River Awareness Kit: supporting capacity development;
  - 13.11 Multi-riparian basins: the Salween River;
  - 13.12 Community-based approaches to conflict management: groundwater in the Umatilla Basin.
- Project Development and Implementation
  - 13.13 Gabčíkovo-Nagymaros Project: a test case for international water law?;
  - 13.14 Benefit sharing and the Lesotho Highlands Water Project;
  - 13.15 Cooperation over the Senegal River.

Some of the cases have been contributed by chapter authors and several of them are based on work first developed by Aaron Wolf and Joshua Newton, but all have been edited by Virginia Hooper and Michael McWilliams into a standardized format.

13.1 Changing nature of bargaining power in the hydropolitical relations in the Nile River Basin

Ana Cascão and Mark Zeitoon

Key points

- Bargaining power is a key element of any counter-hegemonic strategy, and effective if used collectively. Negotiating positions can be improved by leveraging bargaining power, as is the case of the upstream riparian states in the Nile River Basin.
- Not all power is equal. Increases in bargaining power for upstream riparians relative to Egypt have come without any shift (yet) in the other pillars of hydro-hegemony – ideational and material power.

Introduction

This study applies some of the critical hydropolitics and hegemony theory of Chapter 3 to the upstream–downstream dynamic relations over the Nile River. We showed there how recent political developments in the Nile Basin may reflect significant changes in the balance of bargaining power among the riparian states. There is evidence that over the last decade the upstream riparians have made increasing use of bargaining tools to influence negotiations at the Nile Basin Initiative (NBI).[1] Our focus is on the most recent period, that is the ongoing negotiations between the nine Nile riparian states for a cooperative framework agreement (CFA).[2]

Theoretical context

Power plays a significant role in influencing transboundary water relations and allocative outcomes, and must therefore be incorporated into any analysis. In this approach, hydropolitics are also considered to be characterized by hegemonic configurations, wherein the most powerful riparian states have an advantage over their riparian neighbours to influence the allocation of the resources. Notably, the power available to the 'basin hegemon' assumes different forms – material, bargaining and ideational.

Bargaining power refers to the capability of actors to control the rules of the game and set agendas, in the sense of their ability to define the political parameters of an agenda (Bachrat and Baratz, 1962; Lukes, 2005). Importantly, however, bargaining power is not the exclusive possession of the hegemon. It is bargaining power that makes the weaker actors in a given basin not as weak as they may be perceived (Daoudy, 2005). By leveraging bargaining power, the non-hegemons can in theory improve their negotiating position vis-à-vis the hegemon(s), counterbalance their weaknesses in other fields of power, and eventually contribute to change the hydropolitical configuration. As such, bargaining power is a key element of any counter-hegemonic strategy (Cascão, 2008a).

A history of asymmetric bargaining power in the Nile Basin

Whether in terms of material, ideational or bargaining power, Egypt has traditionally been the strongest of the ten Nile riparian states. The construction of the Aswan High Dam in 1971, which provided a large over-year storage capacity, significantly strengthened Egypt's control over water resources. Subsequent Egyptian governments have since succeeded in establishing the parameters of the regional agenda and negotiations according to their own interests.

For the last five decades Egypt has benefited from steady entitlement to more than three-quarters of the river's flow. This has been possible because the allocation was agreed to by Egyptian and Sudanese governments in the 1959 bilateral Nile Water Agreement, which did not include any entitlement for the upstream riparians. Coupled with the physical command the dam provides, the 1959 Agreement has reinforced control over the flows by providing an inflexible starting point upon which future negotiations may be based. Indeed, Egyptian control of negotiating platforms has been almost as extensive as its control of the flows.

Egypt has demonstrated the stronger capacity to influence the negotiations by imposing the terms of bilateral agreements (e.g. 1929 and 1959 Agreements with Sudan), as well as strictly refusing to negotiate with the upstream riparians, at least until the 1980s. Since the mid-1990s, Egypt has agreed to negotiations for a multilateral and all-inclusive cooperative framework agreement – but on its own terms. The intent of such positioning seems by all accounts to be perpetuated through Egyptian attempts to continue to impose 'red lines' on, or at least to stall, the negotiations. Thus far, the *status quo* continues to reflect the power relations in the basin in favour of the basin hegemon.

In terms of bargaining power, the upstream Nile riparians have historically been weaker than their downstream neighbours. The disadvantage is attributed to a number of factors, including a lack of internal capacity to establish coherent water policies, discourse and agendas, as well as an absence of coherent water negotiations strategy. In any case, the upstream riparians have often been weak in projecting power in the regional or basin arena, and in promoting their interest onto the hydropolitical agenda. Up to the 1990s, the upstream bargaining strategies were mainly made up of scattered nationalist-type discourses about water rights, claims of territorial sovereignty (the 'Harmon Doctrine') and reactive diplomacy. The upstream states have not been able to elaborate collective and/or proactive strategies to contest the downstream quasi-monopoly of the Nile waters, or to bring them to the negotiating table and possibly reach a new agreement on the Nile. The

situation began to change, however, around the mid-1990s.

Changing bargaining power – cooperation and negotiations

The tipping point in the evolution of the bargaining power relations in the Nile Basin can be traced back to the beginning of the 1990s, when the riparian states and third parties put forward the idea of establishing an all-inclusive cooperative institution in the basin – the NBI. For this to succeed, it would have had to be through full membership of all riparian states. This included Ethiopia, which along with other upstream riparians had consistently refused previous cooperative attempts on the grounds that they did not address the problematic legal issues (Tamrat, 1995; Collins, 2000).

Aware of the relevance of Ethiopia's participation to the success of the NBI, and perhaps equally aware of potential funding for hydraulic projects in the Blue Nile Basin that may derive from it, the government of Ethiopia decided to engage. It leveraged its bargaining power to begin to influence the rules of the game (Lemma, 2001; Arsano, 2004). Ethiopia's participation in the NBI was predicated upon the single condition that negotiations for a new and multilateral legal framework were to be held in parallel to the NBI (Amare, 1997). One immediate outcome of this bargaining strategy was that all riparians, including Egypt and Sudan, agreed with the Ethiopian conditionality and negotiations for a new legal and institutional agreement – the Cooperative Framework Agreement (CFA). This was initiated in 1997.

The more enduring outcomes of the Ethiopian bargaining strategy are twofold. First, for perhaps the first time in Nile history all riparians began negotiations on a more equal footing. According to interviews with several negotiators, the legal and water experts from upstream and downstream riparians had, working together, been able over the last decade to deliberate the details of the specific Nile agreement, propose alternative wordings and promote trade-offs. This is considered as a considerable achievement by the upstream negotiators, particularly in comparison with previous decades

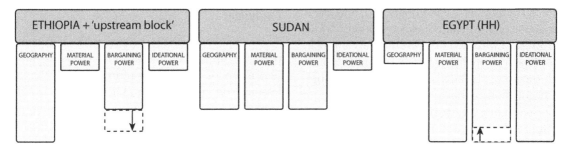

Figure 13.1 Redistribution of (bargaining) power in the Nile River Basin

when downstream riparians would not even consider talks for a multilateral agreement.

Second, Ethiopia and the other upstream riparians had been able to put the legal issues back in the centre of the Basin's hydropolitical agenda and to consistently highlight the urgent need for an all-inclusive negotiated water agreement (Kasimbazi, 2000; Arsano and Tamrat, 2005). The discussions paved the way for the first multilateral water agreement in the Basin and – possibly – for the establishment of a Nile River Basin commission (NBI, 2007).

Were the CFA to be adopted through due process, the Nile Basin states will have the merit of a permanent, multilateral and sustainable cooperative river basin organization (the Nile Basin Commission). The adoption of a new legal agreement means that the upstream riparians will achieve what until recently was not available to them: access to international funding, development of infrastructure projects within their national territories, and socioeconomic benefits derived from these projects. It remains to be seen, of course, whether the CFA will be agreed to, and what role Ethiopian bargaining power plays in determining the terms or indeed successful conclusion of the agreement.

The influence of collective bargaining power

Analysis of the evolution of the negotiations, contents and details of the CFA provides a good example of how bargaining power has been used by both upstream and downstream riparians. Here, evidence of the influence of the bargaining power of upstream states is provided through CFA negotiation events of 2007 and 2009.

By June 2007, the Nile riparians had all but concluded negotiations on a final draft of the CFA with the exception of one article: Article 14b.[3] This article is about 'water security' but indirectly implies the once taboo and excluded subject of reallocation of Nile waters and the 1959 Sudan–Egypt Water Agreement. Allegedly the wording of the article was ambiguous enough to satisfy both upstream and downstream parties, and to move towards the signature and ratification of the agreement. 'Water security' appeared as a pragmatic tool to promote (at least apparent) consensus among the parties (see Cascão 2008b). The upstream riparians voted in favour (as a block), and were ready to adopt the agreement that would lead to the Nile Basin Commission. The Egyptian and Sudanese sides had reservations about the formulation of the article and proposed an alternative.[4] The upstream riparians interpreted the changed wording as a way to legitimize the past agreement instead of moving away from it, and rejected it as a whole (see *The Standard*, 29 June 2007; *The New Vision*, 30 July 2007).

The attempt by Egypt and Sudan to maintain their positions on the supremacy of the 1959 Water Agreement should not come as a surprise. What is of greater interest in this case, is the possible erosion of the basin hegemon's ability to generate consensus to its preferred outcome. Taken from the hydro-hegemony perspective of Chapter 3, the platform created by the CFA negotiations enabled the upstream riparians to substantially increase their collective bargaining power *vis-à-vis* their downstream neighbours. The effects of this leveraging of the 'weaker' riparians' position became all the more clear when they voted as a block in favour of the agreement as originally drafted. The clear contestation of Egypt and Sudan's position is by far the strongest expression of collective bargaining in recent Nile hydropolitics. There are limits to how much the power balance can be pushed through CFA negotiations alone, however. In the absence of consensus among all riparian states, the technical and ministerial negotiations were closed in June 2007, and the process was referred to the Heads of State of the Nile riparians to resolve the outstanding article. The process evolved into a political deadlock (*The New Vision*, 9 November 2008). By mid-2009, the Heads of States had not reached any deliberation on Article 14b, nor even met to discuss the issue at the highest political level. This can be considered as an 'active stalling' strategy (term accredited to Anton Earle) employed by the downstream riparians, which consists of deliberate attempts to delay the end of the negotiations as much as possible in order to preserve the *status quo*.[5]

This prolonged deadlock prompted the upstream riparians to once again resort to collective bargaining power. In May 2009, the 'upstream block' decided to put an end to the more powerful side's 'stalling strategy' by proceeding with ratification of the CFA, despite the lack of agreement to this by the two downstream riparians (Nile-COM, 2009; see also *The New Times*, 24 May 2009). According to the terms of the CFA, the ratifications of a two-thirds majority would be enough to establish the Nile Basin Commission. The action provoked some consternation in the downstream riparians and the donor community (cf. *Al-Ahram Weekly*, 28 May 2009; *The East African*, 10 August 2009).

This unique decision in the hydropolitical history of the Nile Basin reveals the new vocal attitude and determination of the upstream riparians to move to the next step of the cooperation process. These riparians are increasingly and systematically making use of their bargaining power, namely collective, to impose changes in the rules of the hydropolitical game and discourse in the Nile Basin. As Figure 13.1 exemplifies, the current situation is contributing to a redistribution of power between riparians, as upstream riparians are increasing bargaining power at the expense of downstream neighbours.

Conclusions

The extent to which an increase in the bargaining power of upstream riparians may erode Egyptian hegemony on the Nile reinforces one of the key lessons contributed by water managers to the analytical approach of hydro-hegemony: that 'not all power is equal' (see Chapter 3). The increase in bargaining power has come without any shift (yet) in the other pillars of hydro-hegemony – ideational and material power. The relatively level playing field offered by the CFA negotiations reinforce the suggested revision that 'strong asymmetries in terms of access and production of data, information and knowledge weaken the non-hegemonic riparians in their ability to influence the political agenda or to bargaining at the negotiations table' (see Chapter 3).

But we may speculate that well-executed strategies employing bargaining power may yet prove to be a form of 'tipping point' in Nile hydropolitical relations. By increasing their collective bargaining power, the upstream riparians may be on the right path to decrease the asymmetries in the other dimensions of power and to begin to level the power imbalance with the downstream riparians. The will to leverage bargaining power may indeed be the driver behind the upstream block's insistence in the adoption of the CFA and the establishment of a permanent river basin organization.

The particular question to draw from this case is – *how might adoption of the Cooperative Framework Agreement contribute to changing the balance of power in the Nile Basin?* If the strategy proves successful, and the CFA is adopted, the Nile Basin Commission will be established in place of the transitional cooperative mechanism (the NBI). With a permanent Commission in place, the Nile riparians would in theory be able to benefit from donor support for the construction of several projects currently under preparation by the NBI, including infrastructure both for hydropower and agriculture purposes. The majority of these projects are expected to bring a number of economic benefits to the riparians, in particular to the upstreamers, who have largely been unable to leverage such 'development'. The process may in turn contribute to increase substantially their material power. In terms of ideational power, although the asymmetries between upstream and downstream riparians are very wide and not expected to be overcome so soon, it is anticipated that such an institutionalized type of cooperation, as already exemplified by the NBI, can contribute to decreasing the existing 'strong asymmetries in terms of access and production of data, information and knowledge', while generating sound, independent and legitimated knowledge about the Nile water resources and its optimal utilization across the Basin. This 'new' knowledge can eventually increase the ability of the upstream riparians to shape perceptions of their downstream neighbours and the international water community. The jury is still out, however, on whether such increases in material and ideational power driven by the shift in bargaining power will be able to overcome the asymmetries in wealth and political allies.

Notes

1   The NBI was officially established in 1999, but the process for its establishment started at the beginning of the 1990s. In 1993, an informal mechanism for riparian dialogue and involving third parties was initiated. In 1995, the Nile River Basin Action Plan was prepared by the Nile riparian states and presented to the donors' community. It was then agreed that cooperation on the Nile would follow two parallel tracks: 1) negotiations for a legal and institutional framework (CFA); and 2) multilateral programmes and projects (NBI). The CFA negotiations were launched in 1997. In 1999, the Nile Basin Initiative was formally established. The two tracks had been run in parallel until the start of 2010.

2   Eritrea only plays an observer role at the NBI. The nine countries included in the NBI include Egypt, Sudan, Ethiopia, Uganda, Kenya, Tanzania, Burundi, Democratic Republic of Congo and Rwanda.

3   Article 14b states that 'the Nile Basin States therefore agree, in a spirit of cooperation, to work together to ensure that all states achieve and sustain water security and *not to significantly affect the water security* of any other Nile Basin State' (*East African Business Week*, 20 August 2007; emphasis added).

4   The alternative wording would be less ambiguous in terms of protecting the earlier agreements: '... in a spirit of cooperation, to work together to ensure that all states achieve and sustain water security and *not to adversely affect the water security and current uses and rights* of any other Nile Basin State' (emphasis added).

5   The term has been applied first to transboundary waters by Anton Earle, following discussion of the Nile at a roundtable discussion hosted by the University of Tokyo's Hydropolitics Research Group, February 2006.

References

Al-Ahram Weekly (2009) 'A source of disagreement', *Al-Ahram Weekly*, Egypt, 28 May–3 June 2009

Amare, G. (1997) 'The imperative need for negotiations on the utilization of the Nile waters to avert potential crises', *Proceedings of the 5th Nile 2002 Conferences*, Ethiopian Ministry of Water Resources, Addis Ababa, Ethiopia, 24–28 February 1997, pp287–297

Arsano, Y. (2004) 'Ethiopia and the Nile: dilemma of national and regional hydro-politics', PhD thesis, Center for Security Studies, Swiss Federal Institute of Technology, Zurich

Arsano, Y. and Tamrat, I. (2005) 'Ethiopia and the Eastern Nile Basin', *Aquatic Sciences*, vol 67, no 1, pp15–27

Bachrat, P. and Baratz, M. S. (1962) 'The two faces of power', *American Political Science Review*, vol 56, pp941–952

Cascão, A. E. (2008a) 'Ethiopia – challenges to Egyptian hegemony in the Nile Basin', *Water Policy*, vol 10 (S2), pp13–28

Cascão, A. E. (2008b) 'Ambiguity as strategy in transboundary river negotiations: the case of the Nile River Basin', *Proceedings of II Nile Basin Development Forum*, Nile Basin Initiative/Sudanese Ministry of Irrigation and Water Resources, Khartoum, Sudan, 17–19 November 2008, pp569–579

Collins, R. O. (2000) 'In search of the Nile waters, 1900–2000', in H. Erlich and I. Gershoni, (eds) *The Nile: Histories, Cultures, Myths*, Lynne Rienner Publishers, Boulder, CO, pp245–267

Daoudy, M. (2005) *Turkey and the Region: Testing the Links Between Power Assymetry and Hydro-Hegemony*, Presentation given at First Workshop on Hydro-Hegemony, 21/22 May 2005, King's College London, London, UK, London Water Research Group

The East African (2009) 'Donors back Egypt, Sudan on Nile water pact', *The East African*, 10 August 2009

East African Business Week (2007) 'River Nile agreements – no change for poorer downstream countries', *East African Business Week* (Uganda), 20 August 2007

Kasimbazi, E. (2000) 'Challenges of negotiating an agreement on a transboundary river basin: lessons for the Nile Basin', in *Proceedings of the 8th Nile 2002 Conferences*, Ethiopian Ministry of Water Resources, Addis Ababa, Ethiopia, 26–29 June 2000, pp557–563

Lemma, S. (2001) 'Cooperation on the Nile is not a zero-sum game', *United Nations Chronicle*, vol 38, no 3, www.un.org/Pubs/chronicle/2001/issue3/0103p65.html, accessed 15 May 2010

Lukes, S. (2005 [1974]) *Power: A Radical View*, 2nd edition, Palgrave MacMillan, Hampshire, UK

NBI (2007) 'Ministers agree a cooperative framework for the Nile Basin', 26 June 2007, www.nilebasin.org/index.php?option=com_content&task=view&id=50&Itemid=84, accessed 10 July 2007

The New Times (2009) 'Nile pact finally adopted', *The New Times*, 24 May 2009

The New Vision (2007) 'Country blocked from tapping Nile waters', *The New Vision* (Uganda), 30 July 2007

The New Vision (2008) 'River Nile treaty talks hit deadlock', *The New Vision* (Uganda), 9 November 2008

Nile-COM (2009) *Minutes of Nile Basin Initiative Extraordinary Nile Council of Ministers' Meeting May 22, 2009*, Nile Basin Initiative, Kinshasa, Democratic Republic of Congo

The Standard (2007) 'Egypt, Sudan against equitable sharing of Nile water', *The Standard* (Kenya), 29 June 2007

Tamrat, I. (1995) 'Constraints and opportunities for basin-wide cooperation in the Nile: a legal perspective', in J. A. Allan and C. Mallat (eds) *Water in the Middle East: Legal, Political and Commercial Implications*, I. B. Tauris, London, pp177–188

13.2 The Indus Treaty: conflicting riparians and the role of a third party

Aaron T. Wolf and Joshua T. Newton

Key points

- Active participation of third parties to mediate negotiations, provide strong leadership and even raise funds from the international community can be crucial to reaching consensus.
- Technically sub-optimal but politically feasible solutions may be necessary in cases where relationships between riparians are poor and disagreements are entrenched.
- Separation of water from broader issues of contention can speed up negotiations.

Introduction

The Indus flows through China, Afghanistan, and India and drains into the Arabian Sea. At more than 3000km in length it is one of the most important rivers in South Asia as it supports substantial agriculture in India and Pakistan. Irrigation has been practised in the Indus River basin for centuries and water from its network of tributaries irrigates some of the most agriculturally significant areas of the subcontinent. The major tributaries of the Indus, of which five have their upper reaches in the Indian-administered state of Jammu and Kashmir, are fed by glacial melt and the seasonal monsoon which gives them a variable runoff regime.

The Indus Water Treaty was signed by India and Pakistan in 1960 after the World Bank intervened as a third party. It emerged from a long-standing dispute over water for irrigation and hydropower between these upstream and downstream riparians. The conflict became an international dispute after international political boundaries were redrawn on the basis of population distribution and religion following independence from the British in 1947. Little thought was given to water resources and, as a consequence, India was given control of the upper reaches of the Indus tributaries in addition to much of the irrigation infrastructure constructed by the British. In effect, irrigation canals in Pakistan became dependent on water supplies from India.

Initially, an interim agreement mandated that India would continue supplying water to Pakistan. Once the agreement lapsed, India stopped water supplies, thus forcing Pakistan to negotiate over water allocations.

Key lessons for Transboundary Water Management

The Indus Water Treaty functions in the context of wider conflict and tension between India and Pakistan. Since partition, the two countries have fought three wars and in 2002 the relationship between them deteriorated to the extent that observers predicted a fourth conflict was imminent. Perhaps the most significant source of tension is the control of Kashmir, and therefore control of the upper reaches of the rivers of the Indus plain including irrigation infrastructure built under British rule (Clemens, 2005).

The Indus Treaty sought to turn water disputes between the two countries into a functional discussion based on technical and engineering data rather politics. This was achieved through the mediation of the World Bank, which crucially facilitated negotiations between the countries as a neutral third party. Despite the efforts of the World Bank, this case is widely viewed as an example of successful negotiations which led to independent water resource management rather than an example of successful transboundary management.

The following key features of the negotiating process and resulting treaty provide a number of lessons for Transboundary Water Management (TWM) in the context of hostile riparian relationships (Alam, 2002).

The effect of shifting political boundaries

Conflict over water allocation for irrigation had been a feature of the Indus Basin long before the partition of India and Pakistan in 1947. Prior to independence this was an inter-province dispute between Sindh and Punjab. Before a resolution to the dispute could be found, India and Pakistan gained independence and the changing political boundaries resulted in an international disagreement over water.

Speeding up negotiations

After unsuccessful negotiations between India and Pakistan, the World Bank entered into discussions as a neutral third party facilitator, and this enabled the negotiation process to move forward. An example of the role played by the Bank was the breaking of a stalemate when the two sides could not agree on a plan for the future development of the basin. The Bank suggested that each country should submit its own plan which could then be compared with the other to identify common ground.

The following factors also contributed to successful negotiations. First, although critical discussions occurred at a political level, protracted negotiations took place between senior engineers from both sides (Salman and Uprety, 2002). Given the technical complexity of the scheme, this allowed negotiations to move at a faster place. Second, sensitivity towards the particular hydrologic concerns of each party aided the process, and finally the explicit understanding that the process would not set precedents for other areas accelerated the decision process. For example, in an early meeting (May 1952), both sides agreed that any data may be used without committing either side to its 'relevance or materiality', thereby precluding delays to negotiations over data discrepancies.

Sub-optimal solutions

The Indus Water Treaty highlights the need for a pragmatic approach and recognition that a sub-optimal solution may be the only achievable solution in certain circumstances. In the case of the Indus, initial suggestions envisaged cooperative management and joint development of water resources in an integrated fashion. However, issues of political sovereignty were not compatible with this approach (Salman and Uprety, 2002). The World Bank quickly recognized that dividing the water resources to allow each state to control its own share was the best possible solution given the wider political circumstances. The final solution required that the river basin be split, with the eastern tributaries (the Ravi, Beas and Sutlej) being exclusively controlled by India, and the western tributaries (the Indus, Jhelum and Chenab) to be controlled by Pakistan.

Disregarding the principle of integrated water management and recognizing that the most important issue is control by each state of its own resource, precludes the benefits that can be derived from co-management of water resources. Nevertheless there are areas of the world, including parts of the Middle East, where such an approach, designed to allow some level of mutually acceptable agreement, may be successfully implemented.

The role of funding for river basin development

The solution of dividing the water resources of the river basin needed considerable financial investment in order to build storage and interlinking infrastructure. However, the cost of this infrastructure could not be borne by either India or Pakistan. The crucial role of the Bank, in this case, was in raising the finances to support development of the storage and physical interlinkages. By doing so, the bank neutralized and eliminated a significant obstacle for ratification of the treaty.

Dispute resolution mechanisms

The treaty contains inbuilt dispute resolution mechanisms such as the appointment of a neutral expert if the two sides failed to reach agreement. If the neutral expert fail to resolve the dispute, the mechanisms call for the appointment of negotiators and convening of a Court of Arbitration. The treaty also calls for both sides to notify the other of any further plans to undertake engineering works on any of the tributaries.

Power inequalities as a disincentive for agreement

India was clearly the dominant and more powerful party to this agreement and was also the upstream riparian. This significant power asymmetry undoubtedly contributed towards distrust from Pakistan and may have acted as a disincentive for agreement.

Conclusions

The Indus Water Treaty allowed India and Pakistan to control their own water resources within the context of wider conflict between the two states. Political sovereignty over water resources prevented joint development and integrated management of the basin but, through the mediation of a third party, a successful water-sharing agreement was negotiated.

Adapted from Aaron T. Wolf and Joshua T. Newton (2009) 'Case studies of transboundary dispute resolution', Appendix C in J. Delli Priscoli and A. T. Wolf (eds) *Managing and Transforming Water Conflicts*, Cambridge University Press, Cambridge

References

Alam, U. (2002) 'Questioning the water wars rational: a case study of the Indus Waters Treaty', *The Geographical Journal*, vol 168, no 4, pp354–364

Clemens, J. (2005) 'Peace through transboundary water management? The Indus Waters Treaty between India and Pakistan', http://rural21typo3.medianet.de/uploads/media/ELR_engl_60-63.pdf, accessed 9 November 2009

Salman, S. M. A. and Uprety, K. (2002) *Conflict and Cooperation on South Asia's International Rivers: A Legal Perspective*, World Bank Publications, Washington, DC

13.3 The International Joint Commission: a successful treaty between Canada and the USA

Aaron T. Wolf and Joshua T. Newton

Key points

- General guiding principles allow the International Joint Commission to adapt flexibly to an evolving agenda resulting from new environmental concerns, new information and new situations.
- Independent commissioners are able to seek the best long-term solutions in the interests of both countries.
- Joint fact finding and research provides a platform for consensus building.

Introduction

The International Joint Commission (IJC) was established in 1909 by the Boundary Waters Treaty between the United States and Canada. The treaty aims to prevent and resolve disputes over the extensive transboundary waters shared by the two countries along their approximately 9000km border (IJC, 2008). These water bodies include the Great Lakes–St Lawrence System which contains one-fifth of the world's freshwater.

This shared resource is essential for economic development and historically has been beset by heavy pollution from industry and agriculture. As a result, water quality concerns are central to the work of the IJC and considered important by both sides. The two countries made a commitment not to pollute the waters to an extent that would cause injury to health or property in the other country (IJC, 2008). Further concerns are the management of water for competing sectors such as hydropower, navigation and the environment. Over the past 100 years, over 100 potential disputes have been resolved by the IJC. The success of this institution and the Boundary Waters Treaty is attributed to its clarity and the flexibility provided by the implementation of general guiding principles rather than a detailed prescriptive approach.

The IJC is made up of six independent members or commissioners (three appointed by the United Stated and three appointed by Canada) and numerous advisory boards. In addition to working

in the best long-term interests of both countries, the IJC studies and recommends solutions to transboundary water problems referred to them by the governments of either country and works on approving and overseeing dam projects. Since 1909, the treaty has been supplemented by further agreements such as the 1972 'Great Lakes Water Quality Agreement' and more recently, the 1991 'Agreement of Air Quality.'

Key lessons for transboundary management

Through the IJC, transboundary waters between the US and Canada are managed effectively. However, there remain areas of criticism, most notably regarding the limited authority of the treaty. Here we will examine the factors contributing both to the success of the IJC and where lessons can be learned.

Sovereignty and the limited authority of the IJC

A significant criticism of the treaty lies with its limited authority. As a result of concerns about sovereignty over water resources, the remit of the treaty does not extend to river basins. This means that tributaries are excluded from consideration by the IJC. Observers note that this exception prevents the IJC from adopting an ecosystems approach to water management in the shared basins as advocated in the 1987 Protocol (which resulted from the 1972 'Great Lakes Water

Quality Agreement'). This Protocol expanded the Commission's authorities and activities with respect to water quality, and specifically called for a review of remedial action plans, the development of lake-wide management plans and, as previously mentioned, the implementation of an ecosystems approach.

A further criticism is that the IJC can only act on federal referrals and not those from Canadian provinces, US states, or municipalities on either side of the border. This effectively limits the ability of the IJC to act on known problems until a federal referral is made. As a result, federal governments may be tempted to resolve problems politically rather than referring cases to the IJC, thereby limiting the authority of this institution (Institute for United States Policy Research, 2007).

Public engagement and participation

Since the inception of the 1909 Boundary Waters Treaty, Canada and the United States, and all stakeholders within the Great Lakes Basin, have worked together and have been brought together as a community as a result of the commitment in preserving the shared waters of the two countries.

Specifically, as required by Article 12 of the treaty, the IJC takes proactive steps to engage all stakeholders including the public (Clamens, 2005). Public engagement has been shown to raise awareness and is also used successfully to gauge public reaction regarding proposals made by the Commission.

Joint research

In common with many of the case studies provided in this volume, joint research and data collection has been shown to enhance collaborative working and trust. In this case, joint fact finding undertaken by the various boards appointed by the IJC, and therefore by both signatories, provides a basis for consensus and an integrated approach. This is an important feature of IJC work and is the basis for dispute resolution and a balanced focus on the best interests of both parties (Clamens, 2005).

Independence

The IJC acts to achieve the best long-term interests of the US and Canada regarding their transboundary waters. The members appointed to the Commission are independent of instruction or management by their respective governments. This independence is thought to contribute to the effectiveness of the IJC.

Conclusions

Despite criticisms resulting from its limited formal authority, the Boundary Waters Treaty has, through the work of the IJC, peacefully resolved over 100 transboundary water issues and provided early warnings about potential future problems. It has also evolved to meet the changing agenda of both countries over the decades in response to new environmental pressures. It remains a good example of an institution designed to manage transboundary water for the benefit of both parties.

Adapted from Aaron T. Wolf and Joshua T. Newton (2009) 'Case studies of transboundary dispute resolution', Appendix C in J. Delli Priscoli and A. T. Wolf (eds) *Managing and Transforming Water Conflicts*, Cambridge University Press, Cambridge

References

Clamens, M. (2005) 'The international joint commission', *VertigO – la revue électronique en sciences de l'environnement*, hors-série 2, septembre 2005, http://vertigo.revues.org/index1885.html, accessed 10 November 2009

Institute for United States Policy Research (2007) *Transboundary Water Policy Issues: The Western North American Region*, Conference Report, Institute for United States Policy Research, University of Calgary, www.nnasc-renac.ca/documents/ TransboundaryWaterPolicyReport.doc, accessed 10 November 2009

IJC (2008) *Annual Report for 2008, Boundary Waters Treaty Centennial Edition*, www.ijc.org/php/publications/ pdf/ID1629.pdf, accessed 9 November 2009

13.4 La Plata River Basin Treaty

Aaron T. Wolf and Joshua T. Newton

Key points

- The lack of a supra-legal body to manage treaty provisions results in delays to project implementation, because for each project there is a requirement to go through each individual country's legal processes.
- The proliferation of sub-treaties and associated institutions can hinder the achievement of a coherent overall vision for a river basin.
- The prospects for joint management are better if cooperation is sought before conflicts arise.

Introduction

The La Plata River Basin drains an area of 3 million km² taking in parts of Argentina, Bolivia, Brazil, Paraguay and Uruguay. It is one of the world's largest transboundary river basins, covering over 15 per cent of the land area in South America (OAS, 2005). The major waterways in the basin are important trading routes.

The five riparian states have historically cooperated over water, and in 1969 signed the La Plata River Basin Treaty to provide a broad framework for joint development and integration of river basin management. Subsequent multilateral and bilateral treaties outline the specifics of economic investment, hydroelectric development and transportation enhancement. The main objectives are to develop the river basin to ensure sustained growth, to exploit the potential for hydropower and to improve navigation. Pollution control and mitigation are also noteworthy concerns, whereas the availability of water is a secondary issue given the relative abundance in the catchment.

In order to achieve these objectives, basin states agree to identify and prioritize cooperative projects, and to provide the technical and legal structure to see to their implementation. This has resulted in 130 dams along the Paraná, the construction of the world's largest hydroelectric project, Itaipu, and successive development, infrastructure and transportation projects.

The Itaipu Project was particularly significant due to the contested rights to the watercourse at the location of the dam. Both Brazil and Paraguay had claims to this region and wanted to develop hydropower. The outcome of negotiations led to cooperation and the joint development of the hydropower facility with both countries deriving electrical power from the project.

Another significant project in the basin is the proposed 'Hydrovia' waterway. This will be the largest scheme to be developed by the Comité Intergubernamental Coordinador de los Países de la Cuenca del Plata (CIC, the Intergovernmental Coordinating Committee of the La Plata Basin) and will involve all five countries. Many feel that this will be the most important test of the institutional capability of CIC and the La Plata Treaty. This transportation and navigation project will involve dredging and straightening sections of the Paraná and Paraguay Rivers. It will allow year-round barge access from ports in Argentina and Uruguay through to Bolivia and Paraguay. Controversially, some of the work will be undertaken in the Pantanal wetlands, an area of rich biodiversity. Environmentalists have serious concerns about the impact of such a project on the ecosystem and functioning of the Pantanal.

More recently, a dispute has arisen between Uruguay and Argentina over the construction and operation of paper mills that Argentina claims are polluting the River Uruguay. This shared watercourse acts as the border between the two countries and is the subject of the 1975 Treaty of the River Uruguay. Argentina has taken the case to the International Court of Justice and a ruling is expected in 2010. This bilateral dispute between two signatories of the La Plata Treaty demonstrates the limited ability of the CIC to resolve issues arising between national governments.

Key lessons for transboundary water managers

The river basin organization derived from the La Plata Treaty comprises the Conference of Foreign Affairs Ministers, the CIC and the Secretariat. The CIC is the main tool for delivering the objectives of the treaty which requires open transportation and communication along the river and its tributaries, and prescribes cooperation in education, health and management of 'non-water' resources (e.g. soil, forest, flora and fauna). International organizations, such as the Global Environment Facility (GEF), have also supported the effort towards integrated management by providing financial and technical assistance.

The La Plata River Basin Treaty is generally regarded as successful, particularly with respect to providing an overarching framework for joint development of the basin and cooperative work between states. Cooperation between states commenced before major projects were planned and this has effectively alleviated the potential for future conflict over water in the La Plata Basin. The treaty has also been praised for its ability to resolve inter-sectoral disputes. Nevertheless some aspects of the treaty and subsequent water management strategies have drawn criticism. The fate of the Hydrovia project will be a significant test and will determine the extent to which basin states can effectively work together on a large project which will bring not only benefits but also potential environmental degradation.

Scale and fragmentation

The sheer size of the area covered by the river basin creates complexity for joint water management. It has an impact both on the efficacy of management decisions but also for coordination of a 'vision for basin development'. This is compounded by the non-exclusive nature of the treaty which has given rise to a host of bilateral and multilateral supplemental treaties and agreements. More than 20 agreements or contracts between two or more of the countries within the basin have been signed, in addition to the La Plata Treaty. While this proliferation of agreements allows many different perspectives, the high level of fragmentation hinders the adoption of a coherent strategy. Indeed the numerous institutions created by these agreements frequently do not communicate with each other directly or through the CIC (GEF, 2007).

Some, for example Pochat (2008), have suggested that management at a sub-basin scale and over a more manageable area may provide better results. Whether this would contribute to further fragmentation of the coherent basin vision remains to be seen.

Lack of supra-legal power

There has been some concern over the inherent institutional weakness and the lack of a supra-legal power within the La Plata Treaty. This weakness often allows lengthy delays in both the decision-making processes and the implementation of agreements for projects in the basin. In the absence of supra-legal powers, decisions taken by the CIC have to be passed through the national legal system of the respective basin states before action can be taken.

Role of public participation

Despite criticisms of its top-down approach, there is a role within the framework of the La Plata Basin Treaty for civil society organizations and participation (GEF, 2007). Given the lack of supra-legal power and the variations in the national legal structures, the process of participation is not the same for each country. In order for the CIC to implement integrated water resource management (IWRM), the participation of all stakeholders should be ensured. To meet this objective, the countries have agreed to undertakings such as the Digital Map Project to increase external participation in an accessible manner. This project is hosted on the CIC website (www.cicplata.org/?id=md&s=b7dc0d720b2766 da8244555085744c0f, accessed 15 May 2010).

Conclusion

The La Plata Basin Treaty is a successful over-arching agreement for the river basin and effective cooperation between its signatories. The numerous dams that have been developed in the basin are a testament to this cooperation. Some criticisms of the treaty and the institutional arrangements point to the proliferation of individual and separate river basin institutions spawned by the numerous treaties that supplement the La Plata agreement. Another criticism is the lack of supra-legal power and the impact this has on the efficiency of the decision-making process. A crucial test for the CIC will be the management of the Hydrovia project.

Adapted from Aaron T. Wolf and Joshua T. Newton (2009) 'Case studies of transboundary dispute resolution', Appendix C in J. Delli Priscoli and A. T. Wolf (eds) *Managing and Transforming Water Conflicts*, Cambridge University Press, Cambridge

References

GEF (2007) *Annex D. SAP for the La Plata Basin. Public Participation Plan (PPP)*, www.gefweb.org/.../05-09-07%20Revised-Plata%20FP%20-%20Final%20Compendium%20of%20Annexes.pdf, accessed 10 January 2010

OAS (Organization of American States) (2005) *La Plata River Basin*, Water Project Series, no 6, www.oas.org/dsd/Events/english/Documents/OSDE_6LaPlata.pdf, accessed 14 November 2009

Pochat, V. (2008) 'How to establish coordination mechanisms among stakeholders (La Plata River Basin)', presentation at *World Water Week, 2008*, www.worldwaterweek.org/documents/WWW_PDF/2008/thursday/K11/IRBM/5_Mr_V_Pochat_IBRM_seminar.pdf, accessed 14 November 2009

13.5 Lessons from the Johnston negotiations 1953–1955

Aaron T. Wolf and Joshua T. Newton

Key points

- The exclusion of the wider political context from water negotiations can preclude consensus between hostile riparians.
- The influence (both positive and negative) of regional powers outside a river basin should be recognized during the negotiating process.
- The process of negotiation can lead to informal adherence to a plan even in the absence of agreement.

Introduction

The Johnston negotiations of the early 1950s were an attempt to allocate water between the five riparians of the Jordan River Basin. While the hydrological and technical aspects of the plan were considered to be practicable (Phillips et al, 2007), the wider political environment was a barrier to the ratification of the treaty. This case study highlights the convergence of water and politics and shows that allocation of water resources cannot be undertaken in isolation. While the Johnston Plan was never ratified, there have been subsequent regular meetings between Jordan and Israel, known as the 'Picnic Table' talks, to discuss the allocation of the waters of the Yarmouk tributary.

Context

The Jordan River is shared by Israel, Palestine (the West Bank), Lebanon, Syria and Jordan. It rises in a mountainous area on the borders of Israel, Lebanon and Syria, and flows through an increasingly arid landscape before discharging into the Dead Sea. One of its most significant tributaries, the Yarmouk, originates in Syria and Jordan. These water resources are essential for riparian states given the levels of water scarcity in the region and the dry climate. In addition to the Jordan River, there are also strategically important groundwater reserves that supplement surface water supplies in the catchment.

The states sharing the Jordan River were formerly part of the Ottoman Empire. Its fragmentation and the redrawing of state boundaries that ensued, coupled with the emergence of the new Jewish State of Israel, fuelled international disputes and tension which included the subject of water allocation. In order to promote peace and to mitigate against the proliferation of unilateral water resource development plans, the United States sent Eric Johnston as an ambassador to the region in 1953. His aim was to devise a water allocation plan to share the water equitably between riparian states. This was not the first attempt at dividing the water between the riparian states, and it came to be known as the Unified Plan, because it was essentially a compromise between earlier plans.

The underlying objective was to ensure water allocation that would be accepted by all parties and bring stability to the region, while providing enough water for Israel to develop as a viable state. The plan approached water allocation on the basis of irrigable area in each country; made provision for regulatory works such as dams and canals; and included steps towards joint management of the river. Broadly, the plan allocated much of the water to Jordan and Israel (although Israel was given less than the volume it had originally requested), with less water for Syria and Lebanon.

Johnston undertook these negotiations in a series of visits to the Middle East between 1953 and 1955. Despite eventual agreement on the technical aspects of the plan, the deterioration of the political relationships in the region precluded finalization of the agreement. Syria in particular proved to be a blocker (through the Council of the Arab League) as the Syrian government was unwilling to sign an agreement which recognized the legitimacy of Israel.

Key lessons for transboundary water management

Even though Johnston's first proposal, known as the Main Plan, was rejected, the Unified Plan was more successful. After careful negotiation, its technical aspects were largely accepted by all sides, despite concerns about out-of-basin transfers for Israel. Nevertheless, the political environment and the deteriorating relations between riparians prevented ratification, regardless of the proposal's rational basis.

Separation of water from political realities

The failure of the Johnston Plan demonstrates the importance of addressing political barriers in parallel with questions of water resources. While it is true that during negotiations between hostile riparians the separation of water and politics can allow more rapid progress, for example through the collection and sharing of water data to build a platform of mutual trust, questions of politics cannot be ignored altogether. The complete separation of the technical aspects of water management from the wider political agenda is unlikely to lead to a long-standing agreement.

It has been acknowledged that despite presenting an achievable plan to the nation states, the Johnston Plan was never likely to be accepted due to political considerations. For example, Syria, Lebanon and Jordan were not likely to sign up to an agreement that implicitly recognized the State of Israel, especially in light of outstanding concerns with regard to borders and the question of Palestinian refugees moving into Jordan.

National sovereignty

The case also demonstrates the difficulties inherent in overcoming questions of national sovereignty. Where riparians have hostile relationships, sub-optimal solutions may be the only course of action. In the Jordan River Basin, the co-management of resources was not possible and the division of resources between basin states was the only viable option. For example, the question of storing winter floods in the Sea of Galilee became a contentious issue as the Sea is situated in Israel.

The influence of regional powers

The case studies in this volume have highlighted the role that the international community can play in supporting and facilitating water resource negotiations and agreements. Here, the importance of regional powers is demonstrated. The inclusion of Egypt, a regional power, in the negotiations was thought to have brought negotiations closer to fruition. The influence that Egypt was able to exert, and the leverage it brought to bear on proceedings, shows that regional parties may have more influence than more distant international allies. Equally important, the exclusion of regional powers such as Iraq and Saudi Arabia was significant as it allowed them to pressure Lebanon and Syria into refusing to ratify the agreement.

The United States also attempted to exert influence through the conditional provision of funds. It promised financial support in the event that the riparian states cooperated with each other.

Inclusion of all water resources and water quality issues

Unlike more recent water-sharing agreements, the Johnston Plan did not adequately encompass all the resources within the basin. The absence of a plan for the division of groundwater resources or questions of water quality is notable.

Implicit cooperation

The failure of the Johnston negotiations prevented the ratification of an explicit water allocation agreement, and yet some degree of implicit cooperation between Israel and Jordan has been possible, resulting in fairly high stability, if also sub-optimum water management.

Israel and Jordan have been able to meet to discuss water allocations several times a year at the confluence of the Jordan and Yarmuk Rivers, resulting in decreased tensions between the two states. These 'Picnic Table' talks have allowed a venue for some level of technical agreement, and an outlet for minor disputes, for more than 40 years.

Conclusions

The failures of the Johnston negotiations demonstrate the difficulties of achieving consensus between hostile riparians. These problems are exacerbated by political and ideological differences. The separation of water issues from the politics of borders and refugees contributed to the failure of the riparians of the Jordan River Basin to reach agreement. The decades following the Johnston negotiations witnessed armed conflict and further tension over water. This eventually resulted in peace talks in the 1990s and the establishment of the Multi-Lateral Working Group on Water Resources.

Adapted from Aaron T. Wolf and Joshua T. Newton (2009) 'Case studies of transboundary dispute resolution', Appendix C in J. Delli Priscoli and A. T. Wolf (eds) *Managing and Transforming Water Conflicts*, **Cambridge University Press, Cambridge**

References

Phillips, D. J. H., Attili, S., and Murray, J. S. (2007) 'The Jordan River Basin: 1. Clarification of the allocations in the Johnston Plan', *Water International*, vol 32, no 1, pp16–38, www.thirdworldcentre.org/phillips1.pdf, accessed 22 November 2009

13.6 The Danube River Basin: joint responsibility for river basin management

Aaron T. Wolf and Joshua T. Newton

Key points

- Participation at all levels of river basin management contributes to cost and time savings.
- Alignment of national legislation with the objectives of transboundary river basin organizations can ensure more rapid progress towards integrated river basin management.
- International river basin organizations can aid implementation of supra-national legislation such as the Water Framework Directive (WFD).

Introduction

One of the largest rivers in Europe, the Danube rises in Germany and flows east to its mouth in the Black Sea. Nineteen countries share this river basin and over 80 million people live in the catchment, making it one of the most culturally and politically diverse river basins in the world.

The international body that manages the Danube River Basin, the International Commission for the Protection of the Danube River (ICPDR), consists of Austria, Bosnia and Herzegovina, Bulgaria, Croatia, Czech Republic, Germany, Hungary, Moldova, Montenegro, Romania, Slovakia, Slovenia, Serbia, Ukraine and the European Union. It implements the 1994 Danube River Protection Convention and ensures 'the sustainable and equitable use of waters and freshwater resources within the Danube Basin' (ICPDR, 2008). The ICPDR is regarded as a positive example for transboundary water management (TWM) because of the emphasis it places on participation, the processes it adopts to oversee that the national legislation of contracting parties aligns with the objectives of the ICPDR, and its implementation of the European Water Framework Directive.

Challenges in the Danube Basin: pollution and navigation

Poor water quality and balancing environmental needs with river navigation are the most pressing concerns of the ICPDR. Improving water quality is complex because the sources of pollution are varied. Three categories of pollution are of most concern: nutrient, organic and hazardous substances. These come from point and diffuse sources as a result of agricultural runoff, industrial effluent and urbanization. The concentration of urban areas along the Danube causes particular problems: the Danube flows through or is adjacent to some 60 large cities including 4 national capitals: Vienna, Bratislava, Budapest and Belgrade.

Reducing water pollution requires coordinated strategies and monitoring to gauge progress towards targets. To accomplish this, the ICPDR has put in place a transnational monitoring network for collecting environmental information across and along the basin. This, and related strategies, have achieved some improvements in water quality, for example lower levels of ammonium have been recorded in the upper stretches of the Danube and a reduced phosphate load along the joint Slovak–Hungarian Border (ICPDR website, www.icpdr.org, 2009).

Navigation is a contentious issue as the ICPDR is required to balance the needs of the environment with economic growth and industrial activity. Proposals to increase navigation on the river are the subjects of much debate, the ultimate aim being a strategy which prevents the basin's essential navigation requirements from degrading the environment.

Key lessons for transboundary water management

Cooperation has been a feature of relations between basin states for some time and has evolved as political relationships have changed, new states have formed and as EU legislation for water policy has become more stringent. This transition from cooperation to coordinated river basin management provides a positive example for other basins around the world, particularly in highlighting the importance of participation at all levels in increasing the efficiency of decision making and the implementation of strategies. The mutual reinforcement between the Danube River Convention and the European Water Framework Directive is also of interest as it demonstrates the interaction between supra-national legislation and transboundary water management in the Danube Basin.

Participation

Participation at all levels in the work of the ICPDR has been taken seriously, and this has had two effects. The first is to foster further cooperation between riparian states and the second is a reduction in the time taken to implement ICPDR strategies.

A participatory approach was adopted from the early stages of planning and management. For example, the public was involved in the development of strategic action plans (SAPs)[1] – the first instance where public participation was expressly included in an international water management plan. This assured support from those who would be directly affected by the implementation of the plan, and therefore shortened the time taken for approval of the SAP. The degree of cooperation among representatives of participating governments, and the importance given to public participation in developing the SAP, marked signif-

icant achievements in promoting regional cooperation in water resources management. A second example is the inclusion of observer organizations in the ICPDR. This mechanism ensures transparency and, to date, 19 organizations have achieved observer status, including NGOs and representatives from the private sector.

Efforts at awareness-raising and the promotion of joint responsibility, such as the annual Danube Day, further demonstrate the ICPDR's inclusive approach. These events generate high levels of publicity and, through education and outreach, encourage an understanding of how individual actions can contribute to the health of the river.

The role of European legislation: The Water Framework Directive

The Water Framework Directive (WFD) was enacted shortly after the ICPDR came into existence. The WFD uses policy to improve water management, with a focus on achieving water quality objectives. Implementation of the WFD is now one of the highest priorities of the ICPDR, despite the fact that not all members of the ICPDR are part of the EU. Nevertheless, all contracting parties have agreed to work towards compliance with the WFD.

Countries are now working jointly on the Draft River Basin Management Plan for the Danube. This is a key requirement for the WFD and will help countries to improve water quality by the target date of 2015. More broadly, the WFD plays an important role in supporting integrated water resources management (IWRM). The requirement for member states to comply with European legislation helps to overcome national interest and provides incentives for improved river basin management within the Danube Basin.

The ICPDR and national legislation

Coordination and alignment between the objectives of a transboundary water management (TWM) organization and the national legislation of member states is aimed at achieving timely implementation of strategies and projects. As described in a number of case studies in this volume, in the absence of supra-legal powers or processes to assure that national legislation reflects the policies of international river basin organizations, progress towards coordinated management is slow. The work of the ICPDR and its contracting parties towards the alignment of national law with the ICPDR provides a good example for transboundary water managers.

Conclusions

The ICPDR is regarded as a successful model for TWM and demonstrates that a large number of countries can successfully be brought together to manage a river basin despite different national agendas. Much of this work is made possible by the levels of cooperation brought about by the emphasis the ICPDR places on participation at all stages and levels of its work. This emphasis on participation explicitly recognizes the vital link between internal politics among different sectors and political constituents within a nation on the one hand, and the strength and resilience of an agreement reached in the international realm on the other. The focus on joint responsibility and awareness-raising is also a good example for TWM.

Adapted from Aaron T. Wolf and Joshua T. Newton (2009) 'Case studies of transboundary dispute resolution', Appendix C in J. Delli Priscoli and A. T. Wolf (eds) *Managing and Transforming Water Conflicts*, Cambridge University Press, Cambridge

Notes

1   The SAP described a framework for regional action, implemented through national action plans. It contained four goals for the environment of the Danube River basin: strategic directions, including priority sectors and policies; a series of targets within a timeframe; and a phased programme of actions to meet these targets. These goals can only be achieved by means of integrated and sustainable management of the waters of the Danube River basin.

References

ICPDR (International Commission for the Protection of the Danube River) (2008) *15 Years of Managing the Danube Basin*, www.icpdr.org/icpdr-pages/15_years_managing_danube_basin.htm, accessed 15 November 2009

13.7 Guaraní Aquifer: the development of a coordinated management approach

Aaron T. Wolf and Joshua T. Newton

Key points

- Groundwater management should be integrated into regional water management strategies.
- Preventive action and early cooperation that pre-empts conflict can minimize the potential for future disputes.
- Knowledge and reliable data are essential for building confidence and assisting management of a hidden resource such as an aquifer.

Introduction

The Guaraní Aquifer is a vast transboundary groundwater resource shared by Brazil, Paraguay, Uruguay and Argentina. It contains approximately 45,000km^3 of water and is an important strategic supply which, due to the thermal characteristics of parts of the aquifer, is also a potential energy source. Approximately 15 million people live above the aquifer and the demand on its water resources has risen significantly in recent years due to increased urbanization and the expansion of irrigated agriculture. The aquifer is also vulnerable to pollution from agriculture, such as the intensive cultivation of sugarcane, and from untreated municipal wastewater. At present, this pollution is limited to the local scale.

There is no current conflict or dispute over the aquifer, but in the absence of an aquifer management strategy there is growing concern that despite modest levels of exploitation, excessive levels of abstraction and pollution in the future could raise problems, particularly in vulnerable areas. With this in mind the basin states, together with the Global Environment Facility (GEF), undertook the Project for the Environmental Protection and Sustainable Development of the Guaraní Aquifer. The project, which was funded jointly by the countries sharing the aquifer and GEF, ended in early 2009 having achieved progress towards developing an effective strategy for transboundary aquifer management. The

project was tasked with expanding the knowledge base of the aquifer, providing measures to combat non-point source pollution and increasing public participation. In conjunction with these tasks, pilot projects were also undertaken at potential hot spots. The work of GEF demonstrates the importance of support from the wider international community for the development of transboundary water management (TWM) agreements and implementing institutions.

There is a long-standing history of cooperation in the region, as demonstrated by the MERCOSUR trade agreements and more specifically, the La Plata Basin Treaty. This overarching agreement to cooperate over water resources was signed in 1967 and is explained in further depth in Case Study 13.4. Nevertheless, a more detailed strategy for managing the Guaraní Aquifer is required because the La Plata Basin Treaty has very limited scope with regard to groundwater and does not provide an effective framework for aquifer management. Ideally, an approach taken by the riparian states to transboundary aquifer management should be institutionalized by integrating the management framework of the Guaraní Aquifer into the wider La Plata Basin Treaty.

Key lessons for transboundary water management

The decision taken by the governments of Brazil, Argentina, Uruguay and Paraguay to develop an aquifer management strategy has resulted in new institutional and regulatory processes to oversee the use of groundwater resources. This is particularly important at the local level where there is often limited knowledge about groundwater and the practices to manage its use effectively. In the light of the recent (December 2008) United Nations General Assembly Resolution on the use and management of transboundary aquifers, the Guaraní Aquifer illustrates how the UN resolution may be implemented.

A key lesson is that early intervention pre-empts potential conflicts and minimizes the risk of future disputes. This was possible because the countries had a common interest in protecting their water resources from pollution and using water sustainably. The following section examines some of the important aspects for transboundary groundwater management which are exemplified by the Guaraní Project.

Developing a knowledge base

Groundwater management is difficult because aquifers are unlikely to behave in a predictable manner across large (or even small) areas. They vary in depth, composition and structure, and all these factors affect how the water within them behaves and flows. For example, the Guaraní Aquifer consists of two sub-basins separated by structural highs which determine the water flow regime (World Bank, 2006). This, taken together with the 'hidden' character of groundwater creates the need for a comprehensive knowledge base. Such a knowledge base will help managers better understand how water moves around the aquifer, where recharge occurs, and which areas of high vulnerability and risk need more protection.

One of the aims of the GEF project was to improve monitoring of the aquifer across national boundaries to better understand its characteristics. This shared knowledge base is crucial to build cooperation between riparians and to develop effective aquifer management strategies.

The complexity of these groundwater systems also means that effective stakeholder communication is essential. Each stakeholder group may have a different view of the condition and behaviour of the aquifer. Dissemination of knowledge in a manner that is appropriate to the target audience is likely to prevent unfounded perceptions of the state of the aquifer.

Harmonizing national legislation

The harmonization of national legislation in the countries that share the Guaraní System is a critical foundation for an effective management strategy. However, achieving such harmonization is challenging because Brazil and Argentina are federal countries where the responsibility for water is at the state or province level, whereas Uruguay and Paraguay are unitary states. These differences resulted in different institutional advances with respect to groundwater management in each country. In Argentina, the National Government created the Inter-Ministerial Committee for Groundwater, and provincial governments have now included groundwater management into their water management agendas. In Brazil, increased budgetary allocations were made to the Guaraní Aquifer System, and state governments included groundwater in their water agendas. Paraguay, which previously lacked a framework for groundwater management, found that the project provided stimulus for a new Water Law and the creation of a Guaraní Aquifer Unit in the Ministry of the Environment. Finally, in Uruguay, parliament is considering a law which contains provision for the creation of a new Guaraní Aquifer Unit (World Bank, 2009).

Local management practices replicated at an aquifer scale

Through four pilot studies, the GEF project demonstrated the need for action and intervention at the local scale as a best practice for aquifer management. However, these local actions need to be replicated across the aquifer and across international boundaries. This requires the development of local capacity and local institutions to put these strategies into effect (World Bank, 2006).

Conclusions

Through the support of the GEF, Brazil, Argentina, Uruguay and Paraguay have sought to develop a management strategy for the Guaraní Aquifer before the emergence of widespread problems over groundwater exploration and pollution. A more sustainable approach is now evolving through harmonization of national legislation, along with the creation of management tools and investment in the collection and analysis of data for this large aquifer system.

Adapted from Aaron T. Wolf and Joshua T. Newton (2009) 'Case studies of transboundary dispute resolution', Appendix C in J. Delli Priscoli and A. T. Wolf (eds) *Managing and Transforming Water Conflicts*, Cambridge University Press, Cambridge

References

The World Bank (2006) *The Guaraní Aquifer Initiative for Transboundary Groundwater Management*, http://siteresources.worldbank.org/EXTWAT/Resources/4602122-1210186345144/GWMATE_English_CP_09.pdf, accessed 25 November 2009

The World Bank (2009) *Environmental Protection and Sustainable Development of the Guaraní Aquifer System Project*, Implementation Completion and Results Report no ICR00001198, www-wds.worldbank.org/external/default/WDSContentServer/WDSP/IB/2009/08/13/000333037_20090813231853/Rendered/PDF/ICR11980P068121IC0disclosed08112191.pdf, accessed 10 January 2010

13.8 The Mekong River Commission

Aaron T. Wolf and Joshua T. Newton

Key points

- The effectiveness of the Mekong River Commission is compromised by the lack of binding involvement of upstream riparians including China.
- The support of the international community is directly linked to the pace of development; when there is greater involvement, development occurs more quickly.
- The collection of data prior to the implementation of projects to develop the river basin allows for greater planning efficiency and promotes cooperation between states.
- Lack of alignment between national legislation and the goals of the MRC restricts its ability to meet objectives.

Introduction

The Mekong River runs through China, Myanmar, Thailand, Laos, Cambodia and Vietnam. At over 4000km, it is one of the world's longest and most important rivers. The riparians of the Lower Mekong Basin (Thailand, Laos, Cambodia and Vietnam) are members of the Mekong River Commission (MRC), a transboundary water organization established by the Agreement on the Cooperation for the Sustainable Development of the Mekong River Basin, signed in 1995. This framework agreement establishes the MRC as an intergovernmental organization supporting joint basin-wide planning and a host of other initiatives including fisheries, safe navigation, irrigation, environmental monitoring, flood management and the development of hydropower options (MRC, 2009). The framework also ensures 'reasonable and equitable use' of the water resources, thereby incorporating a key principle of international water law (Sadoff et al, 2008).

China and Myanmar, the upper riparians, are not members of the MRC but are dialogue partners to the agreement. Upstream development by China in particular may have a significant impact on the proposals of the downstream countries and would diminish the efficacy of the MRC.

The history of the MRC can be traced back to the early 1950s when the United Nations Economic Commission for Asia and the Far East noted that there was potential to develop hydropower and irrigation in the Mekong Basin. Later reports recommended that the development of the basin should be undertaken cooperatively by the nations sharing the water resources and that an international body should be established to undertake this coordination. Over time, this organization has evolved in response to geopolitics and varying levels of support (both financial and otherwise) from the international community. Before the MRC emerged in its present form there were two earlier international river basin organizations:

- The Mekong Committee (1957–1978) which included Cambodia, Laos, Thailand and Vietnam;
- The Interim Mekong Committee (1978–1995) which included Laos, Thailand and Vietnam, but not Cambodia.

In its current form, the MRC consists of three permanent bodies: the Council, the Joint Committee and the Secretariat (the location of which moves every five years, but from June 2010 will be permanently co-hosted by Lao PDR and Cambodia).

Key lessons for transboundary water managers

Since the inception of an international committee to oversee joint coordination of the water resources in the Mekong Basin, the MRC has successfully completed a considerable amount of important work. This includes the collection of information and data and the blasting of the river channel to allow enhanced navigation. The MRC has also raised stakeholder awareness about water issues in the Mekong region.

Despite these advances, there has been little progress with regard to development in the main river channel. Critics have cited a variety of reasons for this inaction, which provide useful lessons for transboundary water managers.

Reconciling domestic national water policy with the MRC

A criticism of the 1995 Mekong Agreement is the lack of ratification or a mandate that key aspects of the agreement be incorporated into national legislation of the MRC signatories (Sadoff et al, 2008). In effect, this means that there is limited alignment between national water policies and the goals of the MRC. This is compounded by the disparity observed between the national policies of national states and international standards. Taken together, this restricts the extent to which states can work towards implementing development in the Mekong Basin and has led to calls for greater focus on governance and capacity-building in the region. Nevertheless, there has been collaboration between the Secretariat and the National Committees of the signatory nations, to develop policies on a range of areas including navigation, flood control and fishing.

Joint collection of data provides a platform for cooperation

The joint collection of data in the Mekong River Basin has provided a basis for cooperation and discussion. This is despite the relative lack of development in the main river channel. The first task of the Mekong Committee was to develop a system for the collection of essential data and information for the basin. This joint effort in data gathering allowed basin states to work together and in the process built trust and fostered cooperation between nations.

During the course of data collection, a significant shift was observed in the type of data requested by the Mekong Committee. Social and economic studies were undertaken early, and a mission sponsored by the Ford Foundation in 1961 found that the maximum benefits of development in the basin could only be derived if there was extensive training of the local population. This observation and the subsequent collection of social and economic data indicate a transition to a more holistic approach to river basin management.

Riparian relationships

A critical reason for the cooperation of basin states is the earliness of the agreement between them. The riparians were brought together by the United Nations prior to the development of plans for large projects on the Mekong and its tributaries. The work of the committee has also reduced suspicion between riparians, meaning that regardless of prior conflict, the riparians in the Lower Mekong Basin have largely managed to continue with the water-related work.

One of the most glaring problems with the MRC is that neither China nor Myanmar is a member and that China is developing unilateral plans to dam the Mekong River. This will have a significant impact on the water supply for downstream countries. While the communication channels with China are more open than they have been in previous decades, the rate of development in China and the pressing need for energy security and irrigation means that unilateral Chinese development of the Mekong is very likely.

Importance of international support and finance

The MRC is funded by contributions from the four member countries and from aid donors (MRC, 2009). The level of financial aid and international support given to the MRC has been correlated with the level of cooperation and progress in the basin, as can be seen from the history of the MRC and its predecessors since the 1950s. Early accomplishments were impressive, impelled in part by strong UN support and a

'Mekong spirit' on the part of the 'Mekong Club' of donors. By the 1970s, the pace of cooperative development began to slacken, partly the result of decreasing involvement by an international community daunted by political obstacles and the size of planned projects.

Conclusions

The MRC has enabled the basin states of the Lower Mekong Basin to coordinate management of the river, yet any unilateral action by China may jeopardize its effectiveness and progress made to date. Despite advances in data collection and the implementation of smaller projects, the MRC has demonstrated limited capacity to move from data collection to project implementation on the main river. A key reason for this is the lack of alignment between national legislation and the objectives of the MRC, and the need for greater focus on governance and capacity-building

Adapted from Aaron T. Wolf and Joshua T. Newton (2009) 'Case studies of transboundary dispute resolution', Appendix C in J. Delli Priscoli and A. T. Wolf (eds) *Managing and Transforming Water Conflicts*, Cambridge University Press, Cambridge

References

MRC (Mekong River Commission) (2009) *Mekong River Commission Website*, www.mrcmekong.org, accessed 20 November 2009, www.mrcmekong.org/about_mrc.htm, accessed 15 May 2010

Sadoff, C., Greiber, T., Smith, M. and Bergkamp, G. (2008) *Share – Managing Water Across Boundaries*, IUCN, Gland, Switzerland

13.9 Stakeholder participation in the Orange-Senqu River Basin, Southern Africa

Daniel Malzbender and Nicole Kranz

Key points

- A strong political commitment from all basin states is essential to initiate and maintain stakeholder involvement processes.
- Effective stakeholder participation in a transboundary setting requires well-established (and progressively strengthened) institutional structures at basin level for coordination and monitoring of activities.
- Stakeholder involvement activities in a transboundary setting need to be mindful of different legal and cultural settings between countries and regions, and tailored accordingly.

Introduction

The Orange-Senqu River Basin is shared by the four southern African countries Botswana, Lesotho, Namibia and South Africa. At over 1 million square kilometres the Orange-Senqu River Basin is the largest basin south of the Zambezi (Earle et al, 2005). It is also the most developed transboundary river basin in the southern African region, with a variety of water transfer schemes to supply water to municipalities, industries and farms in and outside of the basin (Earle et al, 2005). One of the major catchments, the Vaal, is particularly highly populated and urbanized, with 48 per cent of the population of South Africa living in the catchment and relying on its water (Heyns, 2004). The basin is characterized by a high diversity of stakeholders, ranging from large-scale industry, mining, commercial agriculture, hydropower producers and large water utilities, to domestic users, small-scale and subsistence agriculture. The basin is faced with a wide range of management issues, such as declining water quantity and quality, climate change and socioeconomic problems (UNDP/GEF, 2008).

A number of bilateral commissions dealing with different management aspects of the basin (primarily large-scale infrastructure projects such as the Lesotho Highlands Water Project) have existed for some time. In 2000, the four basin countries established the Orange-Senqu River Commission (ORASECOM),[1] the first commission in the basin comprised of representatives from all basin states. The objective of the Commission is to serve as technical advisor to the Parties (the four countries) on matters relating to the development, utilization and conservation of the water resources in the river system. Article 5 of the Agreement further specifies technical areas in which the Commission is to provide advice, and among them lists, in Article 5.2.4, issues concerning stakeholder involvement in the management of the basin.

The development of stakeholder participation

In May 2005 the Water Ministers of the four basin states mandated the Commission to develop a strategy for the Commission to progressively increase the engagement with basin stakeholders. A 'Roadmap towards Stakeholder Participation' was subsequently developed through a facilitated process[2] over 18 months involving the Commission (through its different organs) and local and international experts. The Roadmap was presented to the ORASECOM Council (the highest decision-making body of the Commission) in April 2007 and endorsed by the Commission.

The Roadmap actively seeks to facilitate the input of all stakeholders and explicitly opened the process to multiple interactions, aiming for the creation of strong relations among different stakeholder groups. It covered the short to medium term (five to ten years), while highlighting the need for (future) alignment between the stakeholder involvement strategy and ORASECOM's 'vision' as an institution.

The implementation of the Roadmap is loosely structured by a 'Framework of Activities', in order to allow stakeholders to determine the pace of the implementation process. The Roadmap recognized that ORASECOM 'has relatively limited capacity for the implementation of many of the processes'. It is envisioned that most of the actions will be carried out by other organizations, such as NGOs, community-based organizations, academic institutions, development partners, etc. To reduce overlaps and gaps that often occur when different development partners work in a region, Terms of Engagement have been drawn up for development partners wishing to work on transboundary stakeholder projects in the basin.

Despite the strong political commitment and overall positive experiences with public participation, the practical challenges facing stakeholder participation in the Orange Basin are still significant. For example, most stakeholders in rural communities are not organized, or if organized lack the awareness to become involved with water management issues. This makes it very difficult to reach all relevant stakeholders and to develop joint action. In order to facilitate awareness creation, knowledge exchange and better engagement among the diverse groups of stakeholders, the Roadmap, following the example of the Okavango River Basin, proposes the establishment of a Basin-Wide Forum (BWF). The BWF would coordinate with already existing forums at the sub-basin level in the individual countries that have been (or are currently being) set up in the basin states, as part of ongoing integrated water resources management (IWRM)-inspired water law reform efforts, which include the devolution of management responsibilities to local levels.

It needs to be understood that the Roadmap is not a directly implementable blueprint for stakeholder involvement in the basin. Instead its key objectives are:

- to develop and strengthen institutional mechanisms for effective stakeholder participation;
- to build and strengthen capacity in basin forums;
- to develop and maintain horizontal and vertical communication between and among the structures of ORASECOM and basin stakeholders.

In this way it can form guiding principles for the development of practical stakeholder-involvement activities during future implementation phases.

Subsequent to the Roadmap development, an initial stakeholder analysis was conducted in 2008 as part of the Transboundary Diagnostic Analysis carried out by a UNDP/GEF project. In addition to identifying the main stakeholder groups in the basin, the analysis aimed at gauging the respective interests and opinions of different stakeholders about pertinent water management challenges. The findings of this analysis can serve as a complementary input to the development of an appropriate engagement mechanism in future stakeholder involvement plans.

In support and preparation of planned participative efforts in the basin, a River Awareness Kit (RAK) was developed in 2009 in order to support the training of and communication with relevant stakeholders throughout the basin. The web-based tool provides easily accessible information about relevant water management issues in the basin.[3] The RAK provides comprehensive, easily accessible and understandable information about the physical characteristics of the river basin as well as socioeconomic factors, governance structures and the specific water-related challenges to be addressed (see following case study 13.10 on the RAK).

At present a number of ICP-supported projects in the Orange-Senqu River Basin are ongoing.[4] Through coordination by ORASECOM and Deutsche Gesellschaft für Technische Zusammenarbeit (GTZ) as joint international cooperating partner (ICP) coordinator for water in the Southern African Development Community (SADC), these are coordinated as an overall basin programme, addressing key management challenges in the basin. All of these projects contain elements of stakeholder involvement, which will aim, in time, to implement in practice the principles set forth in the Roadmap.

Conclusions

The case of the Orange-Senqu Basin demonstrates that best-practice principles as set out in the Roadmap still need to be translated to the specific situation in the river basin, such as the respective legal frameworks of basin states, different participative cultures in various parts of the basin, and between different stakeholder groups or the different capacity levels between stakeholder groups. During the implementation of practical stakeholder involvement activities ORASECOM and its support projects became faced with the challenge of leveraging buy-in from stakeholders and communities on the ground. This requires consideration of the issue of scope and scale. Which issues are addressed at which level? Who is organizing which participatory processes and how is the involvement and ownership of stakeholders maintained?

With ORASECOM providing a solid institutional structure at transboundary level to progressively coordinate management activities in the basin, including the involvement of stakeholders, it is likely that the Orange-Senqu River Basin will remain an interesting case to follow as far as transboundary stakeholder involvement activities are concerned. The Roadmap process initiated in 2005, as well the ongoing, coordinated practical implementation of stakeholder involvement activities, might provide valuable lessons for other transboundary basins in the years to come.

Notes

1   Agreement between the governments of the Republic of Botswana, the Kingdom of Lesotho, the Republic of Namibia and the Republic of South Africa on the Establishment of the Orange-Senqu River Commission.

2   The Stakeholder Roadmap development was financially supported by InWEnt (Germany) and facilitated by the African Centre for Water Research (www.acwr.co.za).

3   The Orange-Senqu River Awareness Kit is accessible at www.orangesenqurak.org and was developed by Hatfield Consultants with financial support of GTZ.

4   The main projects currently are supported by GTZ/DFID, UNDP/GEF and the EU.

References

Earle, A., Malzbender, D., Turton, A. R. and Manzungu, E. (2005) *A Preliminary Basin Profile of the Orange/Senqu River*, AWIRU, University of Pretoria, Cape Town, South Africa

Heyns, P. (2004) 'Achievements of the Orange-Senqu River Commission in integrated transboundary water resource management', presented at the *General Assembly of the International Network of Basin Organizations*, Martinique, 24–28 January 2004

UNDP GEF (2008) Preliminary Transboundary Diagnostic Analysis for the Orange-Senqu River Basin, UNDP GEF, New York

13.10 Orange-Senqu River Awareness Kit: supporting capacity development

Simon Hughes, Erin Johnston and Grant Bruce; Horst Vogel,
Peter Qwist-Hoffmann and Bertrand Meinier

Key points

- Tools such as the Orange-Senqu River Awareness Kit (Orange-Senqu RAK), a strategic information, knowledge management and communication tool, support capacity development through centralized access to data, information and knowledge related to the management of transboundary waters.
- User-driven tool design and incorporation of stakeholders in the development enables ownership and trust in the Orange-Senqu RAK and promotes cooperation between the various stakeholders and member countries.
- The approach used to develop the Orange-Senqu RAK has important policy implications in the Southern African Development Community (SADC) region in terms of information and knowledge management and capacity development.

Context

Over the years, many shared watercourse institutions in the SADC region have benefitted from technical and financial support from international cooperating partners (ICPs). In addition, various national governments monitor hydrology and climate and also collect socioeconomic information. The net result is that river basin organizations (RBOs) and riparian states are custodians of significant repositories of data and information related to the various river basins; however, this information is often distributed among a number of organizations and individuals, and is not widely or uniformly accessible to all stakeholders. This information is often presented at the national or state level, which limits the relevance for river basin managers who require information at the river basin scale. Furthermore, data are rarely presented in ways that can be easily employed by stakeholders, often requiring substantial knowledge to extract key messages for environmental education and awareness-raising activities.

In its mandated role as regional coordinator of ICP development assistance for transboundary water management projects in the SADC region,

GTZ identified the opportunity to address pressing needs for access to information through the implementation of RAKs, an approach that has been successfully pioneered in other parts of the world. The Orange-Senqu RAK is the first such tool developed in the SADC region. Through its support to the SADC Water Division, GTZ is promoting the use of this innovative information and knowledge management tool to support capacity development and enhance cooperation in transboundary water management (TWM).

Designed to foster RBOs and national focal point access to socioeconomic, hydrological and environmental data needed for informed decision making in TWM, a River Awareness Kit is one of the fundamental building blocks of the multi-dimensional capacity development approach that GTZ has developed to guide its interventions in the region, as illustrated by the concept of the *Transwater Capacity Cube*.

Background

The Orange-Senqu River Basin is the most developed river basin in Africa in terms of water infrastructure. The basin covers all of Lesotho, a large portion of South Africa and southern

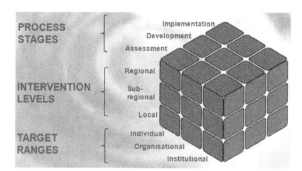

Figure 13.2 GTZ Transwater Capacity Cube

regions of Botswana and Namibia. The climate, landscape and cultures of the river basin vary significantly from the source of the Senqu River in the highlands of Lesotho, across semi-arid central and northern South Africa and south-eastern Botswana, to desert conditions where the river forms the border between Namibia and South Africa before flowing into the Atlantic Ocean. The river basin includes the Vaal River, which passes through the megalopolis Johannesburg–Pretoria (Gauteng Province), the economic powerhouse of Africa. This variation poses many challenges to the management of the water resources of the basin, under growing pressure from increasing populations and economic activities and the uncertainties of climate change.

Established in 2000, the Orange-Senqu River Commission (ORASECOM) promotes the equitable and sustainable development of the resources of the Orange-Senqu River. ORASECOM is the first commission established following the agreement by all member states of the SADC Revised Protocol on Shared Watercourses in 2000, which after ratification came into force in 2003.

River Basin Awareness Kits:
lessons for transboundary water management

Challenges related to information access and availability, coupled with the amount and quality of available information, prevents many stakeholders from turning data and information into the knowledge necessary to improve TWM. An RAK is designed to address these key challenges through the development of a knowledge and information base for a river basin. This central 'hub' of information and knowledge, which may be updated regularly, can then be used to support informed decision making and enhance understanding regarding TWM, as open and transparent access to information has the potential to significantly improve development cooperation.

A secondary, but equally important, potential benefit of developing an RAK is the value derived from the stakeholder engagement process itself. The participatory process used to develop and maintain an RAK helps create a forum for stakeholders to identify and discuss key issues. This forum helps to create an enhanced understanding of cross-cutting themes and identify knowledge and capacity gaps by the riparian stakeholders. The end result of the stakeholder engagement process is consensus on the information needs required to support river basin and transboundary water management.

The Orange-Senqu River Awareness Kit

The goal of the Orange-Senqu RAK is to address challenges related to information accessibility, availability and presentation within the basin in order to support the mandate of ORASECOM. As a relatively new organization facing significant environmental and water management challenges, ORASECOM was limited by the lack of consolidated information about the basin and a lack of a common vision for basin management by the various stakeholder groups.

Approach

The Orange-Senqu RAK was developed using participatory techniques designed to capture the needs and requirements of a wide range of stakeholders. The stakeholder engagement process is intended to operate within existing institutional mechanisms and be conducted in partnership with RBO representatives at the national and regional levels. This approach leverages existing meetings and forums to promote sustainability and ownership by the RBO.

The core content of the RAK is divided into four main themes:

1   The *River Basin* provides an overview of the physical aspects of the basin.

2   *People and the River* covers the social and economic aspects of the basin.

3   *Governance* discusses environmental governance and transboundary integrated water resources management.

4   *Meeting the Water Challenge* explores the physical management of water in the Orange-Senqu River basin.

The table of contents for the Orange-Senqu RAK was defined through a series of participatory user needs assessment workshops, involving a wide variety of stakeholder groups within the Orange-Senqu River Basin, providing expert insights into different aspects of water resources management. Content developed included information sourced from over 400 scientific papers, technical reports, documents and interviews. The material gathered was supported by 83 maps, 300 photographs and 20 interactive components explaining key messages. The topics covered included the basin landscape and geography, an introduction to integrated water resources management (IWRM), a summary of ORASECOM and animated diagrams explaining issues such as water management and the distribution of infrastructure.

Stakeholders are provided with mechanisms for input and review throughout the RAK development process to ensure that the tools and information created match their requirements. The end product is a website and CD-ROM that addresses different aspects of transboundary water resource management in clear, non-technical language. Self-learning resources are supported by interactive tools, maps and videos. The tools are designed so that they can be updated and modified by the stakeholders and users as required.

Sustainability is promoted through capacity development workshops at the regional and national levels, focused on developing the capacity of stakeholders to use the RAK. Training-of-the-trainer workshops were conducted with national focal points from each riparian nation, to transfer the knowledge and resources to deliver the RAK as a training course to other stakeholders. This ensures that the tool is widely disseminated through existing institutional mechanisms.

Emerging impacts

The Orange-Senqu RAK was developed over a period of 18 months and was officially launched in October 2009. The impacts have already started to emerge. The Orange-Senqu RAK is considered a platform for common understanding among key stakeholders on the major issues and challenges in the Orange-Senqu River Basin. As an electronic tool, the information included in the RAK is accessible through the internet, and by CD-ROM in areas with poor internet connection. The ORASECOM Secretariat has observed that since its launch, it has been receiving an increasing number of requests for information about the Orange-Senqu River, from a wide range of stakeholders outside the water sector. Furthermore the Secretariat is receiving very positive feedback from all over the world, mainly on the comprehensive nature of the information included in the RAK. Learning from these experiences, it is expected that the development of RAKs for the Limpopo and Kunene rivers will create broader river basin stakeholder platforms in the SADC region, as well as ensure broader participation in the implementation of the Revised SADC Protocol on Shared Water Courses.

Distribution

The Orange-Senqu RAK is available via the internet (www.orangesenqurak.org) and as a CD-ROM, and addresses different aspects of transboundary water resource management for a variety of stakeholder groups.

It was initiated and is being supported by GTZ through the 'Transboundary Water Management in SADC' regional cooperation programme. Funding was provided by the Bundesministerium für wirtschaftliche Zusammenarbeit und Entwicklung (BMZ, the German Federal Ministry for Economic Cooperation and Development), in partnership with the UK Department for International Development (DFID). The Orange-Senqu RAK was developed and implemented by

Hatfield Consultants, in conjunction with Strata360, both from Canada. River Awareness Kits for the Limpopo and Kunene river basins are currently under development, also supported by the GTZ through the 'Transboundary Water Management in SADC' Programme.

13.11 Multi-riparian basins: the Salween River

Aaron T. Wolf and Joshua T. Newton

Key points

- In the absence of a transboundary water management (TWM) agreement, riparian stakeholders may proceed with unilateral initiatives that make eventual agreements even more difficult to implement.
- Developing a single point of cooperation for TWM – such as hydropower – may provide the foundation for more comprehensive frameworks encompassing multiple issues.
- When the poopulation of a transboundary river basin is not included in the decision-making process, local conflict may arise despite cooperation between national governments.

Context

The Salween River is shared by China, Myanmar and Thailand. At almost 2500km, it is the longest undammed river in mainland Southeast Asia. More than 10 million people from this ethnically diverse area depend on the Salween river basin for their livelihoods. The watershed also supports a rich biodiversity, and the Nujiang, the section of the Salween that flows through China, is part of the Three Parallel Rivers area which is designated as a UNESCO World Heritage Site.

For a significant part of its length, the Salween flows between deep gorges, providing a number of locations for hydropower development. Each of the three countries sharing the river is developing proposals to exploit this hydropower potential and these projects are likely to have significant impact on downstream water users, local communities and the environment.

TWM between China, Thailand and Myanmar is inextricably linked with wider political concerns. The military junta ruling Myanmar is regarded as an oppressive regime with a poor human rights record. However, China is a significant donor of aid to Myanmar and is already collaborating with the government to develop dams inside the country's borders (McNally et al, 2009). Thailand also has agreements to develop the Salween in Myanmar as a result of domestic pressure to supplement national water supplies and generate more electricity. As much as 10 per cent of the Salween's flow could be diverted to augment freshwater supplies to Thailand if water-sharing agreements with Myanmar come into force. National sovereignty to protect water resources has, in this case, outweighed international pressure against working with the military regime in Myanmar.

To date, there have been a number of proposals and plans to build dams on the main river but, as yet, no major development has been completed. In 1989 Thailand and Myanmar established a joint technical committee to pursue feasibility studies. In 2002, discussions about the construction of large dams took place between the electricity boards of the two countries. In 2003, China announced plans to build 13 dams on the Nujiang in China. These were met with protests from environmental and social groups in Myanmar and Thailand, international pressure groups such as International Rivers, and from within China. The plans were officially suspended by Premier Wen Jiabao in 2004 while a review was undertaken. In 2006, a committee organized by the National Environmental Protection Agency and the National Development and Reform Commission concluded that construction of four of the dams could proceed (Brown and Xu, 2010). Pressure to increase electricity supply is likely to lead to the eventual development of more than the 13 dams originally proposed.

Key issues for transboundary water management

Water sharing agreements and institutions

The Salween watershed has a number of characteristics that exemplify common issues encountered when managing water across international borders. The most important of these is that there is no working TWM agreement for the river between all three riparians – China, Myanmar and Thailand. In practice, this means that the three riparians are making unilateral but incompatible plans to develop the basin. Currently both China and Thailand have independently entered into cooperative agreements with Myanmar, and China is proceeding with its unilateral plans to develop a number of large dams. If unchecked, this unsustainable arrangement will lead to inequitable water allocation between the three countries, setting the stage for regional conflict over water resources.

The absence of an agreement or an institution for TWM in the Salween Basin places all three countries in a vulnerable position. The rapid political and economic reforms in China (McNally et al, 2009) and uncertainty regarding the intentions of the military junta in Myanmar, make the prospects for effective water sharing agreements questionable at best. Given the desire by all parties to develop hydropower in the area, it is envisaged that a transboundary water organization will evolve in the future to allow these developments to proceed.

Any TWM agreement or organization in the basin will need to be resilient and flexible enough to manage the uncertainty both with respect to the changing political situation but also the expected variations in river flow due to climate change. The Salween is fed from glacial melt, and flows in the river are vulnerable to changes in the rate of glacial melting.

Power asymmetry

The Salween Basin is characterized by power asymmetry. China is the upstream riparian and is the dominant economic and political power in the watershed. This imbalance is likely to affect negotiations for use of the water and China may not feel it necessary to participate with its downstream neighbours.

Benefit sharing

The interests of all three countries converge and may lead to 'benefit sharing' due to their respective domestic requirements to develop hydropower capability. This greater energy security will enhance economic stability for each of the riparian countries (McNally et al, 2009). An alignment in interests over hydropower lends itself well to interstate negotiations, and provides a common purpose on which to base future discussions and agreements extending into other areas of mutual benefit. It also limits the likely levels of conflict, as both Thailand and Myanmar may benefit from hydropower development in the upstream portion of the Salween (in China), particularly if electricity can be exported.

Participation

Given the drive by all three countries to develop their hydropower resources, trilateral participation is crucial. This is because the construction of dams for hydropower has a dramatic social and environmental impact. If local communities are not consulted, this may lead to greater intrastate instability. This is especially important for Myanmar, where there is high resistance by local populations in some areas of the Salween River Basin, and the government does not exert full control over the region. Thai and Myanmar officials have been working together discreetly in an insurgent area where the Myanmar army has persecuted the Shan civilian population. With an already large number of Shan people being forced from the region, environmental groups and local populations are worried that a dam project will only exacerbate the problem. Whereas Thailand and Myanmar may work cooperatively to avoid conflict, large-scale projects such as the development of hydropower resources may create or exacerbate intrastate conflicts in regions where local populations have not been included in decision-making processes.

Transparency and feasibility studies

The recent events regarding the opposition to China's proposals for dams on the Upper Salween or Nujiang demonstrates the effect that greater levels of transparency can play in basin development. Changes in the Chinese legal system have enabled social organizations to challenge development projects (Magee and Kelley, 2009) and the results of this can clearly be seen. Questions do remain, however, regarding how much of the results from environmental impact assessments and related studies will be released into the public domain. Despite these reservations, it should be noted that all three riparian countries have conducted feasibility studies. This is a positive step toward cooperation over development of the river and away from potential conflict.

Conclusions and lessons learned

A number of lessons and conclusions can be drawn from the example of the Salween River. The relationship between Myanmar and both China and Thailand shows that cooperation over water for economic development can supersede working with an oppressive regime. The importance of participation at all levels to prevent intrastate conflict has also been highlighted. This is particularly the case for hydropower development which has a significant impact on local communities and ecosystems. Despite the early stage of development in the watershed, best-practice examples can already be found. For example, feasibility studies, investigations and management discussions are proceeding prior to the construction of any major development. This paves the way for integrated management of water resources.

Finally, the current absence of water-sharing agreements or water management institutions provides a baseline to track how a transboundary organization may evolve, and to see if lessons from established institutions such as the Mekong River Commission will have an impact in this watershed.

Adapted from Aaron T. Wolf and Joshua T. Newton (2009) 'Case studies of transboundary dispute resolution', Appendix C in J. Delli Priscoli and A. T. Wolf (eds) *Managing and Transforming Water Conflicts*, Cambridge University Press, Cambridge

References

Brown, Philip H. and Yilin, Xu (2010) 'Hydropower Development and Resettlement Policy on China's Nu River', forthcoming, Journal of Contemporary China, www.colby.edu/economics/faculty/phbrown/papers/BrownXuJCC2009.pdf, accessed 1 July 2010

Magee, D. and Kelley, S. (2009) 'Damming the Salween River', in F. Molle, T. Foran and M. Käkönen (eds) *Contested Waterscapes in the Mekong Region: Hydropower, Livelihoods and Governance*, Earthscan, London, pp115–140

McNally, A., Magee, D. and Wolf, A. (2009) 'Hydropower and sustainability: resilience and vulnerability in China's powersheds', *Journal of Environmental Management*, vol 90, ppS286–S293

13.12 Community-based approaches to conflict management: groundwater in the Umatilla Basin

Todd Jarvis

Key points

- It takes time and patience to see tangible benefits from negotiations over a 'hidden' resource such as groundwater, as there are many types of knowledge about groundwater that must be shared between the stakeholders.
- The collaborative learning approach matters, specifically as it relates to the public education and outreach programme, by increasing public awarness and building trust.
- Communications about water resources must extend beyond the traditional methods of technical journal articles and presentations at conferences, to the use of other media such as documentary film.

Introduction

Falling groundwater tables in the basalt aquifers of the Umatilla River Basin in the US state of Oregon prompted the formation of the Umatilla County Critical Groundwater Task Force (UCGT). Its mandate was to develop and recommend solutions to both long- and short-term water problems in the river basin. The UCGT used a community-based, collaborative learning approach and has made good progress to date. Its cost-effective methods provide a basis for conflict management, and show that shared groundwater resources can be managed more effectively through education, awareness raising and joint undertakings. This case study examines the development of this approach and highlights key aspects that are applicable to transboundary water management (TWM).

The history of the Umatilla County Critical Groundwater Task Force

The Umatilla River is a tributary of the Columbia River, and groundwater in this area has been exploited for over 50 years. The river basin is shared by the states of Washington and Oregon, and by the Confederated Tribes of the Umatilla Indian Reservation, for whom maintaining salmon fisheries is culturally important.

Irrigated agriculture places the greatest demand on the basin's water supplies. Farmers use groundwater to supplement surface water when supplies run short. Over recent years, due to the expansion of cultivated areas and changing crop choices, increased groundwater pumping has caused the water table adjacent to the Umatilla River to fall dramatically. This has had a number of consequences and one of the most culturally sensitive is the impact on river flows.

The Umatilla Indian Reservation was established by a Treaty with the US Government in 1855. This treaty conferred historic fishing and consumptive rights regarding water use for Indian Tribes in the Umatilla Basin. Sensitive salmon fisheries are dependent on sufficient flows in the Columbia and Umatilla Rivers, and suffer when water levels decline. The water rights of the tribes therefore conflict with the needs of the farming community, which extracts water for irrigation. This conflict adds complexity to groundwater management in the basin. The impetus for the creation of the UCGT was the state-designated Critical Groundwater Areas (CGAs). These had been established to conserve the quantity and quality of groundwater for future generations of farmers starting in 1976 and ending in 1991, thus precluding additional wells for public drinking water and irrigation supplies (Oregon Water

Resources Department, 2003). After struggling for over eight years in dealing with a state mandate to address water quality and quantity declines in western Umatilla County, the Umatilla County Planning Commission (Planning Commission) held a hearing in Hermiston, Oregon, in 2003 to consider implementing a land-use overlay zone within the state-designated CGAs.

Over 540 Umatilla County citizens, including irrigators, rural residents, city residents, scientists and consultants, attended the Planning Commission hearing to voice their opposition to the proposed overlay zone. 20 of the 25 individuals who testified in opposition to the proposed overlay requested that Umatilla County establish a local group of citizens to address the wide-ranging issues surrounding the water problems in lieu of adopting an overlay zone only to limit a handful of new domestic wells. Following these stakeholder-based recommendations, the Planning Commission and Umatilla County Board of Commissioners appointed the 20-member UCGT to develop and recommend solutions to short- and long-term water quantity issues in Umatilla County, especially within the CGAs.

The UCGT is made up of representatives from stakeholders who represent the interests of water users from across the basin including state and federal government, tribal groups, the agricultural sector, the media and Oregon State University. It has recently released a draft version of the Water Management Plan for the Umatilla Sub-Basin up to 2050 in which it proposes a number of strategies to manage falling water tables.

Key lessons for transboundary water management

The successes of the UCGT are attributed to its community-based collaborative learning approach. This facilitates the sharing of knowledge between stakeholders and also raises awareness in the local community of the complex issues associated with groundwater management. There are many lessons to be learned, particularly that this approach takes considerable time and is not a 'quick fix' for solving the water problems in Umatilla. These aspects are examined in more detail below.

Collaborative learning and participation

Schlager (2007) indicates that the difficult physical, social and economic challenges facing the Umatilla Basin groundwater users and governments required addressing the issues collaboratively. The adopted approach was based roughly on the principles of collaborative learning developed by Daniels and Walker (2001), which have been used in many different environmental situations. The process was supported by students from Oregon State University (OSU) who compiled a single document containing essential technical information about the Umatilla Basin. This was used as a basis for negotiation.

The UCGT recognized the value of public participation so that the public could choose the level of risks rather than serving simply as recipients of risk, as described by Delli Priscoli (2004). Regular monthly meetings were held which were open to the public to ensure that the development of the basin plan was transparent and open. Through these regular meetings, public awareness was increased and greater trust was fostered with the local community. This trust facilitated more rapid political decision making, such as the passing of new legislation required to move the management process forward. The Oregon State Senate passed Senate Bill 1069 in 2008 dedicating US$750,000 for the purpose of conducting the Umatilla Basin regional aquifer recovery assessment. During the 2009 legislative session, House Bill 3369 was passed and signed into law by the Governor of Oregon in August 2009.

While the UCGT has been successful at achieving consensus, an important lesson from the Umatilla case is that collaborative learning takes time. It is not a quick fix and is something which evolves continuously. This is especially true when negotiating over a 'hidden resource' such as groundwater.

Beyond traditional methods of communication

Closely linked to the collaborative learning process is the use of experts to provide advice to the UCGT and to support the community outreach programmes on water. These programmes demonstrated the benefits of communication beyond the traditional methods of technical journal articles and presentations at conferences.

The UCGT used specialists in water sciences, policy and management from OSU to train community volunteers for speaking engagements at public schools, newspaper and radio advertising, brochures, radio interviews, telephone interviews and staffing booths at regional agricultural industry events. Telephone polling by the Soil and Water Conservation District revealed that 75 per cent of the respondents had knowledge of a groundwater problem. OSU and the outreach subcommittee addressed over 200 schoolchildren from kindergarten to grade 12 in a two-county area, as well as at the Confederated Tribes of the Umatilla Indian Reservation (CTUIR) School. A 30-minute documentary film of the groundwater situation, *Water Before Anything*, was developed by an OSU graduate student to document the collective action by the community.

Successful negotiations and low cost of community approaches

The community-based approach adopted in Umatilla has proved to be cost-effective. This was particularly important given the limited funds available from other sources such as federal, state and tribal budgets.

Conclusions

Collaborative learning provides a low-cost approach to conflict management between differ-ent water user groups. The UCGT has managed to produce its plan for the Basin until 2050 and has been instrumental in the passing of new legislation, notably House Bill 3369.

The key lesson is that in order to achieve a long-term solution, time must be devoted to community learning. Building trust and raising public awareness are excellent additional benefits derived from this approach.

References

Daniels, S. E. and Walker, G. B. (2001) *Working through Environmental Conflict: The Collaborative Learning Approach*, Prager, Westport, CT

Delli Priscoli, J. (2004) 'What is public participation in water resources management and why is it important?', *Water International*, vol 29, no 2, pp221–227

Oregon Water Resources Department (2003) *Ground Water Supplies in the Umatilla Basin*, http://www1.wrd.state.or.us/pdfs/UmatillaGWWkshpRptApril2003.pdf, accessed 14 August 2007

Schlager, E. (2007) 'Community management of groundwater', in M. Giordano and K. Villholth (eds) *The Agricultural Groundwater Revolution Opportunities and Threats to Development*, Comprehensive Assessment of Water Management in Agriculture Series no 13, CABI, Wallingford, UK, pp131–152

UCGT (Umatilla County Critical Groundwater Task Force) (2008) *Umatilla Sub-Basin 2050 Water Management Plan*, www.co.umatilla.or.us/planning/pdf/2050%20Plan%20Final.pdf, accessed 7 November 2009

13.13 Gabčíkovo-Nagymaros Project: a test case for international water law?

Owen McIntyre

Key points

- Unilateral action on water resources causes tension and hinders future cooperation.
- Third parties such as the European Union can enable cooperation in the face of conflict.
- Joint environmental monitoring provides a politically neutral platform for negotiation.

Introduction

The dispute between Hungary and Czechoslovakia (now Slovakia) over the Gabčíkovo-Nagymaros Project on the Danube River was the first international water conflict to be taken to the International Court of Justice (ICJ). At the root of the argument was Hungary's decision to halt a joint development project because of concerns about its environmental impact. The judgement from the ICJ was passed in 1997 and vindicated neither side. Both countries have continued negotiating to find a mutually acceptable decision, but with limited success. The need to find a solution is now more pressing than ever due to the requirements of the Water Framework Directive, an important piece of European legislation.

The disagreement between Hungary and Slovakia has particular significance because it was viewed as a test case for the 1997 UN Watercourses Convention. Ultimately, the ICJ did not base its decision on international water law, and instead used a more conventional approach focusing on the particulars of treaty law. Nevertheless, its judgement has been interpreted as an endorsement of the UN Watercourses Convention because it makes reference to the principle of 'reasonable and equitable' use.

The history of the conflict

The disagreement is based on negotiations dating back to the 1950s which focused on plans to construct hydroelectric plants on the Danube River. In 1977, both countries signed a treaty on the construction and operation of two dams and systems of locks, one on Slovak territory at Gabčíkovo and one on Hungarian territory at Nagymaros.

Shortly after construction work began, the Hungarian government, under pressure from environmental lobbies, including the influential Danube Circle, began to express concern about the impacts of the project on the environment. Their concerns focused on the impact of pollution and reduced recharge on aquifers linked to the Danube, and the potential damage to Szigetkoz, an ecologically important wetland area.

The Hungarian government eventually decided to suspend work until the environmental effects of the project were fully assessed. Soon after, the Slovak authorities decided to proceed unilaterally with an alternative solution called 'Variant C'. This effectively diverted the Danube into Slovak territory and kept the development entirely within its borders. This scheme dramatically reduced the amount of water flowing into Hungary and had a significant impact on that nation's water supply and environment. As a result, the Hungarian government terminated the 1977 Treaty.

After further negotiations failed, Hungary and Slovakia submitted the dispute to the International Court of Justice (ICJ) and signed a Special Agreement to this effect. Pending the Court's judgement, the parties agreed to establish and implement a temporary water-management regime for the Danube.

The eventual verdict from the ICJ vindicated neither country completely. It found that Hungary was in violation of the 1977 Treaty and that Slovakia's implementation of Variant C was illegal.

Key issues for transboundary water management

There have been numerous attempts to reach an acceptable solution between Hungary and Slovakia but, to date, there has been no resolution. This can be attributed to factors such as the limited scope of the ICJ to undertake effective dispute resolution, and the extent of unilateral action by both parties, which is now a stumbling block to cooperation. These themes are explored in more detail below.

The role of the ICJ in international water disputes

This case highlights a number of important issues for transboundary water management and the role of the ICJ. During the course of the Gabčíkovo-Nagymaros case it became apparent that the court was not a particularly effective tool for dispute resolution. Indeed, as a result of the ruling, there have been many subsequent rounds of negotiations which have attempted to establish the way forward, but with only limited success. Nevertheless, the court was able to push both countries to discuss future cooperation rather than dwell on historical conflicts. Observers note that this is an important function of the ICJ. By legitimizing the *status quo*, the ICJ has compelled both parties to focus their efforts on solutions rather than devoting resources to arguing over past events (Jansky et al, 2004a).

More widely, this case sets an important precedent for international water law because the judgement of illegality against 'Variant C' endorses the principle of 'reasonable and equitable use' of water resources. This is a central principle of the UN Watercourses Convention, which has subsequently been enshrined in international transboundary agreements and institutions, such as the Mekong River Commission.

Third parties: the role of the European Union

Negotiations between Hungary and Slovakia demonstrate the importance of third parties and the influence of the international community. Here, the European Union was able to use political leverage because both Hungary and Slovakia were pursuing accession to the EU.

The EU used its influence to persuade Hungary and Slovakia to cooperate by submitting their dispute to the ICJ. Upon achieving EU membership in 2003, they both began the mandated process of harmonizing their environmental policies and practices to those of the EU. It remains to be seen whether this, their membership of the International Commission for the Protection of the Danube River, or the requirements of the Water Framework Directive, will bring about a solution to the dispute.

Joint environmental monitoring

Despite the lack of progress in the search for an agreement, Hungary and Slovakia have managed to implement joint environmental monitoring systems. This was undertaken as part of a joint agreement established in the interim period before the ruling from the ICJ was delivered (Jansky et al, 2004b). Joint monitoring was deemed to be helpful because it served to separate the political and technical aspects of the water dispute and provided a basis for negotiation. The use of joint data collection, and the isolation of technical data as a neutral basis for negotiation, is a common theme which runs through many of the case studies presented in this volume.

Consequences of unilateral action on future cooperation

Unilateral action taken by both the Hungarian and Slovakian governments has been a significant factor in the failure of negotiations to date. Both sides have invested significant time and finance in their respective unilateral plans, creating barriers to compromise and cooperation. The construction of infrastructure is a particular problem; once investments have been made into developing infrastructure, countries can be unwilling to reconsider their options, and backtracking becomes politically difficult.

Conclusions

The Gabčíkovo-Nagymaros case illustrates how unilateral action can hinder future cooperation between riparian states and hamper efforts to resolve disputes; in this case negotiations between the two parties remain ongoing. Notwithstanding the difficulties associated with these protracted negotiations, a programme of joint monitoring has provided a platform for closer collaboration on environmental issues. Finally, this case study clearly demonstrates that entrenched differences can require considerable time and resources to reconcile; despite repeated rounds of negotiation, a ruling from the ICJ, and membership of the European Union, there has been limited resolution to this case.

Adapted from Owen McIntyre (1998) 'Environmental protection of international rivers: case concerning the Gabčíkovo-Nagymaros Project (Hungary/Slovakia)', *Journal of Environmental Law,* **vol 10, no 1, pp79–91**

References

Jansky, L., Pachova, N. and Murakami, M. (2004a) 'The Danube: a case study of sharing international water', *Global Environmental Change*, vol 14, pp39–49

Jansky, L., Pachova, N. and Murakami, M. (2004b) *The Danube: Environmental Monitoring of an International River*, The United Nations University Press, Tokyo

13.14 Benefit sharing and the Lesotho Highlands Water Project

Aaron T. Wolf and Joshua T. Newton

Key points

- The potential for economic benefit can overcome difficulties posed by power disparities between riparians when negotiating over water resources.
- Environmental and social impacts should be thoroughly assessed before construction of large infrastructure. Failure to do so is likely to result in increased costs at a later date.
- External parties, such as international donor agencies, can balance asymmetries between riparians by providing both technical and financial support.

Context

The Lesotho Highlands Water Project is an ambitious transboundary infrastructure project between the small enclave state of Lesotho and its much larger neighbour, South Africa. The scheme is an illustration of the benefits that can be derived when states collaborate over water resources. In this case, the project simultaneously generates income and electricity for Lesotho while increasing water supply to South Africa via water transfers to the Vaal River Basin.

South Africa is a relatively water-scarce country and therefore has sought to increase its water availability to support industrial and population centres. Lesotho, by contrast, is a much less economically developed country, but has abundant water resources derived from the Senqu/Orange River. This originates in Lesotho's mountainous highlands and despite the small proportion of the catchment lying in the country, the area generates over half of the annual runoff of the river system (LHDA, 2009). In years of normal rainfall these resources exceed both the current and future requirements of its population.

The Lesotho Highlands Water Project consists of a series of dams on the Senqu/Orange River together with tunnels for the delivery of water to South Africa. Phase 1 has been completed and comprises, in addition to a extensive supporting infrastructure, the Katse and Mohale Dams, the Mtsoku Weir, connecting tunnels to transfer water to South Africa, and a hydropower plant. The original treaty envisaged further phases – indeed, Phase 2 is due for completion in 2017 (Lesotho Government Online, 2009) – but their implementation is dependent on feasibility studies, environmental assessments and finance.

The history of the project is closely tied to the shifting political relationship between South Africa and Lesotho. In 1966, South Africa officially proposed the joint water project to Lesotho and a Joint Technical Committee was eventually formed in 1978. This was mandated to gather information and undertake studies to assess the feasibility of the project. In order to move forward to design and construction, a treaty was required. Negotiations proceeded, and in 1986 the 'Treaty on the Lesotho Highlands Water Project between the Government of the Kingdom of Lesotho and the Government of the Republic of South Africa' was signed.

Under the terms of the treaty, the water transfer infrastructure was to be financed by South Africa, in addition to payments for the actual water delivered. Meanwhile, Lesotho was responsible for financing the hydropower and development components. To assist with this financial burden, Lesotho received aid from a number of donor agencies, including the World Bank.

Key lessons for transboundary water managers

This collaborative project provides a useful example of benefit sharing as both Lesotho and South Africa gained from the allocation of resources. Despite this success, there are aspects of the agreement that have drawn criticism. For example, the emphasis on technical feasibility at the negotiation and planning stages contributed to the environmental and social impacts being overlooked. It is also worth noting that serious corruption overshadowed the construction of the project, and eventually led to the prosecution of various multinational companies by the Government of Lesotho. However, this issue will not be examined further in this case study.

The following section details the important aspects and lessons that can be taken from this case.

Benefit sharing

The Lesotho Highlands Water Project is a practical example of sharing the benefits derived from a transboundary watercourse. Lesotho gained financially through foreign exchange and obtained energy from hydropower, and South Africa assured its supply of water for further economic growth. This demonstrates the potential of an integrated approach to negotiating the allocation of a 'basket' of resources, by providing specific benefits to both riparians.

South Africa now receives cost-effective water from high-altitude storage. This is a more efficient delivery mechanism for water, much of which would have eventually entered its territory prior to the agreement. South Africa saves on the cost of pumping water from the Orange-Vaal Transfer Scheme – an alternative source of water for industrial growth (Earle et al, 2005) – and also benefits from reduced evaporative losses; meanwhile Lesotho receives revenue and hydropower for its own development.

Renegotiation clauses and flexibility

In spite of the dramatic political shifts in South Africa (as a result of the fall of the apartheid regime), the treaty has never been significantly altered or renegotiated. This reflects an inherent flexibility which allows for adaptation of the treaty to changes in conditions and project terms. This is especially important given the phased implementation of the project, the long timescales, and the need for continued feasibility assessment and environmental management.

Context of wider political relationships

Understanding the influence of wider political relationships between riparians allows water negotiations to be viewed in the correct context. Initially, there was a strong relationship between the two states, particularly due to Lesotho's reliance on South Africa for trade and employment. At this early stage, progress towards working on a joint project occurred quickly.

This relationship then cooled as a result of South Africa's continued apartheid policies. This effectively stalled cooperation over water resources. The Treaty between the two countries was finally signed in 1986 after a change of government in Lesotho. This history highlights the influence of political relationships on water treaty negotiations.

The role of power asymmetry and the balancing effect of external parties

The differences in power and resources between South Africa and Lesotho show how such asymmetries can influence negotiations over water. South Africa used its financial and technical dominance to provide Lesotho with incentives for greater cooperation. With South Africa's considerable history in water resource development, it was able to make available both the technical and financial resources to undertake feasibility studies before it commenced negotiations. This allowed South Africa to set the agenda for these early discussions (Kistin, 2009).

Nevertheless, Lesotho managed to leverage its position by bringing in external experts from the World Bank to augment its technical water capability and to help engage negotiations from a position of greater capacity and knowledge. This intervention was instrumental in the finalization of the agreement between the two states.

Assessment of environmental impacts

Building dams and transferring water out of a river basin has a considerable impact upon the environment due to the submergence of land behind the dam and changes in the flow regime of the river downstream of the development. It also has a large social impact because it requires relocation of communities and settlements away from flood waters.

Notwithstanding the involvement of the World Bank, relatively little attention was directed to its requirements for environmental and social assessments. Far greater attention was given to the engineering challenges of the scheme and the feasibility of the water delivery infrastructure. In fact, the environmental action plan (EAP) for Phase 1A was only undertaken once construction had already begun. Evidence from the Phase 1B EAP contributed to the formulation of an Instream Flow Requirement Policy which was implemented in 2002. Conducting EAPs after design and construction can raise project costs and is not best practice.

Conclusions

The Lesotho Highlands Water Project illustrates the possible benefits that can be derived from collaborative water projects. There are many points of best practice that can be drawn from this scheme, such as the renegotiation clauses in the treaty, and the involvement of a third party to reduce the effect of technical resource asymmetries between states. Lessons can also be learnt with regard to the need for environmental impacts to be fully assessed prior to the design and construction of large infrastructure projects.

Adapted from Aaron T. Wolf and Joshua T. Newton (2009) 'Case studies of transboundary dispute resolution', Appendix C in J. Delli Priscoli and A. T. Wolf (eds) *Managing and Transforming Water Conflicts*, **Cambridge University Press, Cambridge**

References

Earle, A., Malzbender, D., Turton, A. and Manzungu, E. (2005) 'Preliminary profile of the Orange/Senqu River', www.acwr.co.za/pdf_files/05.pdf, accessed 10 January 2010

Kistin, E. J. (2009) 'Explaining output: the formation of the Orange Senqu water governance regime', paper presented at the *2009 World Water Week*, LWRG/UPTW/UNESCO Session on 'Cooperation as Conflict? Towards Effective Transboundary Interaction', Stockholm, Sweden, www.worldwaterweek.org/documents/WWW_PDF/Resources/2009_19wed/E_Kistin_Formation_of_the_Orange_Senqu_Water_Governance_Regime.pdf, accessed 10 January 2010

Lesotho Government Online (2009) *Phase II of Lesotho Highlands Water Project On the Go*, www.lesotho.gov.ls/articles/2009/PHASE_II_LESOTHO_HIGHLANDS.php, accessed 15 November 2009

LHDA (2009) 'Lesotho Highlands Water Development Project', www.lhwp.org.ls/overview/default.htm, accessed 15 November 2009

13.15 Cooperation over the Senegal River

Aaron T. Wolf and Joshua T. Newton

Key points

- Cooperation on the development of water resources can positively impact wider political and economic relationships.
- Public participation in decision-making processes, particularly with communities affected by potential environmental degradation, is important to reduce economic costs and frustration.
- Sharing the costs of development in proportion to benefits derived is an equitable system which enables greater cooperation.

Introduction

The Senegal River Development Organization (OMVS) was formed by Mali, Mauritania and Senegal in 1972. The agreement between these nations enabled dam construction in the Senegal River Basin, resulting in shared benefits from the transboundary water but also shared problems as a result of environmental degradation. The OMVS has successfully facilitated co-management of the basin despite the fact that until 2006 Guinea, an upstream riparian, was not a member.

The establishment of an international river basin organization was prompted by a number of factors including the regional climate, a rapidly growing population and the need for more rapid economic development. Drought, which is a common occurrence in this part of sub-Saharan Africa, increases the vulnerability of basin states and effectively constrains agricultural and industrial growth. Flows in the river depend on the variable amounts of rain that fall in the upper reaches of the basin in Guinea. Low rainfall in this area leads to water shortages for downstream states. As a result of this vulnerability to drought, the countries were happy to come together to build storage infrastructure and to regulate the flow of the river. A severe drought between 1968 and 1973 added impetus to this agreement.

Given this context, one of the aims of the OMVS was the development of projects that would allow more effective management of limited resources. Two noteworthy projects have resulted: the Manantali hydropower dam in Mali, which has

been in operation since 2001, and the Diama anti-salt dam in the Senegal River Delta. These allow the provision of year-round freshwater, and increase the resilience of the basin states to drought, by reducing water scarcity; however, the dams have resulted in serious ecosystem degradation.

Key lessons for transboundary water management

There has been culture of cooperation between the river basin states for over 30 years. Soon after independence, a series of conventions was signed which recognized the Senegal as an international river, and created management institutions. Unfortunately, owing to political tensions, these organizations gradually ceased to function. Finally the OMVS was created in 1972. This was mandated to not only manage the river basin, but also had the wider objective of the political and economic integration of the basin states (UN/WWAP, 2003).

The OMVS provides an interesting example by virtue of the levels of cooperation it engendered among the riparian states. This can be attributed to the broadly aligned interests of Mali, Mauritania, Senegal and, more recently, Guinea. The impetus to manage the river basin collaboratively results from the vulnerability to drought, complementary interests in hydropower, navigation and irrigation (althoug the trade-offs between these goals need to be managed), and the availability of international aid. Despite these levels of cooperation there are a number of less positive aspects which are examined in the following sections.

Guinea and the OMVS

Until 2006, one obvious problem for the OMVS was that not all riparian states were party to the agreement. While Guinea had been a member of transboundary organizations formed prior to the OMVS – for example the Organisation of Boundary States of the Senegal River – political instability had made joining the OMVS difficult. Prior to joining in 2006, Guinea held observer status to the treaty and crucially did not object to the development undertaken by other states. Nevertheless, given Guinea's position as an upstream riparian, its membership is the preferred solution for co-management of the river. Now the focus is on integrating Guinea fully into the organization and adjusting the various benefit sharing mechanisms to reflect the needs of the fourth riparian.

Environmental degradation

Dam construction on the Senegal River has provided electricity and irrigation, but unforeseen environmental impacts of the dams have produced a number of negative side-effects. These include increases in waterborne disease, pollution from irrigated agriculture, degradation of ecosystems and a reduction in the natural recharge area of important aquifers in the basin (UN/WWAP, 2003). This prompted the formation of an Environmental Impact Mitigation and Monitoring Programme to help develop strategies to mitigate the effects of these problems.

The importance of participation

During the design and construction of the Manantali and Diama dams there was limited participation by the various stakeholder groups involved, particularly the local population. This resulted in frustration and even economic losses among stakeholders. More recently, local coordination committees (LCCs) have enjoyed an increased role as the OMVS has altered its institutional framework in response to the environment impact of the dams. Greater emphasis has now been placed on the participation of the communi-

ties who reap benefits from the dams but who are also impacted by related environmental degradation (Sylla, 2006).

Benefit-sharing mechanism

The OMVS uses an equitable procedure to apportion benefits derived from irrigation, hydropower and navigation. Benefits are divided between the states using a formula based on the levels of investment respective countries have contributed. This design promotes equality between the states with regard to development of the river basin. The OMVS also uses an original mechanism for the allocation of water which is described below.

Management of inter-sectoral conflicting demands: optimal distribution

The OMVS has created an effective strategy to manage competing claims for water. Resources are not allocated on the basis of volume to a country but rather the use to which the water can be put by sector (UN/WWAP, 2003). The objective of such an approach is to ensure that benefits reach local populations and support sustainable development in the basin.

The approach centres on inventories of need produced by OMVS National Committees. These are sent to the OMVS High Commission which produces recommendations for the Council of Ministers. A decision is made by the Council on the basis of the following criteria (UN/WWAP, 2003):

- reasonable and fair use of the river water;
- obligation to preserve the basin's environment;
- obligation to negotiate in cases of water-use disagreement/conflict;
- obligation of each riparian state to inform the others before undertaking any action or project that could affect water availability.

This method of water allocation is detailed in the 2002 Senegal River Charter and provides an example of an innovative mechanism to maximize the potential benefits of sharing water between riparians and different sectors.

The role of international aid

Consistent with the experience of transboundary organizations in other parts of the world, the role of the international community in providing financial and technical support has proven to be critical for OMVS. In this case, a large consortium of donors raised the finance necessary to undertake the construction of water infrastructure by the OMVS.

Conclusions

The OMVS embodies a flexible framework for the management of the Senegal River Basin by all four of the riparian states. Cooperation over water resources is underpinned by common goals including sustainable growth and greater resilience in the river basin. The joint funding of the OMVS, with countries re-paying loans from donors to a level proportional with the likely benefit accrued is an equitable system and one which has worked well. Nevertheless, it will need to adjust to reflect the inclusion of Guinea. This example of benefit sharing demonstrates a number of positive aspects of cooperation for other transboundary river basins, and one which will be strengthened if some of the negative environmental effects of water resource development in the region can be mitigated.

Adapted from Aaron T. Wolf and Joshua T. Newton (2009) 'Case studies of transboundary dispute resolution', Appendix C in J. Delli Priscoli and A. T. Wolf (eds) *Managing and Transforming Water Conflicts***, Cambridge University Press, Cambridge**

References

Sylla, M. M. (2006) 'The role of basic community organizations in the management of the natural resources of a transboundary water basin – the example of the local coordination committees of the Senegal River Development Organization', in A. Earle and D. Malzbender (eds) *Stakeholder Participation In Transboundary Water Management – Selected Case Studies*, www.acwr.co.za/pdf_files/06.pdf, accessed 22 November 2009

UN/WWAP (United Nations/World Water Assessment Programme) (2003) *1st UN World Water Development Report: Water for People, Water for Life*, UNESCO and Berghahn Books, Paris, New York and Oxford

14

Towards a Conceptual Framework for Transboundary Water Management

Joakim Öjendal, Anton Earle and Anders Jägerskog

Introduction

As described in the introductory chapter of this volume, the three major groups involved in transboundary water management (TWM) are the water resource community (including water managers from government, as well as water users from the private sector and civil society), the research community (including academics, international financial institutions (IFIs) and development partners) and politicians (at various levels of scale). The politicians, as the prime movers in TWM systems, project and use the four forms of power of the state (geographical power, material power, bargaining power and ideational power) in accordance with the range of domestic and international pressures exerted on them (see Figure 1.1 in Chapter 1). The figure of the three interconnected cogs is incomplete in that it does not include the cogs representing other sectors in a political economy, which are driven by factors such as energy security, food security, cultural and identity issues. A model, as a simplified representation of reality, is by nature incomplete, but for a better understanding of how power is used in various TWM situations there needs to be aware-

ness that there are influences and forces lying outside the three groupings mentioned above.

In the chapters of this book we are introduced to five basic yet crucial points about TWM:

1   Improved TWM is one key for responding to the global water crisis, given the current and future distribution of available water resources.

2   To impact the allocation of these resources – hence the quality of water governance – is a complicated process with several sub-fields to consider, in particular knowledge (research), interests (politics) and practice (water community experience), which cannot be left out.

3   To change any *status quo* (and sometimes to *not* do that), is a conflictual process with many interwoven stakes and stakeholders, with complex issues and trade-offs involved.

4   Among the water communities there are attempts at cooperation and development of better practices bringing, if not solutions, at least a certain momentum to the field, consequently making it receptive to further input.

5   In spite of the vast need and a certain amount of goodwill, the established knowledge on

TWM is limited, fragmented, and, often, case-specific, providing poor guidance for the future governance of the field. Some would say 'scientific knowledge' is so poor that there is evidence that it can hardly be researched at all in a distinct way (Allan and Mirumachi, this volume Chapter 2).

These are seemingly straightforward points; and yet so crucial. The final point in particular – or rather to counter it – is what ultimately motivates this volume. The editors have aimed to achieve three things. The *first* would be to identify the knowledge frontier, through the contribution of cutting-edge research carried out by some of the prime researchers in the field. The *second* is to collect, in a systematic way, experiences from the practitioners within particular niches that are deemed to be of key significance, as well as to make an inventory of the current challenges and dilemmas in TWM. *Third*, we aim to compile this in a comprehensive way, including future challenges, and make it accessible to a broad audience. We have also given room for a more speculative, forward-looking section, where we try to encourage authors to look 'around the corner', find solutions, and make the best possible use of existing knowledge. All these ambitions will, to varying degrees, be reflected in this concluding chapter.

Below, we will recollect some of the key insights the contributors have presented, and distinguish what, in our interpretation as editors, the key analytical aspects are in the different sections, before we finish with a section where we assess what we have learned from this endeavour, and, indeed, what one *can* learn from a volume like this. Finally, we reflect on how best knowledge can and should be distributed and applied, given our particular understandings of the dynamics in this field.

Contributors' Insights

The contributions to this volume are by necessity from a varied background and presented in different traditions, spanning a wide field, capturing both practitioner experience and recent research findings, and presenting these in a handbook fashion. At the outset the editors developed three categories of chapter for the book – the first aimed at developing the theoretical framework for TWM, the second related to the practice of TWM and the final section identifying emergent trends. Here follows some review, divided into the key categories that have structured the volume.

The first section, *Analytical Approaches to Transboundary Water Management*, is described in the introduction to this volume as aiming 'to explain why wars over water are unlikely, explain some of the interactions between states from a hydropolitical perspective and describe how expanding the focus of negotiations to include benefits associated with a water system can ease possible disputes over shared waters'. The core message here is that at the interstate level, where we may expect conflict over water to be most acute, there are in fact a variety of mitigating mechanisms, preventing water scarcity disputes leading to larger or more intense conflicts. However, the initial optimism shown by some commentators on TWM processes has more recently been tempered by a nuanced and critical understanding of the limits of cooperation, and its relationship with power.

As is shown in several chapters, not all power is equal, benevolent or desirable (for all); and likewise, not all cooperation is equal. As a consequence of asymmetric power relations between watercourse states, the management of transboundary waters in conditions of scarcity becomes increasingly a securitized issue. This makes it especially difficult for outsiders to study the processes leading to different types of cooperation, essentially having to rely on proxy indicators for the quality of cooperative processes. Similar to the thought experiment devised by Erwin Schrödinger, in which a cat in a sealed box is deemed both alive and dead until the box is opened and is seen to be one or the other, TWM processes present a paradox for the research community. The successful cases – whether defined by their outcome, linked to an infrastructure development project for instance, or defined by process – such as the formation of an RBO, are the ones which are transparent enough to access data on. The more intractable cases are opaque, with only limited or partial and inferred observations possible. Securitized issues are, by definition, not in the public domain. The implication for the research community, including IFIs and donors, is that their understanding of, as well as their ability to influence TWM processes, remains limited. Comparisons across cases are difficult to make, ending up being meaningless or possibly inappropriate to the specific situation. This is a pressing methodological dilemma for rigorous research and makes it challenging to accumulate transferable knowledge on TWM. Therefore, we have chosen to complement the system-critical 'from-the-outside' research with a systematic review of experiences 'from-the-inside'.

In an effort to bridge the gap between the theory and the practice of TWM the second section of this book, *Transboundary Water Management Polity and Practice*, presents an experience-based inventory of issues typically associated with the governance of transboundary waters. Recognizing the above point that each TWM situation is unique and that what works in one geographic or political setting does not necessarily work in another, the authors have nonetheless attempted to identify some common features related to good practice in TWM. There exists broad consensus among the international community on the core principles of international water law, notwithstanding dissenting voices. These principles can be adhered to or ignored at the discretion of each state – there is no international police to enforce their implementation. However, irrespective of what stance a specific state takes in relation to customary international legal principles it is important for water professionals tasked with TWM to develop a robust understanding of these principles. Likewise the degree to which the environment is recognized as a legitimate 'user' of water in its own right will differ from state to state and from one watercourse to the next.

There are, however, international norms emerging on the methodologies associated with understanding environmental water requirements. A common understanding of key issues for professionals to base their decisions upon is needed, and not only for professionals. With emphasis now placed on involving a broader array of stakeholders in TWM there is an even greater need for mechanisms which can be used to develop common understanding on a range of environmental, social and economic issues related to the use, management, development and conservation of transboundary watercourses, widening the idea of an 'epistemic community' (Haas, 1992) to its broadest possible extent. This second section of the book thus attempts to share with the reader some of the successful approaches taken to TWM globally over the past two decades. Although each watercourse is unique, it remains possible to learn from successes as well as failures, and in so doing become more proactive in developing appropriate solutions. Some of these failures have been linked to individual cases where, for instance, due consideration was not given to environmental or social concerns. Others are more systemic, a good example being the limited attention placed on the management of transboundary aquifers. For instance, as Puri and Struckmeier describe (Chapter 6), these aquifers supply much of the domestic as well as productive water globally, either as a primary water source or as a buffer during periods of low surface water flows. Yet only recently has an attempt been made to develop international institutions and processes to manage this underground resource.

The final section of the book, *Challenges and Opportunities*, includes 'forward-looking chapters on facing environmental change, and the role of science and education in building capacity to manage effectively the range of changes creating pressure in transboundary basins'. In some sense this has been the most challenging section to compile, as it is more straightforward to describe the current theoretical underpinnings of TWM as well as offering empirical experiences of TWM on the ground. What of the future? Will the world witness an escalation in the number or the intensity of international water disputes? Will environmental change test current TWM institutions to the limit, and beyond? Not if the appropriate investment is made, of time as well as human, financial and political resources, in developing sufficiently robust institutions for TWM. These institutions would also include effective legal frameworks – whether at the global convention scale, for specific watercourses or even bilaterally between states.

Due in part to the complexities of international negotiations, many of the water-sharing and management agreements in place today are not sufficiently flexible to cope with climatic, demographic or economic changes. And yet, the one certainty is that change is inevitable; how do we manage it? Agreements need to recognize and incorporate the water-related impacts of these changes, providing the flexibility which will promote durability. Likewise, watercourse management institutions must be able to grow and develop in tandem with a changing environment, learning lessons from the management of business enterprises where changes in markets, raw materials and technologies are taken for granted. For this to happen, a sound system of knowledge, based on good science, is essential. Investments need to be made now in the education and training of water management professionals of the future to enable them to respond to these challenges.

There are at least two overarching fields of analysis emerging from the broad spectrum of chapters reviewed above. Several of the policy-oriented papers emphasize the 'can do' aspect of TWM, with an understanding that it needs to be done better than it has been to date, whereas the more critical research chapters emphasize the difficulties of managing water efficiently in an international context, pointing at the underlying structural dilemmas of the international system. This is also the crux of the matter in TWM: some established water resource management practices are not adapted to the international system and its particular conditions, and international actors have difficulties assuming experience-based insights mainly developed in the domestic setting. A gap remains between the water management ideals, and the workings of the international system. *Additionally*, there is a disconnect between basin-scale management approaches, where issues are analysed and dealt with inside the basin and by the actors *inside* the basin, and the wider *problematique* needing analysis and action from *outside* the basin (too) and (also) by external actors. In other

words, to what degree should TWM primarily be understood and determined within its geographical extent (watercourse or basin) or by a global water and development community? The latter view is increasingly supported by arguments such as 'virtual water', (Allan and Mirumachi, this volume Chapter 2) 'significance of global climate change' (Falkenmark and Jägerskog this volume Chapter 11), and water management as a 'public good' (Jägerskog et al, 2007).

The management of transboundary watercourses is thus moving away from the state-centric and sovereignty-based model, and towards an expanded and globalized approach. Watercourses supporting internationally important ecosystems, such as wetlands and deltas, form part of a global environmental heritage in much the same way as rainforests and glaciers do. As an illustration of this situation, consider the case of the Okavango Delta in Botswana. The Delta was declared a Ramsar site of international importance; indeed it is the largest Ramsar site in the world, making it the focus of attention for various local and international environmental groups (Turton and Earle, 2003). When the states upstream of Botswana on the Okavango River, Namibia and Angola, started publicizing plans for transferring water out of the basin, they were dissuaded in large part through the pressure from international groups such as the International Union for Conservation of Nature (IUCN). The government of Botswana voiced relatively little concern about the plans – stressing that they would support a full environmental impact assessment and options analysis process prior to reaching a decision (Turton and Earle, 2003). It is no longer sufficient for only the states physically linked to the watercourse to agree on development priorities; the international community in various guises also plays a role. In other words, if human rights breaches are a global concern, and if poverty alleviation is a common goal, then TWM is also a global concern and should be recognized as such. Given the contemporary depth of globalization and the increasing intertwining of societal processes, water governance in a basin can hardly be seen as a concern for the riparians alone.

Key Points for Improved TWM

The chapters in this volume have introduced the reader to the challenges and approaches associated with TWM. On reading each chapter it becomes apparent that the optimal or perfect system does not exist; instead we are introduced to a variety of compromise and volatile situations. The trade-offs which need to be made between economic sectors, the environment, and society at large are neither unique to the transboundary situation, nor to water resource management. Having stated that, and drawing on the review of the chapters above, it seems evident that a few key issues deserve closer attention.

The voice of critical research

Water management could be a *source of conflict* (Starr, 1991). Consequently, as is made clear by Jarvis and Wolf (this volume Chapter 9), water management is intimately linked with conflict management. The need to satisfy an insatiable appetite for water conflicts with the reality of a finite supply. At the transboundary level the propensity for conflict is, apparently, exacerbated by the asymmetric power relations between basin states, and by the lack of an effective international judicial system to mitigate such imbalances. As described by authors in this volume, the seemingly unavoidable zero-sum outcomes associated with the interstate competition for water resources leads to an environment-conflict trap – for the country, basin or the region. It is also pointed out that water is not typically the trigger for intense conflict or war between states; indeed it is rare for states to go to war over any one issue. What emerges from a reading of the chapters is a multifaceted and dynamic flirtation by states with both conflictive as well as cooperative strategies related to sharing water. The concept of a *continuum from conflict to cooperation* is insufficient in describing the variety of positions states choose to take on different water bodies, different issues, with different neighbours at different points in time. *Continuous negotiation* is a more appropriate term, which may constitute a future trend given that many agreements turn obsolete due to (perceptions of, or references to)

global climate change and its impact on rainfall and flow regimes, as well as other demographic or economic change factors.

Allan and Mirumachi (this volume Chapter 2) claim that 'power relations which feature in the politics of transboundary relations, and the invisible processes which feature in the international economy, make researching transboundary water resource allocation and the joint international management of such waters virtually impossible'. In that analysis the degree of transparency of TWM systems is inversely related to the degree of securitization of that situation. As relations between states sharing water resources move from non-politicized to politicized and on to securitized so too does the transparency of decision making and planning decrease. As such, *conflict* and *cooperation* are not static conditions, but subject to the dynamics of *powers* and *interest*; nor are they mutually exclusive. These are underrated factors not taken into account sufficiently.

Hence, transboundary waters are not necessarily shared waters; the former being a chance of topology, the latter a constructed response to promote cooperative management of those waters. As demonstrated by Cascão and Zeitoun (this volume Chapter 3), these management responses are heavily dependent on the power relations between the countries concerned. It is not enough to consider power as a one-dimensional measure of a state's ability to implement its will on others; various dimensions of power are available to states to a greater or a lesser degree at different points in time. The geographical power associated with lying upstream can be either countered or complemented by the *realpolitik* of material power, where countries have neither friends nor enemies – only interests. Thus a state in a relatively powerless geographical position, perhaps a downstream riparian, can exert its will upstream by relying on economic or military muscle. But the concurrent anarchy and interconnectedness of the global political system opens other more abstract avenues to power. Bargaining power, related to the ability to control the rules of the game, and ideational power, the ability to influence the sanctioned discourse, combine to either entrench the two more visible forms of power or militate against them. This more nuanced conceptualization of power dynamics goes some way in explaining why states seem to engage in both cooperative as well as conflictive actions over their transboundary water resources.

Therefore, we claim, '*cooperation*' is elusive in nature. Following from the 'continuous negotiation' thinking, states seem to drift in and out of cooperation and conflict, at various times and in various settings. In addition to the analysis of various contexts, the conflict–cooperation quavering has a lot to do with the normative understanding of the term 'cooperation'. As argued by Cascão and Zeitoun (this volume Chapter 3), cooperation is not by definition good and is not the opposite of conflict. An absence of conflict implies harmony – the issue is not contested. Cascão and Zeitoun build on the thoughts outlined by Keohane (1994) who made a useful distinction in relation to cooperation, conflict, harmony and discord which has been largely left unattended by writers on water conflict and cooperation. He argues that cooperation is sharply distinguished from both harmony and discord. When harmony exists between actors, their respective policies are geared towards attainment of the goal(s) of the other actor(s). When discord prevails the action taken by each actor is effectively a hindrance for the attainment of the other's goal(s). Even if states are in harmony of discord there is scant incentive for them to change their behaviour.

Cooperation (which has often been more or less equated with harmony but in line with Keohane's argument is distinct from it) would therefore require that organizations or institutions change their behaviour, previously not aligned, towards a common position through a process of policy coordination. In this perspective cooperation is dependent on each party changing its behaviour as a reciprocal act. The dichotomy – either water cooperation or water conflict, that has characterized the debate on transboundary waters during the last decade, has not been helpful in unmasking the *nature* of cooperation; indeed it does a disservice to the understanding of the quality of the relations and the available policy options in any particular basin. It is clear that cooperation and conflict coexist (Zeitoun and Mirumachi, 2008) and arguably conflict is a precondition for cooperation to take place. A move by two or more states to bring their policies in line with each other is a move from policies that are in conflict, to adopting policies that are harmonious. Now, with the perspective inspired by Keohane and developed by Cascão and Zeitoun, we are able to pose questions such as:

*   What is the *quality* of the cooperation?
*   *Why* and *under what conditions* is it taking place?
*   Is it sustainable?
*   Will it deepen and lead to tangible socioeconomic development objectives?

As discussed above, this paradox presents the research community, academics specifically, with something of a dilemma. The most researchable TWM cases are those least likely to display innovative approaches to resolving conflicts. The cases where the underlying hydrologic, economic or cultural drivers lead to a greater potential for conflict and hence closer to a state of securitization, are those about which we know the least, and, even worse, are least likely to learn much about (according to Allan and Mirumachi in Chapter 2 in this volume). In these cases it is difficult to assess to what degree the hydrological realities contribute to the possibility of conflict between states, as well as being impossible to judge which solutions were most effective in preventing conflict. The closest researchers get to understanding the situation is through relying on proxy indicators – most famously data on the quantities of virtual water traded between countries. These 'politically silent and economically invisible' (Allan, 2001) exchanges result in a large-scale and effective system of conflict avoidance. A rule of thumb here is that the more securitized a basin is, the more the 'politically silent and economically invisible' indicators matter; and the more a basin can and shall be understood in relation to its regional hydropolitics and its underlying political economy. Hence, understanding – and indeed practising – TWM is adrift between an analysis of its tangible indicators, based on outcomes, and an analysis of regional politics, connected in a web of complexity.

The implication of this safety net – which allows a guarantee of the provision of staple foods to the country by relying on international trade – is that politicians solve a major water management problem by other means, and not through good water management. The actions of the state in relation to managing shared waters thus reflect less the underlying hydro-economic reality of the basin, and have more to do with the strategic polit-ical–economic objectives of the country as a whole. This ability to avoid a certain category of water management problem by tapping into the surrounding political economy provides another 'escape' from the 'proxy' indicators, and calls for yet another type of analysis.

In addition to understanding the interaction of TWM communities, represented by the cogs in Figure 1.1 of the introductory chapter, we need to realize that the 'cog-machinery' operates in three contexts at least: rationalist-proxy, security, and global political-economy, all demanding different types of analysis. The River Rhine in Europe and the Torne River between Sweden and Finland can in general be rationalized from their respective proxy indicators. The Torne River forms the border between two states, but only provides a livelihood for major populations to a limited degree; nor is it securitized. Thus, what can be observed and researched, can also to a large extent form the basis for a deep analysis. At the other extreme, we have a case such as the Jordan River which cannot be understood without assessing the security situation. The geographical and physical facts as well as trade figures – that is, what is in the public domain – are only to a minor degree informing us of the realities of TWM. In this case there is, figuratively, a sealed box hiding securi-tized water issues, based in conceptions of history and the identity of the primary actors involved in TWM.

The Mekong River provides yet another image, because it is neither economically irrele-vant nor highly securitized. It is, however, intertwined in a major political economy with serious implications for a wide range of stakehold-ers. The region is highly dependent on a particular water regime and TWM here cannot be under-stood without grasping the underlying political economy of the region and the options the actors promote. Cases such as the Tigris–Euphrates Rivers, as well as the River Nile, demonstrate a complexity requiring a combined analysis of these three fields. In reality all basins contain elements of, and need to be understood in relation to, all three of these spheres. One of the key tasks of good TWM is to know what weight to give the different approaches. There is no handbook that can predetermine this, but the awareness of the fact that different basins have different underlying processes – with different rationales – is a key insight.

In this volume, besides analysing the aspects of power asymmetries and transboundary waters (Cascão and Zeitoun, Chapter 3), incentives for broadening the perspective to an expanded range of benefits are also analysed (see Daoudy, Chapter 4). While the benefit-sharing discussion has moved forward, it still has to prove itself in practical terms, with no comprehensive water management strategy based on it to date. Nevertheless, the broader perspective offered with the 'benefit-sharing' approach is important, in that it offers potential *incentives* for states to move beyond their narrow interests. These often revolve around securing their equitable share of the water, effec-tively constituting an unsolvable zero-sum game. Benefit-sharing is at the core of the attempts to move 'cooperation' further and deeper, overcom-ing security-based interlocking, and disentangling stalled powers and interests.

It is clear that within policy as well as within academic circles, the understanding of what constitutes cooperation and how that is measured has been underdeveloped and viewed in a rather simplistic way. In many cases development part-ners and academics alike have, in a Kantian/functionalistic manner,[1] assumed that signs of cooperation – no matter how ineffective, shallow or unsustainable – are some sort of measure of success. As emphasized by Cascão and Zeitoun in this volume, that is an incomplete picture. Selby (2003) noted that in the Israeli–Palestinian water conflict, what is often termed cooperation is in fact closer to domination. This leads to the question: what would be the right indicators to analyse and evaluate cooperation over transboundary waters?

It was noted previously that, although useful, the cooperation continuum developed by Sadoff and Grey (2002) is not enough to understand what the quality of cooperation is. Grey et al (2009, p19) have further contributed to developing a deepened understanding of what constitutes 'effective cooperation'. Their definition is: 'Effective cooperation on an international watercourse is any action or set of actions by riparian states that leads to enhanced management or development of the watercourse to their mutual satisfaction.' Thus, mere *coordination* between riparian states, which in some cases has been considered as cooperation, is not enough. It should be to the 'mutual satisfaction' of the riparian states, lead to 'enhanced management', and, ideally, be sustainable. In other words, it needs to be firmly linked to outcomes of the cooperation process. However, from an 'objective' perspective this may not be enough since the definition ignores international water law and its principle of equitable utilization. While equitable utilization is also hard to define, there are legal principles that would help in this case as outlined by McIntyre (this volume Chapter 5), and established in the UN Water Convention.

In the above, the analysis of the nature of the internal conditions in basins and the consequences for TWM has primarily been considered. But, increasingly, demands for a broader analysis of TWM are articulated and conceptualized. This is, however, challenging well-established dogma within TWM. Let us consider this challenge.

Challenging the dogma: 1. Breaking the tyranny of the basin as a basis for management

During the past two decades (and in many cases for longer than this), the dominant paradigm in water management circles has been that the basin should form the unit of management. However, the current debate regarding transboundary waters has moved beyond the basin as unit of analysis. Allan (2001) has succinctly pointed out that the relevant unit of analysis is not the watershed but the 'problemshed'. He describes the problemshed as the operational context in which decision makers and their challenges exist, while the watershed is a naturally determined area in

which variable and limited water resources exist. When a watershed is no longer able to provide the resources for the livelihoods of the people living within it, the solutions are found in the political economy. A prime example of this point is the virtual water argument – when the water within the watershed is not sufficient for production of food, access to the global food market offers a solution to the food security within a watershed.

Furthermore, as noted by Jägerskog (2003), the issue of *linkages* between different issue areas in international affairs and negotiations is likely to affect the outcome in terms of transboundary waters allocation. While the relative powers between states negotiating over transboundary waters are bound to affect water allocation, it is also likely that trade-offs are made between different areas in negotiations. To maintain a basin focus, when decisions affecting water management are often taken outside of the 'cogs' of water professionals (researchers or practitioners), becomes increasingly difficult at the transboundary scale. In other words, scale matters. It makes sense to use the basin as the unit of management on a national level, as the rules of the game are clearer and better established, usually with mechanisms incorporated to protect the interests of less powerful parties. But at the international, transboundary level the situation is different. At the international level power asymmetries are more pronounced, exacerbated by the lack of clear and enforceable legal structures.

While focusing on the basin and what may be produced by a rational use of the water, the benefit-sharing approach could well be seen in a broader light, moving away from the focus on the watershed or the basin. It further involves a range of other factors such as energy, food production, trade, navigation and even peace and regional integration (Sadoff and Grey, 2002). Although still in need of being further substantiated and practically proven, the more holistic benefit-sharing approach clearly moves us away from a narrow basin perspective to a wider regional, as well as a global perspective including political economy considerations. As discussed by Granit (this volume Chapter 10), the regional economic communities (RECs) being constituted and developed in various parts of the world are playing an increasing role in the management of natural resources generally, and water in particular. These organizations are well placed to promote integration across sectors, watercourses and between countries.

As noted by Falkenmark and Jägerskog (this volume Chapter 11), an increasingly important challenge in TWM is connected to mitigating the impacts of climate change. Often, agreements on transboundary waters are based on multi-year averages of flows and fail to account for climate variability or change. Currently this creates political stress and tensions between the signatory states of agreements. With more basins becoming fully allocated, coupled with larger variability in terms of rainfall, runoff and flow, the political stress that agreements will face is bound to increase. From a technical perspective this calls for the renegotiation or amendment of existing agreements, so that greater *flexibility* can be introduced. However, the political feasibility of this, in most cases, remains problematic. As a minimum, though, agreements which are naturally approaching a date for renegotiation, or ones which will be negotiated in the future, should incorporate such flexibility.

This flexibility could be incorporated by relying on relative instead of absolute flow values. Thus each party receives a share of the available runoff – once a minimum allowance has been made for environmental water requirements. Another approach is to link two or more basins under one agreement, as has been done with the Incomaputo Agreement, entered into between South Africa, Swaziland and Mozambique in 2002.[2] The agreement includes provisions for allocating water among the countries and stipulates a schedule of priority for different user groups within the countries during times of drought. Thus, municipal water supply in each country enjoys higher priority than the energy generation sector, which in turn has a higher priority than the agricultural sector (Earle and Malzbender, 2007). Additionally, the agreement in fact covers two hydrologically separate basins located next to each other – the Incomati and the Maputo rivers. The advantage of this approach is that it allows the states to be more flexible over

allocation from one of the basins than over the other – depending on which basin provided water to their most critical water-use sectors.

Challenging the dogma: 2. Breaking the tyranny of water professionals doing TWM

In an increasingly co-dependent world the issue of regional integration of states is becoming more important. Multilateralism is important not only on a global scale but also on the regional level. For example, the European Union is deepening its internal cooperation, and the East African Community (EAC) and ASEAN in Southeast Asia are moving towards free trade areas and increasing cooperation in various sectors. It is clear that the movement towards closer integration in regions opens up opportunities both for increased growth and development and also for joint, and thereby improved, management of common resources. Conversely, it may also be that joint management of resources, such as water, energy sources and transport, opens up opportunities for investments and integration. It is clear that mandates for river basin institutions may vary, depending on political context and region (see Granit, this volume Chapter 10). In some instances, such as with the Lake Victoria Basin Commission (LVBC), the commission has been nested within the relevant REC – the East African Community. This allows for close alignment of water policy and strategy between the five member states. Likewise the Southern African Development Community (SADC) has in force a Protocol on Shared Watercourses, committing member states in that region to the basic principles of international water law (McIntyre, this volume Chapter 5; Earle and Malzbender, 2007). This is significant from a TWM perspective since it provides a political umbrella covering principles for interaction over transboundary waters and the negotiation of basin-based agreements. When a river basin commission (or commission for transboundary aquifer management) is part of a regional structure, its mandate is clearer and it is likely to have greater enforcement capabilities, and therefore better possibilities to create sustainable water management. Whichever form of REC is in place, water can become a vital input to a variety of other sectors and thus no longer the exclusive domain of the water management community. Other cogs are now added to the model – energy, transport, agriculture and health being some of the obvious ones. Although this adds complexity, at the same time it moves away from cooperation for cooperation's sake, and towards a focus on the outcome of that cooperation. These sectors will want effective TWM to deliver a sustainable supply of water in order for them to fulfil their respective mandates. Ideally they would also have the necessary financial resources to invest in developing water resources for this purpose.

While it may be argued that the increased role of RECs in TWM is useful, a number of questions still remain. As noted in many chapters of this volume, political power is central for understanding the old political science question of 'who gets what water, when, where, why and how?' They show that often the more powerful riparians are able to secure the lion's share of the resource for their own benefit. Even if RECs are likely to get more involved in TWM it will nevertheless be useful to keep in mind that although some of the powers may be transferred from states to RECs, most of the power over important decisions will still be vested within the state (politicians). It is important to note that with increasing regional integration taking place within RECs, the incentives for politicians to move in the direction of more cooperation do exist.

Conclusions

A major conclusion in this volume is that a more nuanced understanding of cooperation over transboundary waters is called for. The chapters of this volume as well as recent work by scholars such as Zeitoun, Selby, and Grey and Sadoff clearly show that the current definitions and indicators of success in TWM are not satisfactory. This volume supports this point, arguing that we need a shift of emphasis from the agreements, processes and commissions which are formed through cooperation, and towards the *outcomes* and eventual *impacts* of cooperation. These development outcomes will

be specific for each region. Around economically well-developed regions such as the Baltic Sea or the River Rhine, the emphasis would be on promoting healthy ecosystem functioning. Other parts of the world may view outcomes of successful cooperation as the sustainable development of water storage or transfer infrastructure. Arguably, principles such as equitable utilization need to be at the core of the development of indicators, for clearly measuring early success in promoting cooperation.

Acknowledging the complexity of TWM, and understanding the roles and powers of various actors (researchers, practitioners and politicians) and the way these affect the decisions taken on TWM, is a key insight, and is vividly described in this volume. Many of the decisions guiding outcomes of TWM are not taken by water professionals or practitioners. Rather, the political context in which TWM takes place should be analysed, for a proper understanding of why certain outcomes emerge. In addressing the new challenges, whether in the area of climate change adaptation, or adaptation to demographic changes, this understanding is imperative.

This volume poses two main challenges. *First*, it questions the dominant paradigm that water management should take the basin as its point of departure. It draws on insights by Allan (2001) that put the focus on the problemshed rather than the watershed, as well as the understanding that the inherently political nature of TWM makes it open to *linkages* with other political areas. To return to the initial 'cogs' metaphor it is no longer possible to maintain that we make a comprehensive analysis if we only consider the basin as our point of departure for analysis and planning. Decisions affecting water management in a transboundary basin are to an overwhelming extent taken outside of the basin, and to maintain an exclusive focus on the basin would ignore the political economy in which decisions that most affect TWM are taken.

Second, and following logically from the first, is the role of water professionals. While from a seemingly rational perspective it would make sense to give water professionals a more prominent role, in real-world TWM processes that is not how things typically work now. Politicians, not water professionals, are held responsible for the policies decided upon. From a regional political perspective, increased integration (whether in the EU, EAC, SADC or any other region) is also emphasizing the connection between water and other areas such as energy, transport and food. It is therefore time for the water communities to reach out for an increased understanding of the challenges that lie ahead.

To advance the TWM agenda it is argued that it is time for the various communities involved in water management to intensify work to develop robust indicators of transboundary cooperation, acknowledging legal principles and sustainable and sound management, as well as the interests of the states engaging in cooperation. Once the rules of the game – based on the principles of customary international water law – are established, the next step is to move towards an assessment of the outcomes of that cooperation. This is where the role of the international community – researchers, donors and IFIs – becomes important. By working with governments and other stakeholders it can support the development of strategies and action plans leading to specified development objectives. To date, the role of the international community has been mixed, and sometimes support for narrowly defined cooperation (without enough concern about actual outcomes) has served to entrench the *status quo*, often for the good of the more powerful riparian(s). If the international community will move towards more concern for the rules of the game, as well as become more involved in analysis of underlying power structures, there is an increased likelihood of cooperative outcomes that would be beneficial for those less empowered.

Hence, good TWM is based on an intricate combination of knowledge, politics and communication. We need to deepen understanding of such abstract issues as 'cooperation' and educate more skilled water professionals; and most importantly, we need to enhance the avenues of communication between the groups. Whereas the former are processes in the making, the latter – to make politicians respect water professionals and to make water professionals understand political rationales – possibly requires a deeper rethinking, and may

be the weakest link in this chain. This volume, in its own modest way, represents an attempt to support such a process.

Notes

1   The *functionalist perspective* arose during the inter-war period of the two World Wars, fundamentally in response to what was seen as the failures of the state as a form of organization in international relations. The functionalists focus on common interests and needs that can be shared between states, as opposed to the more self-interested realist perspective. Functionalists include international organizations, corporations and NGOs as important players in the field of international relations, and see global and regional integration as something important and commendable that in turn would diminish the central role that states play in international politics. The roots of functionalism come from Immanuel Kant and his idealist tradition. Functionalism, focusing in integration between states in economic and technical areas, argues that states and their populations would in the process of integration benefit from the integration (aided by international organizations, NGOs, etc.) and thereby support further integration. Basically it assumes that cooperation would feed more cooperation.

2   Tripartite Interim Agreement Between The Republic Of Mozambique And The Republic Of South Africa And The Kingdom Of Swaziland For Co-operation On The Protection And Sustainable Utilisation Of The Water Resources Of The Incomati And Maputo Watercourses.

References

Allan (2001) *The Middle East Water Question: Hydropolitics and the Global Economy*, I. B. Tauris, London and New York, NY

Earle, A. and Malzbender, D. (2007) 'Water and the peaceful, sustainable development of the SADC region', paper for *Towards a Continental Common Position on the Governance of Natural Resources in Africa*, SaferAfrica, Pretoria, South Africa

Grey, D., Sadoff, C. and Connors, G. (2009) 'Effective cooperation on transboundary waters: a practical perspective' in A. Jägerskog and M. Zeitoun (eds) *Getting Transboundary Water Right: Theory and Practice for*

Effective Cooperation', Report no 25, SIWI, Stockholm, Sweden

Haas, P. M. (1992) 'Banning chlorofluorocarbons: epistemic community efforts to protect stratospheric ozone', in P. M. Haas (ed.) *Knowledge, Power, and International Policy Coordination*, University of South Carolina Press, Columbia, SC

Jägerskog, A. (2003) 'Why states cooperate over shared water: the water negotiations in the Jordan River Basin', PhD dissertation, Linköping Studies in Arts and Science, Linköping University, Sweden

Jägerskog, A., Granit, J., Risberg, A. and Yu, W. (2007) *Transboundary Water Management as a Regional Public Good. Financing Development – An Example from the Nile Basin*, Report no 20, SIWI, Stockholm, Sweden

Keohane, R. O. (1994) 'International institutions: two approaches', in F. Kratochwil and E. D. Mansfield (eds) *International Organization: A Reader*, HarperCollins, New York, NY

Sadoff, C. W. and Grey, D. (2002) 'Beyond the river: the benefits of cooperation on international rivers', *Water Policy*, vol 4, pp389–403

Selby, J. (2003). 'Dressing up domination as "co-operation": the case of Israeli-Palestinian water relations', *Review of International Studies*, vol 29, no 1, pp121–138

Starr, J. (1991) 'Water wars', *Foreign Policy*, no 82 (Spring), pp17–36

Turton, A. R. and Earle, A. 2(003) 'Discussion document on the implications of international treaties on the development of a management regime for the Okavango River Basin', deliverable D 6.2 of WERRD (Water and Ecosystem Resources in Rural Development) Project, AWIRU (African Water Issues Research Unit), Pretoria University, South Africa

Zeitoun, M. and Mirumachi, N. (2008) 'Transboundary water interaction I: reconsidering conflict and cooperation', *International Environmental Agreements*, vol 8, pp297–316

Index

Printed and bound by CPI Group (UK) Ltd, Croydon, CR0 4YY

23/10/2024

01777678-0016